QB
461
.H54
1986

Highlights of modern astrophysics

$39.00

DATE			

© THE BAKER & TAYLOR CO.

HIGHLIGHTS OF MODERN ASTROPHYSICS

HIGHLIGHTS OF MODERN ASTROPHYSICS

CONCEPTS AND CONTROVERSIES

Edited by

Stuart L. Shapiro

Saul A. Teukolsky

A WILEY-INTERSCIENCE PUBLICATION

JOHN WILEY & SONS

New York • Chichester • Brisbane • Toronto • Singapore

Copyright © 1986 by John Wiley & Sons, Inc.

All rights reserved. Published simultaneously in Canada.

Reproduction or translation of any part of this work beyond that permitted by Section 107 or 108 of the 1976 United States Copyright Act without the permission of the copyright owner is unlawful. Requests for permission or further information should be addressed to the Permissions Department, John Wiley & Sons, Inc.

Library of Congress Cataloging in Publication Data:

Highlights of modern astrophysics.

"A Wiley-Interscience publication."
Includes index.
1. Astrophysics. I. Shapiro, Stuart L. (Stuart Louis), 1947– . II.Teukolsky, Saul A. (Saul Arno), 1947– .

QB461.H54 1986 523.01 86-1672
ISBN 0-471-82421-6

Printed in the United States of America

10 9 8 7 6 5 4 3 2 1

Contributors

JOHN N. BAHCALL, School of Natural Sciences, Institute for Advanced Study, Princeton, New Jersey

HANS A. BETHE, Nuclear Studies, Cornell University, Ithaca, New York

MARSHALL H. COHEN, Astronomy Department, California Institute of Technology, Pasadena, California

GEORGE B. FIELD, Center for Astrophysics, Cambridge, Massachusetts

WILLIAM A. FOWLER, W. K. Kellogg Radiation Laboratory, California Institute of Technology, Pasadena, California

RICCARDO GIACCONI, Space Telescope Science Institute, Baltimore, Maryland

PHILIP MORRISON, Department of Physics, Massachusetts Institute of Technology, Cambridge, Massachusetts

MARTIN J. REES, Institute of Astronomy, University of Cambridge, Cambridge, England

MALVIN A. RUDERMAN, Physics Department, Columbia University, New York, New York

KIP S. THORNE, Theoretical Astrophysics, California Institute of Technology, Pasadena, California

VERA C. RUBIN, Department of Terrestrial Magnetism, Carnegie Institute of Washington, Washington, D.C.

ROBERT V. WAGONER, Department of Physics, Stanford University, Stanford, California

FREEMAN DYSON, School of Natural Sciences, Institute for Advanced Study, Princeton, New Jersey

THOMAS GOLD, Department of Astronomy, Cornell University, Ithaca, New York

This book is dedicated to
Edwin E. Salpeter—
colleague, teacher, and friend

Edwin E. Salpeter is the James G. White Professor of Physical Sciences at Cornell University. He was born in 1924 in Vienna, Austria. With his family, he moved to Australia in 1939 to escape Nazism. Ed received his undergraduate training at Sydney University and his Ph.D. in theoretical physics at Birmingham University in 1948 under Rudolf Peierls. He came to Cornell in 1949 to work with Hans Bethe, and has stayed there ever since. Currently he is the Director of the Center for Radiophysics and Space Research.

Ed's early work was in the burgeoning fields of quantum electrodynamics and nuclear physics. He became interested in nuclear reactions in stars in the early 1950s, and this led to a lifelong interest in astrophysics. Hardly any area of astrophysics has been left untouched by his contributions. He identified the mechanism by which stars burn helium in stars, thus solving the longstanding problem of how stars produce heavy elements. He derived the stellar birth rate function from Galactic data, a crucial tool in the study of stellar and Galactic evolution. He provided a thorough treatment of the equation of state of cold degenerate matter in white dwarfs and neutron stars, even before neutron stars were known to exist. He has applied the principles of theoretical physics to the interstellar medium, and initiated the field of "solid-state astrophysics" with his work on interstellar grains. He and Zel'dovich independently proposed in 1964 that gas accretion onto black holes would be a powerful source of radiant energy. From first principles, Ed has developed the theory of the structure of the interior of Jupiter. In recent years he has trained himself as a full-fledged radio astronomer in order to study the distribution of neutral hydrogen. He has used the data to answer questions about properties of the interstellar medium, about the "missing mass" in the universe, and about cosmology.

Ed has always chosen to work on problems on the borderlines of physics and astronomy. He is famous for his great physical intuition, which enables him to quickly determine the relevant length scales and nondimensional ratios in any physical problem. He has trained numerous students and influenced countless others in this approach. As a result, he is responsible for much of the progress achieved in theoretical astrophysics in the past 30 years.

Thomas Gold is the John L. Wetherill Professor of Astronomy at Cornell University. He was the founder and for many years Director of the Center for Radiophysics and Space Research at Cornell. He played a leading role in the establishment of the world's largest radiotelescope at Arecibo. His scientific contributions have spanned many areas, from the Steady-State Theory of the Universe to the physics of the lunar surface, and to the identification of pulsars as neutron stars.

Foreword

In 1949 I spent an afternoon walking through the backs of the colleges in Cambridge (the *real* Cambridge, as they call it), with an Austro-Australian postdoc, discussing such matters as the origin of the Universe, the nature of fundamental particles, and the progress of mathematical physics and chatting about many of the personalities of the day involved in those subjects. My visitor had come from Birmingham, where his studies were being guided by the great physicist Rudi Peierls. He was visiting Cambridge to broaden his acquaintance with the British scene in physics, and he expressed a certain interest also in astronomy.

My recollection of this meeting, and of the discussion, is very clear in one respect. Here was a young man who gave me the impression of being a simple Austrian country boy, and yet it became evident from his conversation that he had acquired considerable knowledge and sophistication in physics, and he clearly had gained access to the inner circles of physics in England. I have since come to realize that it is just this unpretentious attitude (without having much to be unpretentious about!) which is the hallmark of Ed Salpeter, and which has made him such a pleasant colleague and collaborator, and such an attractive teacher.

When I met him again, a few years later, he was well established at Cornell. Peierls and Bethe had made a practice of trading their most promising graduate students, and this had brought him there. It was probably Bethe's influence that started Ed on problems of nuclear astrophysics. Major contributions in this field (as discussed by Willy Fowler in this book) brought him into circles of astronomers, including Walter Baade, and he fell victim to Baade's overpowering enthusiasm for astronomy. Now there was no looking back: Astronomy it was to be. His recipe was simple: "Mathematical physics provides the tools to solve innumerable problems that astronomy presents to us. Let's go to it! We'll solve them elegantly, accurately if we can; inelegantly, laboriously, and approximately only if we can't. But in any case we'll get nearer the truth than we were when we started." Whether this attitude stemmed from the influence of Hans Bethe is not known; but the style is unmistakably similar. Salpeter became to astronomy what Bethe is to physics.

Soon the Salpeter School of Astronomy became established, as students flocked to the master craftsman to learn his craft and his style. No boundaries between subjects got in his way. After major contributions to nuclear astrophysics, stellar statistics became established by his work. Then radio reflection from the ionosphere and from plasmas in general greatly helped to attain important results from the ionospheric work at Cornell's great radio instrument at Arecibo, Puerto Rico. The interiors of superdense objects were investigated, and at the other extreme of density, the radio effects of the most tenuous interstellar gas was studied. The first study of space maser action can be found in his writings. The list gets very long and covers nearly every topic studied in astronomy. But even outside of astronomy there are contributions such as those to help with the important biological studies carried out by another Cornell professor, his wife Mika.

On an average he has published 10 serious scientific papers per year in the last 20 years. Some 90 persons collaborated with Ed in this production, and many of those were students learning the trade. The Salpeter School of Astronomy is in full swing now and has shown no signs of slowing down. There cannot be many astronomical centers in the world where Ed does not have a former student or collaborator in residence.

It is not that Ed gives his name lightly to papers written by his graduate students. In every one he has had a major part, often in providing the basic idea and sometimes even doing the hard detailed work. One student at a Ph.D. examination was heard to say that every good idea in this thesis had come from Professor Salpeter.

The unassuming young man from Austria I first met 35 years ago has probably exerted more influence on astronomy than anyone else in the intervening years, both through his own contributions and through those of the large number of students he has instructed and inspired. I am very happy to have the opportunity of wishing him continued success in ramming back the frontiers of ignorance in his own inimitable way.

THOMAS GOLD
Cornell University

Preface

A unique "happening" took place on the campus of Cornell University in October of 1984. Indian summer had just arrived in upstate New York for its brief, annual visit and the Fall colors were in full bloom when over 400 scientists, mostly physicists and astronomers, converged on Ithaca to pay tribute to one of their own. It was quite a celebration that week as scientists from around the globe came together to honor Edwin S. Salpeter on the occasion of his sixtieth birthday. They came to Cornell to express their appreciation and affection for Ed, who has been in the forefront of so many of the important developments in astrophysics for over 30 years. They also came to hear 13 of the world's most talented and prestigious astrophysicists, specially invited for the occasion, discuss and debate the status of modern astrophysics. This book recounts what these 13 scientists had to say.

These physicists and astronomers—the 13 authors of our book—comprise no ordinary group. Some are Nobel Prize winners. Others are directors of leading institutes of astrophysical research. All are distinguished scientists who have made significant and lasting contributions to modern astrophysics. They are pioneers in their fields who have helped shape our current thinking about the physical Universe.

In their respective chapters, the authors review some of the most important and timely issues in astrophysics today. They highlight the accomplishments of observational and theoretical work in these areas and demonstrate how by application of the basic principles of physics the secrets of the distant stars and galaxies can be unlocked. They also pinpoint conflicting views and findings on some important topics, thus revealing the inadequacy of our current understanding in these areas. Finally, they indicate some bold possibilities for future research and help prepare us for the next breakthroughs in the fast-paced field of modern astrophysics.

Black holes and neutron stars. Thermonuclear reactions and supernova explosions. Quasars, radio jets, and the Big Bang. This is the stuff of modern astrophysics! So are Space Telescopes and space satellites and detectors of solar neutrinos. And interstellar blast waves and Dark Matter. These are the pressing issues around which so much research activity is centered today. There are also the issues emphasized in this book.

Who should read this book? Anyone! Anyone, that is, who has ever seriously wondered about the inner workings of our Cosmos. Nonprofessional readers with little technical background but who enjoy the popular literature on astronomy and space science are invited to trace the threads of some of the most exciting nonfiction this side of the Milky Way as recounted by the leading experts in the field. Readers with solid grounding in modern physics will undoubtedly appreciate how seemingly bizarre and puzzling astronomical phenomena can be understood by logical application of well-established physical principles. Caution need be exercised only by students—this book may sufficiently capture your imagination that you may be compelled to dedicate your career to pursuing answers to the riddles that remain!

STUART L. SHAPIRO
SAUL A. TEUKOLSKY

Ithaca, New York
May 1986

Acknowledgments

The production of this book took the cooperation and hard work of many people. We are especially grateful to the other members of the organizing committee of the October 1984 symposium at Cornell that led to this book: T. Gold, K. Gottfried, D. Holcomb, and H. van Horn. Special thanks go to Y. Terzian, committee chairman, for the enormous effort he made to ensure the success of the whole endeavor. Most of the photographs of the authors were taken by M. Cooke, to whom we are very grateful. Financial assistance for this project was generously provided by Cornell University. All the authors were particularly inspired by the opportunity to participate in a book dedicated to Ed Salpeter, for whom they have great admiration and fondness.

S. L. S.
S. A. T.

Contents

HIGHLIGHTS OF MODERN ASTROPHYSICS

NUCLEAR PHYSICS AND STELLAR EVOLUTION

1.

The Synthesis of the Chemical Elements Carbon and Oxygen

WILLIAM A. FOWLER

William A. Fowler is Institute Professor of Physics Emeritus at Caltech. The Kellogg Radiation Laboratory, under his leadership, has made seminal discoveries in nuclear physics. One of the founders of the field of nuclear astrophysics, he has made important contributions to the study of nuclear forces and reaction rates, nuclear spectroscopy, and thermonuclear burning in stars. He was awarded the Nobel Prize in physics in 1983 for his work on the synthesis of elements in stars and supernovae.

I. INTRODUCTION

There are no more fascinating problems in nuclear astrophysics than those involving the synthesis of the chemical elements carbon and oxygen. This chapter starts with a discussion of the historical development of the theory and ends with the current experimental and observational efforts to reach a quantitative solution of the problem.

Why is the synthesis of carbon and oxygen important and interesting? Table 1 gives the answer. The hydrogen and helium that emerged from the "Big Bang" stage of the early expanding universe are still the most abundant elements by mass in the solar system and, for that matter, in the universe. But oxygen (0.85% by mass) and carbon (0.39% by mass) are the next most abundant as a result of the synthesis of the heavier elements from hydrogen and helium in stars. In a sense more important to us as human beings is the fact that oxygen constitutes 65% by mass of the human body and carbon constitutes 18%. Here we cannot dwell on the remarkable chemical, geological, and biological processes that increased the oxygen in our bodies by a factor of 76 over the average in the solar system and the carbon by a factor of 46. We must be content with an attempt to understand the nature of the nuclear processes

TABLE 1. Element Abundance by Mass (%)[a]

Element	Solar	Human Body	Ratio
H	77.2	10.00	0.13
He	20.9	—	0.0
C	0.39	18.0	46
N	0.094	3.0	32
O	0.85	65.0	76
Ne	0.15	—	0
Na	0.004	0.15	38
Mg	0.074	0.05	0.68
Al	0.007	—	0
Si	0.081	—	0
P	0.0006	1.0	1667
S	0.046	0.25	5.4
Cl	0.0001	0.15	1500
K	0.0004	0.35	875
Ar	0.01	—	—
Ca	0.007	1.5	214
Fe	0.15	<0.004	<0.03

[a] Data for solar from Cameron (1982), and for human body from Dyson (1978).

that produced carbon and oxygen in the first place through stellar nucleosynthesis.

II. HISTORICAL DEVELOPMENTS

In his famous paper, Bethe (1939) studied the direct formation of ^{12}C from three ^{4}He nuclei. He concluded that "The process is strongly temperature-dependent but it requires temperatures of $\sim 10^9$ degrees to make it as probable as the proton combination, $H + H \Rightarrow D + e^+$." Such a temperature could only occur in red giants, but he also concluded that, "It seems, however, doubtful whether the energy production in giants is due to nuclear reactions at all."[2] Footnote 2 reads, "G. Gamow, private communication." All of this led Bethe to end his section 6 with the statement:

> The considerations of the last two sections show that there is no way in which nuclei heavier than helium can be produced permanently in the interior of stars under present conditions. We can therefore drop the discussion of the building up of elements entirely and can confine ourselves to the energy production which is, in fact, the only observable process in stars.

It was Edwin E. Salpeter, who came from Cornell to the Kellogg Radiation Laboratory at Caltech in 1951, as the young Lochinvar out of the "east," who saved the day with new insight on the formation of ^{12}C from three ^{4}He nuclei. This leads us to a historical look-back at the instability of ^{8}Be.

III. THE INSTABILITY OF ^{8}BE

Bethe (1939) argued that the instability of ^{8}Be against disintegration had been definitely established by the microchemical experiments of Paneth and Glückauf (1937) and quoted Kirchner and Neuert (1937), who gave the disintegration energy as 40–120 keV on the basis of their studies of the reaction $^{11}B + {}^{1}He \Rightarrow {}^{8}Be + {}^{4}He$. Wheeler (1941) analyzed a number of $\alpha\alpha$-scattering experiments and gave the instability of the ground state of ^{8}Be as 125 keV and of the first excited state of ^{8}Be as 3 MeV. However, his analysis yielded an angular momentum $J = 0$ for the excited state, while $J = 2$ is now known to be the case.

Analysis of a number of the Q-values of reactions involving light nuclei led Allison and his collaborators (1939a,b,c) to the conclusion that ^{8}Be

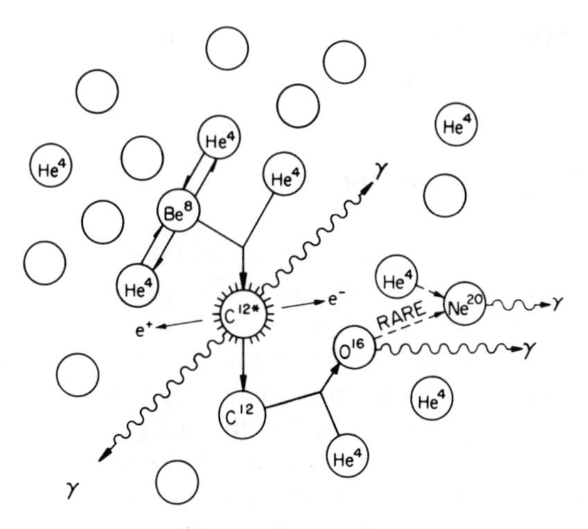

FIGURE 1. Fusion of helium in red giants stars. The density and temperature given are representative of the values in the cores of massive giants ($M \sim 25M_{\odot}$). In red giants in globular clusters ($M \approx 1.2M_{\odot}$) the density is $\sim 10^4$ g cm^{-3} and the temperature is 1.2×10^8 K at the ignition of helium burning at the tip of the red giant branch. The effective energy range given applies to ^{12}C(α,γ)^{16}O(α,γ)^{20}Ne. The $3\alpha \Rightarrow {}^{12}$C process proceeds through resonance at 92 keV in $2\alpha \Leftrightarrow {}^{8}$Be due to the ground state of ^{8}Be (Salpeter 1952) and through resonance at 287 keV in ^{8}Be + $\alpha \Rightarrow {}^{12}$C due to the second excited state of ^{12}C (Hoyle 1954).

was stable against disintegration into two alpha-particles by values ranging from 310 ± 60 keV through 180 ± 160 to 65 ± 180 keV. The uncertainties in these values led to no consensus concerning the stability of ^{8}Be, and the entire matter remained an open question during the hiatus of World War II. It required some time for nuclear laboratories to come into operation after the war, but finally Hemmindinger (1948, 1949) used the photodisintegration process, ^{9}Be(γ,n)^{8}Be, to show that ^{8}Be was definitely unstable with a quoted energy of instability equal to 103 ± 10 keV. This was soon confirmed by Tollestrup, Fowler, and Lauritsen (1949a,b), who used the reaction ^{9}Be(p,d)^{8}Be to establish ^{8}Be $\Rightarrow 2^{4}$He + (89 ± 5) keV. This compares very well with the currently accepted value of 91.89 ± 0.05 keV! Li, Whaling, Fowler, and Lauritsen (1951) used 43 reaction Q-values to determine self-consistent, overdetermined binding energies for 25 light nuclei and assigned an adjusted value of 96 ± 4 keV for the instability of ^{8}Be. Thus when Salpeter arrived in our laboratory in the summer of 1951 there existed the strong conviction that ^{8}Be was unstable with numerical values for the instability good to ~5% to back up this conviction. Salpeter (1952, 1953) used the value 95 ± 5 keV.

Salpeter was able to conclude:

> The nucleus ^{8}Be is unstable to disintegration into two ^{4}He nuclei. But, since an energy of only 95 ± 5 keV, comparable with thermal energies at temperatures over 10^{8} K, is required for its formation, a fraction of about 1 in 10^{10} of the material of the star is kept in the form of ^{8}Be in a state of dynamic equilibrium. The ^{8}Be present then easily absorbs a helium nucleus. Once carbon has been produced, the following reactions also become possible:
>
> $$^{12}C + {}^{4}He \Rightarrow {}^{16}O + \gamma + 7.162 \text{ MeV}$$
>
> $$^{16}O + {}^{4}He \Rightarrow {}^{20}Ne + \gamma + 4.734 \text{ MeV}$$
>
> and so on.

(Modern Q-values are given for the preceding two reactions.) Thus Salpeter laid out the sequence of nuclear reactions that produce and destroy ^{12}C and ^{16}O in helium burning. This sequence is shown in Fig. 1.

IV. RESONANCE IN $^{8}Be(\alpha,\gamma)^{12}C$

Salpeter (1952) correctly treated $2\alpha \Rightarrow {}^{8}Be$ as a resonant process but treated the subsequent reaction $^{8}Be + \alpha \Rightarrow {}^{12}C$ as nonresonant in calculating the overall rate for $3\alpha \Rightarrow {}^{12}C$. He concluded that a temperature $T = 2 \times 10^{8}$ K, would be necessary for the $3\alpha \Rightarrow {}^{12}C$ process to supply sufficient energy to produce the luminosity of red giant stars in Population II globular clusters. However, Sandage and Schwarzschild (1952) concluded from their calculations on red giant models that the temperature was only 1.1×10^{8} K at the tip of the red giant sequence where helium burning was assumed to begin. This was confirmed by Hoyle and Schwarzschild (1955), who found 1.2×10^{8} K.

Hoyle took up the cudgel even before publication of his paper with Schwarzschild. He realized that the $^{8}Be + \alpha \Rightarrow {}^{12}C$ reaction would be speeded up by many orders of magnitude, thus reducing the temperature for its onset, if there existed an excited state of ^{12}C with energy 0.3 MeV in excess of $^{8}Be + \alpha$ at rest and with the angular momentum and parity ($0^{+}, 1^{-}, 2^{+}, 3^{-}, \ldots$) dictated by the selection rules for these quantities. Hoyle came to the Kellogg Laboratory early in 1953 and questioned the staff about the possible existence of his proposed excited state. At that time there was some controversy surrounding the fact that Malm and Buechner (1951) had found no evidence for a state at 7.62 MeV claimed by Holloway and Moore (1940). Most of the Kellogg staff accepted the work of Malm and Buechner (1951) because of their superior detection techniques. However, to make a long story short, Ward Whaling and his visiting associates and graduate students (Dunbar et al., 1953) decided to go into the laboratory and search for the state using the $^{14}N(d,\alpha)^{12}C$ reaction. They found it to be located almost exactly where Hoyle had predicted. It is now known to be at 7.6542 ± 0.0015 MeV excitation in ^{12}C or 0.2877 MeV above $^{8}Be + \alpha$ (Hoyle assumed 0.3 MeV!) and 0.3795 MeV above 3α. Cook et al. (1957) then produced the state in the decay of radioactive ^{12}B and showed it could break up into 3α. They argued that the spin and parity of the state must be 0^{+}, as is now known to be the case.

The $3\alpha \Rightarrow {}^{12}C$ fusion in red giants jumps the mass gaps at 5 and 8. This process could never occur under Big Bang conditions. By the time the ^{4}He was produced in the early expanding universe the subsequent density and temperature were too low for the helium fusion to carbon to occur. Thus George Gamow's hope (see Alpher and Herman 1950) that elements beyond helium could be synthesized in the Big Bang was dashed on the rocks of nuclear theory and experiment.

In contrast, in red giants, after hydrogen conversion to helium during the main sequence state, gravitational contraction of the helium core causes the density and temperature to rise to values where helium fusion is ignited. Hoyle and Whaling showed that conditions in red giants are just right. With his resonance Hoyle (1954) increased the rate of $3\alpha \Rightarrow {}^{12}C$ by a factor 10^{7} over that given by Salpeter, just sufficient to decrease the required temperature by a factor of ~ 2. Salpeter (1957) and Reeves and Salpeter (1959) capped the early era with two papers that treated in a precise quantitative fashion the helium burning and subsequent carbon burning states of stellar nuclear evolution. Salpeter continued to improve our understanding of helium and carbon burning with a series of papers culminating in Deinzer and Salpeter (1964, 1965).

V. THE RATE OF THE $3\alpha \Rightarrow {}^{12}C$ FUSION PROCESS

In this section we confine our attention to the rate of the $3\alpha \Rightarrow {}^{12}C$ fusion process that holds for the full operation of the resonance due to the ground state of ^{8}Be (Salpeter, 1952) and the resonance due to the 7.654 MeV excited state in ^{12}C (Hoyle, 1954). Additional small contributions to the resonant rate, and the rate at temperatures less than 10^8 K, where only the lower-energy tails of the resonances contribute, are given in Caughlan et al. (1985).

Salpeter (1957) showed that the term under discussion in the rate of production of ^{12}C by $3\alpha \Leftrightarrow {}^{12}C^* \Rightarrow {}^{12}C$ + radiation is given by

$$r = n(^{12}C^*)\left(\frac{\Gamma_{rad}}{\hbar}\right) \quad \text{reactions sec}^{-1}\ \text{cm}^{-3} \qquad (1)$$

where $n(^{12}C^*)$ is the number density of the 7.654 MeV excited state of ^{12}C in equilibrium with 3α and $\Gamma_{rad} = \Gamma_\gamma + \Gamma_{pair}$ is the total radiative width of the excited state including the emission of cascade gamma rays and electron–positron pairs. Hoyle's excited state of ^{12}C with spin and parity, $J^\pi = 0^+$, radiates directly to the ground state, $J^\pi = 0^+$, by the emission of electron–positron pairs or cascades by the emission of two gamma rays through the first excited state at 4.4389 MeV with $J^\pi = 2^+$.

The number density of $^{12}C^*$ is calculated straightforwardly from the theory of statistical equilibrium and Eq. (1) yields

$$r = 3^{3/2} n_\alpha^3 \left(\frac{2\pi\hbar^2}{M_\alpha kT}\right)^3 \left(\frac{\Gamma_{rad}}{\hbar}\right) \exp(-E_r/kT) \quad \text{reactions sec}^{-1}\ \text{cm}^{-3} \qquad (2)$$

where M_α is the mass of the alpha particle, n_α is the number density of alpha particles, and E_r is the energy difference between $^{12}C^*$ and three free alpha particles. Although the ground state of ^8Be plays an important role, its mass cancels out, since $E_r = c^2M[^{12}C^* - {}^8Be - {}^4He + ({}^8Be - 2{}^4He)]$. The energy production rate via $3\alpha \Rightarrow {}^{12}C$ is given by rQ ergs $\text{sec}^{-1}\ \text{cm}^{-3}$, where the energy released per reaction is $Q = c^2M[3{}^4He - {}^{12}C]$-7.275 MeV $= 1.1656 \times 10^{-5}$ erg.

Fowler, Caughlan, and Zimmerman (1957, 1975) introduced a standard notation, $N_A^2\langle 012\rangle$ for three-body reactions similar to $N_A\langle 01\rangle \equiv N_A\langle\sigma v\rangle$ for two-body reactions. N_A is Avogadro's number, σ is the cross section, and v is the relative velocity for the relevant reactants numbered 0, 1, 2, and so on. For $3\alpha \Rightarrow {}^{12}C$, $N_A^2\langle 012\rangle = N_A^2\langle\alpha\alpha\alpha\rangle$, which can be used to calculate r/n_α from the following equation:

$$\frac{r}{n_\alpha} = \tfrac{1}{6}n_\alpha^2\langle\alpha\alpha\alpha\rangle = \tfrac{1}{6}(\rho Y_\alpha)^2[N_A^2\langle\alpha\alpha\alpha\rangle] \quad \text{reactions sec}^{-1} \text{ per } \alpha \qquad (3)$$

where the 6 comes from the possible permutations of three identical particles, $n_\alpha = \rho Y_\alpha N_A$, ρ is the density in g cm^{-3}, $Y_\alpha = X_\alpha/4$ is the mole fraction of helium, and X_α is the mass fraction. Equations (2) and (3) yield numerically:

$$N_A^2\langle\alpha\alpha\alpha\rangle = 7.592 \times 10^{-6}\Gamma_{\text{rad}}(\text{eV})T_9^{-3} \times \exp\left[\frac{-11.605E_r(\text{MeV})}{T_9}\right] \text{ reactions sec}^{-1}(\text{mol/cm}^3)^{-2} \quad (4)$$

where $T_9 = T/10^9$ K. This is to be multiplied by $\rho^3 Y_\alpha^3 N_A/6$ to obtain reactions sec^{-1} cm^{-3}.

It will be clear that two parameters, Γ_{rad} and E_r, must be measured experimentally. The measurements have been a real *tour de force* by nuclear experimentalists. The epic story, involving many workers, is summarized in Alburger (1977), Barnes and Nichols (1973), and Nolen and Austin (1976). The major breakthrough came when Schiff (1955) pointed out that the pair width, Γ_{pair}, for the $J = 0^+$ excited state of ^{12}C could be determined from the electric monopole matrix element determined in the excitation of this state in high-energy electron scattering from the ground state. The connection is obvious: by the rules of the game, $^{12}\text{C} + e^- \rightarrow {}^{12}\text{C}^* + e^-$ can be transposed into $^{12}\text{C}^* \rightarrow {}^{12}\text{C} + e^- + e^+$.

Crannell and Griffy (1964) made an accurate experimental determination of the electric monopole matrix element and thus provided an absolute value for Γ_{pair}. It was then necessary to determine the ratio of Γ_{pair} and Γ_{rad} to the total width Γ of the ^{12}C* state, which is mostly due to α-decay. Alburger (1977) summarized all previous work and gave

$$\Gamma_{\text{rad}} = \Gamma_{\text{pair}}\left(\frac{\Gamma}{\Gamma_{\text{pair}}}\right)\left(\frac{\Gamma_{\text{rad}}}{\Gamma}\right) = (3.67 \pm .046) \times 10^{-3} \text{ eV} \quad (5)$$

The experimental value of E_r has been measured directly in $^{12}\text{C}^* \Rightarrow 3\alpha$ by Barnes and Nichols (1973), who found $E_r = 379.6 \pm 2.0$ keV. Proton inelastic scattering $^{12}\text{C}(p,p')^{12}\text{C}^*$ has been used by a number of investigators to determine E_r. Nolen and Austin (1976) made the latest measurements and gave a weighted average of all work as

$$E_r = 379.38 \pm 0.20 \text{ keV} \quad (6)$$

This value is to be used, since it has 10 times the precision of the measurement by Barnes and Nichols (1973).

Using Eqs. (5) and (6), Harris et al. (1983) and Caughlan et al. (1985) were then able to express and tabulate $N_A^2\langle\alpha\alpha\alpha\rangle$ to roughly 15% uncertainty as

$$N_A^2\langle\alpha\alpha\alpha\rangle = 2.79 \times 10^{-8} T_9^{-3} \exp\left(\frac{-4.4027}{T_9}\right) \tag{7}$$

Few reaction rates in nuclear astrophysics are known with comparable accuracy.

VI. THE RATES OF THE $^{12}C(\alpha,\gamma)^{16}O$ AND $^{16}O(\alpha,\gamma)^{20}Ne$ REACTIONS

Unfortunately, the comment at the end of the previous paragraph cannot be applied to the reactions $^{12}C(\alpha,\gamma)^{16}O$ and $^{16}O(\alpha,\gamma)^{20}Ne$ which succeed $3\alpha \Rightarrow {}^{12}C$ in the Salpeter–Hoyle scenario for helium burning. There is a lively controversy about the laboratory cross section for $^{12}C(\alpha,\gamma)^{16}O$ and about its theoretical extrapolation to the low energies (~150 to ~450 keV) at which the reaction effectively operates under astrophysical circumstances of greatest interest (to be discussed later).

The experimental situation is depicted in Fig. 2 taken with minor modifications from the analysis by Langanke and Koonin (1983, 1985) of the Caltech data of Dyer and Barnes (1974) and of the Münster data of Kettner et al. (1982). It will be clear that the measured cross sections differ by a factor of 1.5, well outside the quoted uncertainties, especially near the maxima in the cross sections at 2.3 MeV. With so much riding on the outcome in regard to the relative abundances of carbon and oxygen produced in stellar nucleosynthesis, it will come as no surprise that both laboratories are engaged in checking their absolute values and in extending their measurements to lower energies.

The need for extension of the cross section measurements to lower energies is illustrated in Fig. 3, where the cross-section factor S in MeV-b is shown as a function of center-of-momentum energy. It is appropriate at this point to note that the cross-section factor defined by

$$S \equiv \sigma E \exp\left(\frac{2\pi e^2 Z_0 Z_1}{\hbar v}\right) \quad \text{MeV-b} \tag{8}$$

$$= \sigma E \exp\left[0.9895 Z_0 Z_1 \left(\frac{A}{E}\right)^{1/2}\right] \tag{9}$$

$$\Rightarrow \sigma E \exp\left(\frac{20.568}{E^{1/2}}\right) \text{ for } {}^{12}C(\alpha,\gamma)^{16}O \tag{10}$$

$$\Rightarrow \sigma E \exp\left(\frac{28.323}{E^{1/2}}\right) \text{ for } {}^{16}O(\alpha,\gamma)^{20}Ne \tag{11}$$

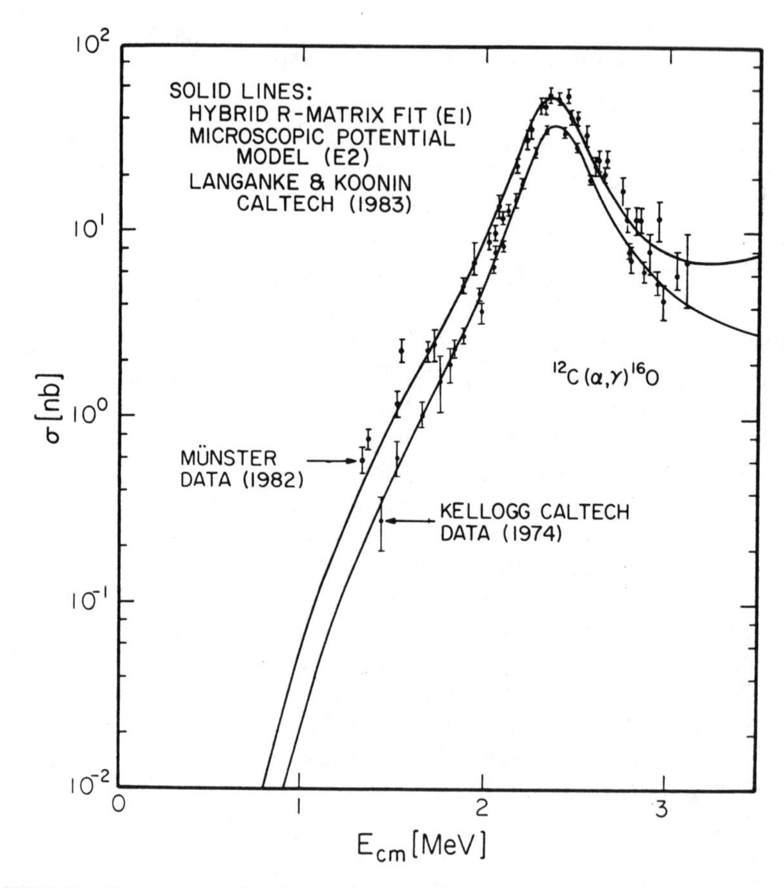

FIGURE 2. The cross section in nanobarns (nb) versus center-of-momentum energy in MeV for $^{12}C(\alpha,\gamma)^{16}O$. The Münster data were obtained by Kettner et al. (1982), and the Kellogg Caltech data were obtained by Dyer and Barnes (1974). The solid lines are theoretical calculations made by Langanke and Koonin (1983).

was introduced into nuclear astrophysics by E. E. Salpeter (1952). It has been used by all workers in the field ever since. In Eqs. (8)–(11) σ is the cross section in barns at center-of-momentum energy, E in MeV, and relative velocity v between two reactants of charge Z_0e and Z_1e and reduced mass A in atomic mass units. The exponential in Eqs. (8) and (9) but with a negative sign in the exponent is the rapidly varying term in the Gamow barrier penetration factor that appears in all cross sections involving charged nuclei interacting off-resonance at low energy. Because of this term the direct extrapolation of cross sections from the lowest

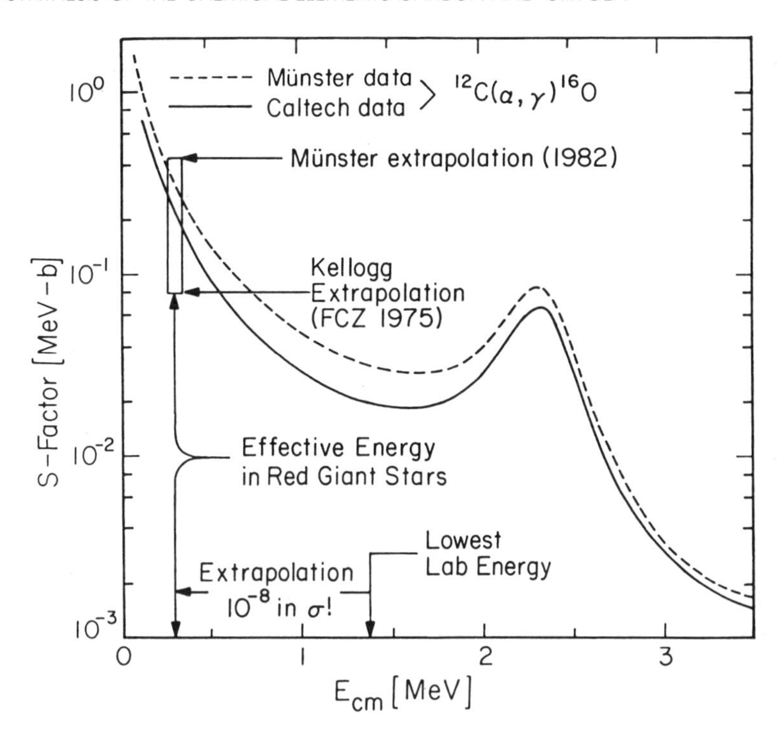

FIGURE 3. The cross-section factor S in MeV-b versus center-of-momentum energy in MeV for $^{12}C(\alpha,\gamma)^{16}O$. The dashed and solid curves are the theoretical extrapolations of the Münster and Kellogg Caltech data, respectively, by Langanke and Koonin (1983). Value for S at 0.3 MeV given by Fowler, Caughlan, and Zimmerman (1975), Kellogg extrapolation, and by Kettner et al. (1982), Münster extrapolation, are also shown.

measured energies to the energies of astrophysical interest is fraught with uncertainty. Salpeter showed that removing this explicitly known term as in Eqs. (8)–(11) yielded the cross-section factor, which depends on intrinsic nuclear factors and is frequently only slowly energy dependent.

Although the low-energy variation of S in Fig. 3 is many orders of magnitude less than that of σ in Fig. 2, S does vary considerably in the energy range 0.15–0.45 MeV, noted at the beginning of this section. The S-factor shows the maximum at 2.3 MeV due to the broad 1^- state in ^{16}O at 9.632 MeV. In addition, there is a rapid increase toward lower energies contributed by the "tails" of the 1^- bound state at 7.117 MeV and the 2^+ bound state at 6.917 MeV in ^{16}O. For s-wave capture (l = 0) of the alpha-particles into these "tails," E1 radiation occurs for the 1^- state and E2 radiation for the 2^+ state.

All of this has been taken into account by Langanke and Koonin (1983,

TABLE 2. Rates of Helium-Burning Reactions

$T_9 = T/10^9$ K	$3\alpha \Rightarrow {}^{12}C$	${}^{12}C(\alpha,\gamma){}^{16}O$		${}^{16}O(\alpha,\gamma){}^{20}Ne$
	$N_A^2\langle\alpha\alpha\alpha\rangle$[a]	$N_A\langle{}^{12}C\alpha\rangle$[b]		$N_A\langle{}^{16}O\alpha\rangle$[b]
0.12	1.88E − 21[c]	3.11E − 19[d]	9.98E − 19[c]	5.07E − 25[c]
0.15	1.48E − 18	2.32E − 17	7.33E − 17	1.45E − 22
0.18	1.14E − 16	6.14E − 16	1.90E − 15	1.08E − 20
0.20	9.60E − 16	3.70E − 15	1.13E − 14	1.31E − 19
0.25	4.01E − 14	1.34E − 13	3.95E − 13	3.46E − 16
0.30	4.37E − 13	2.03E − 12	5.82E − 12	2.56E − 13

[a] Multiply by $(\frac{1}{6})\rho^2 Y_\alpha^3 N_A$ to obtain reactions g^{-1} sec^{-1}.
[b] Multiply by $\rho Y_\alpha^2 N_A$ to obtain reactions g^{-1} sec^{-1}.
[c] Caughlan *et al.* (1985); Langanke and Koonin (1983, 1985).
[d] Fowler, Caughlan, and Zimmerman (1975).

1985), but there is still considerable uncertainty in their theoretical extrapolations to low energy, which compounds the problems raised by the difference in the experimental measurements discussed earlier. Fortunately, the E2 calculation is quite straightforward and the E1 contribution from the 1^- bound state at 7.117 MeV can be experimentally determined from its constructive interference with the tail of the 1^- resonant state at 9.632 MeV. This can be best determined by pushing the measurements to lower energies. That is what is being done at Caltech and Münster. In their latest summation of reaction rates Caughlan et al. (1985) use the Langanke and Koonin (1983, 1985) extrapolation of the Caltech data. Use of the Münster data gives a value approximately a factor of 1.5 higher. That factor is crucial in determining the relative abundance of carbon and oxygen produced in stellar nucleosynthesis.

Langanke (1984) has calculated the rate of the ${}^{16}O(\alpha,\gamma){}^{20}Ne$ reaction at stellar energies. This enhances the rate calculated by Fowler, Caughlan, and Zimmerman (1975) by a factor of 2. Nevertheless this rate is at least two orders of magnitude less than that for ${}^{12}C(\alpha,\gamma){}^{16}O$, so under most circumstances ${}^{16}O$ is the main end product of helium burning.

Table 2 gives a sampling from Caughlan et al. (1985) and Langanke and Koonin (1983, 1985) of the numerical rates of helium-burning reactions over a limited range of temperatures. For the ${}^{12}C(\alpha,\gamma){}^{16}O$-reaction the rate based on the extrapolation of the Dyer and Barnes (1974) data to low energies by Fowler, Caughlan, and Zimmerman (1975) is also given. It will be noted that the newer rates are greater by a factor of 3 even for the extrapolation of the same data. Use of the data of Kettner et al. (1982) would increase this factor to 4.5. The reality of the great uncertainty in

TABLE 3. Primary Terms in Reaction Rates for Helium Burning 1967–1985

$3\alpha \Rightarrow {}^{12}C$ (reactions sec^{-1} $mole^{-2}$ cm^6)	
FCZI (ARA^2 5, 525, 1967)	$2.13 \times 10^{-8}\ T_9^{-3} \exp(-4.294/T_9)$
FCZII (ARA^2 13, 69, 1975)	$2.49 \times 10^{-8}\ T_9^{-3} \exp(-4.4109/T_9)$
HFCZIII (ARA^2 21, 165, 1983)	$2.79 \times 10^{-8}\ T_9^{-3} \exp(-4.4027/T_9)$
CFHZIV (AD & NDT 32, 197, 1985)	DITTO
MULTIPLY BY $\rho^3 Y_\alpha^3 N_A/6$ FOR REACTIONS SEC^{-1} CM^{-3}	
1985 UNCERTAINTY: ± 15%	
${}^{12}C(\alpha,\gamma){}^{16}O$ (reactions sec^{-1} $mole^{-1}$ cm^3)	
FCZII (ARA^2 5, 525, 1967)	$2.19 \times 10^8\ T_9^{-2}(1 + 0.050T_9^{-2/3})^{-2} \times \exp(-32.12T_9^{-1/3})$ $T_9 \leq 6$
FCZII (ARA^2 13, 69, 1975)	$9.03 \times 10^7\ T_9^{-2}(1 + 0.621T_9^{2/3})^2 \times (1 + 0.047T_9^{-2/3})^{-2} \times \exp(-32.120T_9^{-1/3} - (T_9/5.863)^2)$
HFCZIII (ARA^2 21, 165, 1983)	DITTO
CFHZIV (AD & NDT 32, 197, 1985)	$2.93 \times 10^8\ T_9^{-2}(1 + 0.0489T_9^{-2/3})^{-2} \times \exp(-32.120T_9^{-1/3} - (T_9/3.496)^2)$
MULTIPLY $\rho^2 Y_\alpha Y_{12} N_A$ FOR REACTIONS SEC^{-1} CM^{-3}	
1985 UNCERTAINTY: FACTOR OF TWO, MORE TO COME EARLY 1986.	

the rate of the ${}^{12}C(\alpha,\gamma){}^{16}O$ reaction must be borne in mind in reading the next section of this chapter.

To establish the connection between the rates for ${}^{12}C(\alpha,\gamma){}^{16}O$ given in Table 2 and past treatments of the rate of this reaction we append the following comments. In the most recent past, experimental and theoretical results have been expressed in terms of the S-factor for the reaction at 0.3 MeV. The results of Fowler, Caughlan, and Zimmerman (1975) in Table 2 correspond to S (0.3 MeV) = 0.078 MeV-b and those of Caughlan et al. (1985) to S(0.3 MeV) = 0.24 MeV-b. Even earlier, Fowler, Caughlan, and Zimmerman (1967) expressed the S-factor in terms of the reduced alpha-particle width, θ_α^2, of the bound state at 7.117 MeV. The values for this quantity corresponding to the S-factor values just given are θ_α^2 = 0.68 S(0.3 MeV) $\Rightarrow$ 0.053 and 0.16, respectively. For the results of Kettner et al. (1982), S(0.3) MeV = 0.36 MeV-b and θ_α^2 = 0.24. In summary, Table 3 lists the primary analytical expressions for the rates of the $3\alpha \Rightarrow {}^{12}C(\alpha,\gamma){}^{16}O$ reactions which have been recommended for use over the period 1967–1985 (Harris et al., 1983 is an additional source). It must be emphasized that even at moderate temperatures the secondary terms given in the references must not be omitted.

The rates for reactions between positively charged nuclei must be increased because of the screening of the Coulomb barrier between them by the surrounding electrons in astrophysical plasma. Again, everyone in nuclear astrophysics is indebted to Salpeter (1954) for providing the enhancement factors under a variety of circumstances ranging from what he called weak screening to strong screening. In helium burning in stellar cores with $M > 5M_{\odot}$ weak screening applies and the enhancement factor is the same for the $3\alpha \Rightarrow {}^{12}C$ process as for the ${}^{12}C(\alpha,\gamma){}^{16}O$ reaction and amounts to an increase of at most 3%. In less massive stellar cores where the electrons can become degenerate, strong screening applies and the enhancement factors can become quite large.

VII. THE SYNTHESIS OF CARBON AND OXYGEN IN MASSIVE STARS ($M \gtrsim 10M_{\odot}$)

In this final section the relative abundances of oxygen and carbon produced in nucleosynthesis in single massive stars ($M \gtrsim 10M_{\odot}$) will be discussed primarily in connection with the uncertainty in the ${}^{12}C(\alpha,\gamma){}^{16}O$ reaction discussed in the previous section. Comparisons will be made with the solar system ratio, (O/C) = 2.2 by mass or 1.6 by number. This will be done in spite of the fact that the interstellar medium can be enriched in carbon by mass loss from lower mass stars during the asymptotic giant branch stage of stellar evolution. Iben (1984) has summarized the helium shell flashing and carbon dredge-up that occurs during this stage. Other references on helium burning in lower-mass stars that are of general interest in connection with reaction rates are Arnett (1971), Iben (1972), and Nomoto (1982).

Woosley and Weaver (1982) have calculated abundances in the final ejecta of Type II, Population I supernovae for $25M_{\odot}$ stars. Their result, O/C = 3.75 by mass, is 1.7 times the solar system value. They used the ${}^{12}C(\alpha,\gamma){}^{16}O$ rate of Fowler, Caughlan, and Zimmerman (1975). Their large O/C ratio was due mainly to oxygen production in the final supernova explosion. For the heavier elements they found a dip by a factor of almost 10 for the abundances of the elements above Si and below Fe relative to solar system vales. Nomoto, Thielemann, and Wheeler (1984) added an equal contribution from the ejecta of Type I, Population II carbon-detonation supernovae and substantially increased the abundances between Si and Fe but did not change the O/C ratio relative to solar. Recently, Woosley and Weaver (1984) have made calculations with the ${}^{12}C(\alpha,\gamma){}^{16}O$

rate of Fowler, Caughlan, and Zimmerman (1985) increased by a factor of 2.5. They find O/C = 3.9 times solar. Arnett and Thielemann (1984) have calculated quasistatic presupernova nucleosynthesis for $M \approx 20M_{\odot}$ and argue that the final explosion will not substantially change the O/C ratio. Whether or not this is the case, their results are very interesting, in that they made calculations using the Fowler, Caughlan, and Zimmerman (1975) $^{12}C(\alpha,\gamma)^{16}O$ rate × 1, × 3, and × 5. Their factors cover the range of uncertainty discussed in the previous section. They found $^{16}O/^{12}C$ relative to solar equal to 0.50, 1.53, and 2.77, respectively. These values will be decreased by carbon production in Type I supernovae, as will those of Woosley and Weaver (1982, 1984). However, a major point is the fact that nucleosynthesis during the supernova explosive stage may increase the O/C ratio substantially, as indicated by the ratio of the value 1.7 found by Woosley and Weaver (1982) to the value 0.5 found by Arnett and Thielemann (1984) when both used the $^{12}C(\alpha,\gamma)^{16}O$ rate of Fowler, Caughlan, and Zimmerman (1975). It must also be emphasized that the O/C ratio produced in helium burning has a profound influence on nucleosynthesis in subsequent carbon, neon, oxygen, and silicon burning.

It is important to note at this point that helium burning in massive stars occurs at $T \approx 2 \times 10^8$ K and $\rho \approx 10^3$ g cm^{-3}, which are higher and lower, respectively, than the temperatures and densities in globular cluster stars at the tip of the red giant branch that Salpeter (1952) and Hoyle (1954) used in their original treatments of helium burning in stars.

It will be clear that the problem of what determined the solar system abundance ratio for oxygen relative to carbon has not been solved roughly thirty years after Salpeter (1952) and Hoyle (1954) elucidated the theoretical details of helium burning nucleosynthesis.

REFERENCES

Alburger, D. E., 1977, *Phys. Rev.,* **C16,** 2395.

Allison, S. K., Graves, E. R., Skaggs, L. S., and Smith, N. M., 1939a, *Phys. Rev.,* **55,** 107.

Allison S. K., 1939b, *Phys. Rev.,* **55,** 624.

Allison, S. K., Skaggs, L. S., and Smith, N. M., 1939c, *Phys. Rev.,* **56,** 288.

Alpher, R. A., and Herman, R. C., 1950, *Rev. Mod. Phys.,* **22,** 153.

Arnett, W. D., and Thielemann, F.-K., 1984, *Stellar Nucleosynthesis,* C. Chiosi and A. Renzini, eds., Reidel, Dordrecht.

Arnett, W. D., 1981, *Ap. J.,* **170,** L43.

Barnes, C. A., and Nichols, D. B., 1973, *Nucl. Phys.,* **A217,** 125.

Bethe, H. A., 1939, *Phys. Rev.,* **55,** 434.

Cameron, A. G. W., 1982, *Essays in Nuclear Astrophysics,* C. A. Barnes, D. D. Clayton, and D. N. Schramm (eds.) Cambridge University Press, Cambridge.

Caughlan, G. R., Fowler, W. A., Harris, M. J., and Zimmerman, B. A., 1985, *Atomic Data and Nuclear Data Tables*, **32,** 197; see also Nomoto, K., Thielemann, F.-K., and Miyaji, S., 1984, preprint and Langanke, K., Wiescher, M., and Thielemann, F.-K., 1985, preprint.

Cook, C. W., Fowler, W. A., Lauritsen, C. C., and Lauritsen, T., 1957, *Phys. Rev.,* **107,** 508.

Crannell, H. L., and Griffy, T. A., 1964, *Phys. Rev.,* **136,** B1580.

Deinzer, W., and Salpeter, E. E., 1965, *Ap. J.,* **140,** 499; 1965, **142,** 813.

Dunbar, C. N. F., Pixley, R. E., Wenzel, W. A., and Whaling, W., 1953, *Phys. Rev.,* **92,** 64.

Dyer, P., and Barnes, C. A., 1974, *Nucl. Phys.,* **A233,** 495.

Dyson, R. D., 1978, *Cell Biology,* Allyn and Bacon, Inc., Boston.

Fowler, W. A., Caughlan, G. R., and Zimmerman, B. A., 1957, *Ann. Rev. Astron. Astrophys.,* **5,** 25; 1975, **13,** 69.

Harris, M. J., Fowler, W. A., Caughlan, G. R., and Zimmerman, B. A., 1983, *Ann. Rev. Astron. Astrophys.,* **21,** 165.

Hemmindinger, A., 1948, *Phys. Rev.,* **73,** 806; 1949, **75,** 1267.

Holloway, M. E., and Moore, B. L., 1940, *Phys. Rev.,* **58,** 847.

Hoyle, F., 1954, *Ap. J. Suppl.,* **1,** 121.

Hoyle, F., and Schwarzschild, M., 1955, *Ap. J. Suppl.,* **2,** 1.

Iben, I., 1985, *Nucleosynthesis: Challenges and New Developments,* W. D. Arnett and J. W. Truran, eds., University of Chicago Press, Chicago.

Iben, I., 1972, *Ap. J.,* **178,** 433.

Kettner, K. V., Becker, H. W., Buchmann, L., Görres, J., Kräwinkel, H., Rolfs, C., 1982, *Z. Phys.,* **A308,** 73.

Kirchner, F., and Neuert, H., 1937, *Naturwiss,* **25,** 48.

Langanke, K., and Koonin, S. E., 1983, *Nucl. Phys.,* **A410,** 334; 1985, **A439,** 384.

Langanke, K., 1984, *Z. Phys.,* **317,** 325.

Li, C. W., Whaling, W., Fowler, W. A., and Lauritsen, C. C., 1951, *Phys. Rev.,* **83,** 512.

Malm, R., and Buechner, W. W., 1951, *Phys. Rev.,* **81,** 519.

Nolen, J. A., and Austin, S. M., 1976, *Phys. Rev.,* **C13,** 1773.

Nomoto, K., 1982, *Ap. J.,* **253,** 798.

Nomoto, K., Thielemann, F.-K., and Wheeler, J. C., 1984, *Ap. J.,* **279,** L23; 1984, erratum **283,** L25.

Paneth, F. A., and Glüchauf, E., 1937, *Nature,* **139,** 712.

Reeves, H., and Salpeter, E. E., 1959, *Phys. Rev.,* **116,** 1505.

Salpeter, E. E., 1952, *Phys. Rev.,* **88,** 547.

Salpeter, E. E., 1952, *Ap. J.,* **115,** 326.

Salpeter, E. E., 1953, *Ann. Rev. Nucl. Sci.,* **2,** 41.

Salpeter, E. E., 1957, *Phys. Rev.,* **107,** 516.

Salpeter, E. E., 1959, *Ap. J.,* **129,** 608; also see 1955, *Ap. J.,* **121,** 161.

Salpeter, E. E., 1954, *Aust. J. Phys.,* **7,** 373; also see Salpeter, E. E., and Van Horn, H. M., 1969, *Ap. J.,* **155,** 183.

Sandage, A. R., and Schwarzschild, M., 1952, *Ap. J.,* **116,** 463.

Schiff, L. I., 1955, *Phys. Rev.,* **98,** 1281.

Tollestrup, A. V., Fowler, W. A., and Lauritsen, C. C., 1949, *Phys. Rev.,* **75,** 1463; 1949, **76,** 428.

Wheeler, J. A., 1941, *Phys. Rev.,* **59,** 108.

Woosley, S. E., and Weaver, T. A., 1982, *Essays in Nuclear Astrophysics,* C. A. Barnes, D. D. Clayton, and D. N. Schramm (eds.), Cambridge University Press, Cambridge.

Woosley, S. E., and Weaver, T. A., 1984, private communication.

2.

Old and New Neutron Stars

MALVIN A. RUDERMAN

Malvin A. Ruderman is a Professor of Physics at Columbia University. From his early days as an elementary particle physicist, he has gradually shifted more and more toward theoretical astrophysics. He has done pioneering work in the areas of neutron star physics and pulsars, X-ray and gamma-ray astronomy, and other areas of high-energy astrophysics.

I. ANCIENT HISTORY[1]

Six thousand years ago a massive star in our Galaxy came to the end of its life in a gigantic explosion. It hurled its debris into the interstellar void at a speed of over 1000 km sec^{-1}. For a week this expanding blast was more luminous than 10^{10} suns and glowed almost as brightly as our entire Galaxy.

For 5000 years the light from the explosion traveled toward our earth. It arrived here on the 4th of July 1054 AD from the direction in the night sky known as Taurus (the Bull). It then appeared brighter than all other stars in the heavens, brighter even than Jupiter and Venus, and quite visible in the daytime. It surely must have been noticed everywhere on earth where that region of the sky was visible.

The Chinese certainly saw it. The Sung emperor's court astrologer reported to him the meaning of this bright new star:

> Prostrating myself before your majesty I hereby report that a guest star has appeared. . . . If one carefully examines its meaning for the Emperor it is as follows: The fact that the guest star does not trespass into the moon's mansion in Taurus and that it is very bright means that there is a person of great wisdom and virtue in the country. I beg that this notice be given over to the Bureau of Historical Records.
>
> *Yang Wei-te*

With this interpretation both the announcement of the new super guest star and the position of the astrologer were preserved. The Japanese independently observed and recorded this bright new star—a "supernova"—at their medieval capital, Kyoto. In all there are at least five independent accounts of it, four from China and one from Japan. Remarkably, there are no European or Russian records. (It has been suggested that the relevant Russian documents were destroyed in the Mongol conquest of Kiev and that European observers in the otherwise Dark Ages did not record a new star that so conflicted with their Aristotelian notions of an unchanging starry heaven. Comets, which they often did record, were thought to be atmospheric rather than heavenly phenomena.)

There is a very intriguing interpretation of certain old American Indian rock drawings that suggests that these may have been records of the supernova observation of 1054 AD. On the morning of July 5, when it was first visible from western North America, the supernova was adjacent to

[1] Detailed discussions and references are given by Mitton (1978), Clark and Stephenson (1977), and Shklovski (1968).

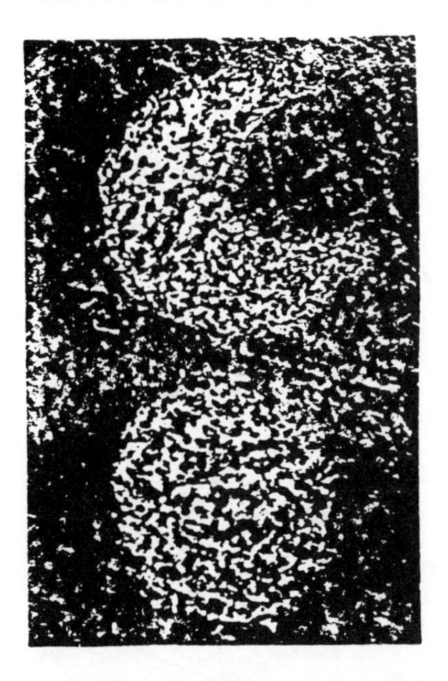

FIGURE 1. Two ancient southwest U.S. petroglyphs that may represent the sudden appearance of the supernova of July 5, 1054 near the crescent moon (Minton 1978; Shklovski 1968).

the crescent moon, a figure of the moon rarely found in ancient American Indian rock art. But several "petroglyphs" in Arizona and New Mexico (Fig. 1) have been discovered that may date from the eleventh century and depict a bright-star crescent-moon juxtaposition.

The exploding guest star of 1054 AD was visible during the day for 23 days, but after two years the supernova could no longer be seen even at night. It disappeared—unseen again until after the invention and development of the telescope in seventeenth- and eighteenth-century Europe. It was first rediscovered in 1731 by an English physician and amateur astronomer, John Bevis, who had an observatory in his house. He produced beautiful copper engraved plates for a planned sky atlas, but his printer went bankrupt as it was being published. Bevis especially noted the exploding supernova remnant (the second "star" in from the top of the Bull's lower horn in Fig. 2) because it appeared as a faint smeared nebulosity rather than as a bright point characteristic of a star. Because the explosion remnant was nebulous instead of pointlike, it could be confused with comets, which at that time were a subject of intense observational interest. The most famous comet of all was Halley's Comet. It

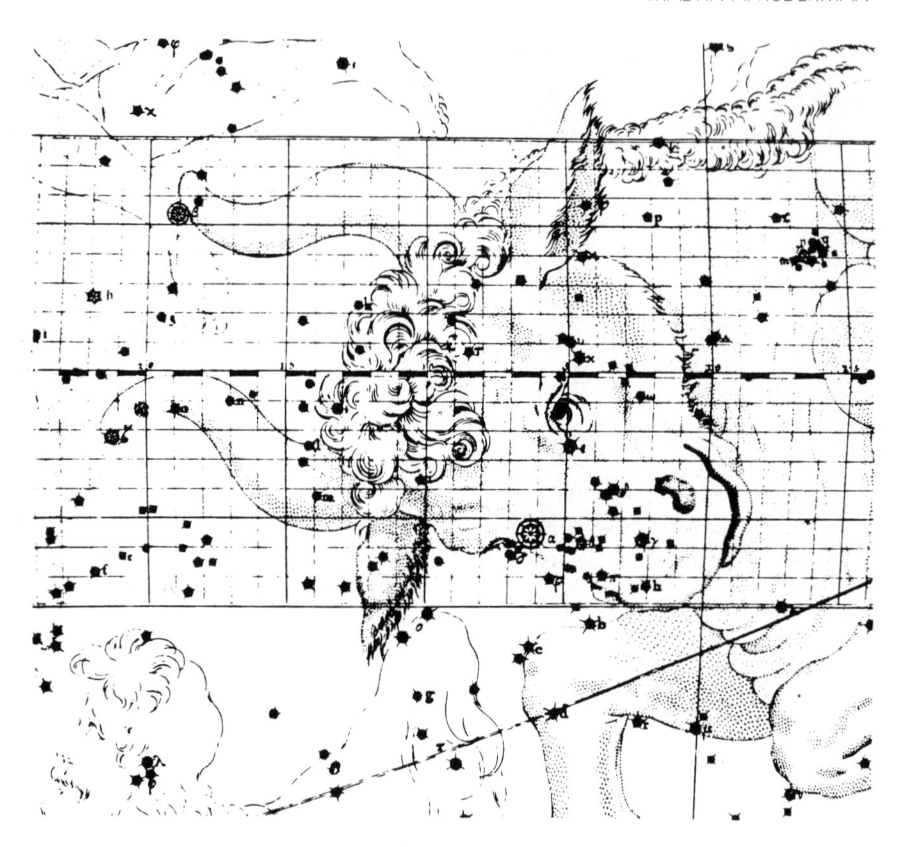

FIGURE 2. The engraving for John Bevis' Sky Atlas, which shows the nebulous supernova remnant as the second star from the tip of the lower horn.

returns every 75 years. (This bright comet was depicted in the Bayeux tapestry representation of William the Conqueror's embarkation for England.) In 1682 Halley observed it. He predicted that the comet would return in 1757, when he expected it to pass through the horns of Taurus. The supernova remnant of 1054 AD first reappeared in astronomical reports when it was misidentified as the missing Halley's comet by the French astronomer Messier in 1758. He later gave it the prime listing in his famous catalogue of nebulous astronomical sources. (The true return of Halley's Comet was in fact first seen four months later than Messier's announcement, not by an astronomer with a telescope but by a French peasant looking heavenward at Christmastime with his naked eye.)

From then on the 1054 supernova remnant became an object frequently observed. The Earl of Rosse observed it through the very large 5-ft-di-

FIGURE 3. Lord Rosse's sketches of the Crab Nebula. The left one was drawn in 1848, the right in 1844.

ameter telescope on his estate in the middle of Ireland. His 1844 and 1848 sketches of it are shown in Fig. 3. It looked like a pineapple in 1844, but he named it, after his 1848 drawing of its shape (which suggested the claw of a Crab), the Crab Nebula. And so it is known today. Rosse and most nineteenth-century astronomers assumed that the nebulosity was an artifact of poor resolution and that the Crab Nebula, like so many others, would ultimately be revealed as a dense cluster of many individual stars.

It was only after the application of photography to astronomical viewing that the nature of the Crab Nebula became clear. That nebula is a cloud of gas 6 light-years across, expanding at 2000 km sec^{-1} with a bright remnant star near the center of the explosion. The light from that star does not look like the light from familiar stars. But this would more easily suggest what the remnant star is not than what it is.

While some scientists looked out at the universe with their telescopes, others explored it only within the confines of their laboratories. In 1932, at the Cavendish Laboratory in Cambridge, England, Chadwick discovered the neutron. The news spread very quickly. On the evening the announcement of that famous discovery reached Copenhagen, Bohr and Landau had a conversation about its implications. Landau, on that same evening, proposed that some stars might be composed almost entirely of neutrons.

In Pasadena, a year later, the astronomers Baade and Zwicky proposed that remnant stars of supernova explosions might be such dense stars of neutrons, and they named them "neutron stars." (The suggestion captured the interest of a local reporter. He thought the Caltech astronomers rather mad. But he gave the proposal considerable play in an article in which he also poked fun at them. So, had you lived in Pasadena 50 years ago, you would already have heard of "neutron stars.") Baade and Minkowski (1942) later explicitly suggested a particular star in the projection of the Crab Nebula supernova explosion as the remnant neutron star.

But how could one verify it? Theorists had estimated the masses and sizes of such stars, but nothing else was known about them. A neutron star should have a mass similar to that of our sun. But most stars have masses in that range and there was no way to weigh the Crab's central star anyway. A neutron star should have a radius of only 10 km, 10^5 times smaller than that of the sun. But whether it was 10^5 smaller or thousands of times larger, the star was so far away that it would still appear as a point to any terrestrial astronomer.

Hoyle, Narlikar, and Wheeler (1964) and Woltjer (1964) suggested that in the collapse to a neutron star, stellar magnetic fields should get hugely amplified. Pacini (1967) argued that a young neutron star formed in that way could also be rapidly rotating and the spinning magnetic dipole could be the power source of the surrounding nebula (see also Kardashev 1965). However, there was no way known at that time to test the neutron star hypothesis. That changed 17 years ago with the discovery of pulsars.

The first spinning radiobeaconed neutron star, a "pulsar," was discovered quite accidentally by Bell and Hewish only in 1968 and immediately interpreted as a spinning neutron star by Gold (1968). Among the earliest pulsars found was the remnant central star of the Crab Nebula. No other interpretation for the object, observed to be pulsing 30 times per second, was reasonable. The star was rotating so rapidly that, in order not to be pulled apart by the familiar centrifugal forces that accompany fast rotation, it would have to have the predicted huge density and small radius of a neutron star. In addition, the input power from the central star needed to account for the emission from the nebula and for the nebula expansion implied that the spinning down central star (Richards and Comella 1969) had a moment of inertia of about 10^{45} g cm^2. For a solar mass star this gives a stellar radius of 10 km, just that which theory insists on for a neutron star (Gold 1969).

Very shortly after the discovery of the radiopulsar in the center of the Crab Nebula it was found that the visible light of that remnant star was also beamed and sweeping across optical telescopes 30 times per second (Staelin and Reifenstein 1968). Apparently, only relatively rare, young, rapidly spinning neutron stars beam strongly enough in visible light to be photographed. (The only other known pulsar—as of 1984—visible by its flashing light, the remnant star in the 10^4-year-old supernova in Vela, is also the weakest optical source ever recorded.)

The Crab Nebula's neutron star is also detected as an emitter of X-rays. Figure 4 is a sketch of the Crab Nebula as seen in X-rays. Its central

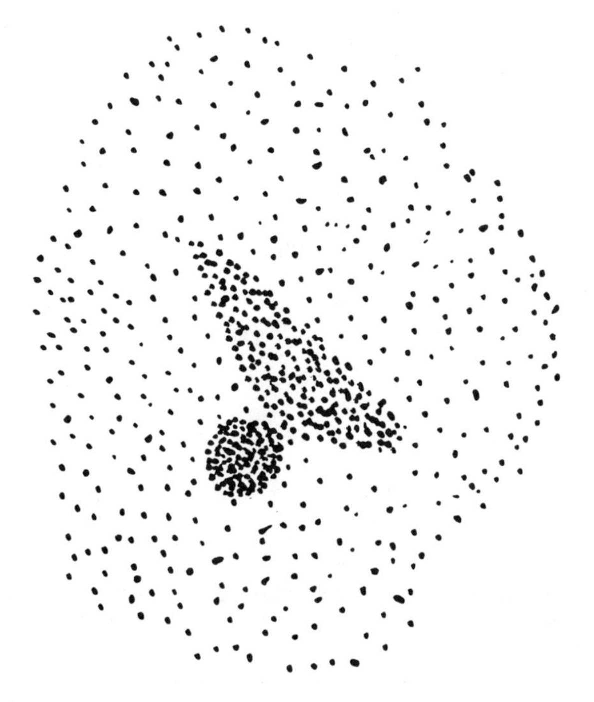

FIGURE 4. A representation of the Crab Nebula and its neutron star as observed in X-rays.

neutron star, although poorly resolved, is now very apparent. As an X-ray source the neutron star is brighter than it is optically, both absolutely and relative to the surrounding expanding nebula. (Had this been our only picture of it, the Crab would probably have been called the Fried Egg Nebula.) As in the optical view of the star the X-ray image flashes on and off 30 times a second as the X-ray beams rotate with the spinning neutron star.

II. YOUNG PULSARS[2]

There are other young supernova remnants, most of whose radiation and expansion are powered by a central, rapidly spinning, strongly magnetized neutron star. One that closely resembles the Crab has been discovered recently in the Large Magellanic Cloud. It has a central X-ray and optical

[2] For reviews and references see Manchester and Taylor (1977) and Michel (1982).

TABLE 1. Properties of the Crab and LMC 0540-69 Pulsars[a,b]

	LMC 0540-69	Crab
Nebula		
Diameter	2 psc	2 psc
Expansion velocity	1600 km sec^{-1}	2000 km sec^{-1}
Mass	$\gg 1\ M_\odot$	$\sim 10\ M_\odot$
X-ray luminosity L_x	9×10^{36} erg sec^{-1}	2×10^{37} erg sec^{-1}
Pulsar		
Period (P)	50 msec	33 msec
Spin-down rate ($\dot{P}$)	480×10^{-15} sec sec^{-1}	430×10^{-15} sec sec^{-1}
Spin-down energy loss rate $\dot{E}$	2×10^{38} erg sec^{-1}	5×10^{38} erg sec^{-1}
Surface magnetic field B	$4 \times 10^{12}\ G$	$3 \times 10^{12}\ G$
"Age" $t \equiv P/\dot{P}$	1660 yr	1200 yr
$L_x/\dot{E}$	0.06	0.05

[a] From Helfand (1984).
[b] The last four lines are inferred from others above. $\dot{E} = (2\pi)^2 I\dot{P}P^{-3}$, where I is the theoretically estimated moment of inertia. B is estimated from the spin-down torque needed to achieve $\dot{E}$ according to Eq. (3).

pulsar (but the X-ray pulse is broad with a large duty cycle, quite different from the double sharply cusped ones of the Crab pulsar). Some properties of the two nebulae and their central pulsars are compared in Table 1.

Helfand (1984) has given arguments that most supernova explosions do not leave the kind of pulsar and nebula now observed in the Crab, a young nebula whose visible and X-ray luminosity are powered by super relativistic electron (and/or positron) fluxes generated by a central rapidly spinning magnetized neutron star:

1. There are fewer than five Crablike nebulae now observable in our Galaxy, suggesting a birth rate of less than 1 per 200 years. The supernova birth rate, on the other hand, is at least several and probably five times greater.
2. Only two of seven historical supernovae produced Crablike nebulae. Of the 36 Galactic supernova remnants within 15,000 light-years searched by the Einstein X-ray observatory satellite, only six give indications of central power sources. The other 30 seem to show only the radiation expected from the expanding shock wave created by the initial supernova explosion, but no activity from a remnant neutron star.

3. In our neighboring galaxy, the Large Magellanic Cloud, only a tenth of the 25 supernova remnants show evidence for a Crablike central source.
4. The expected thermal X-ray flux from hot, newly formed neutron stars has not been detected in most young supernova remnants.

The indicated supernova birth rate for rapidly spinning, highly magnetized neutron stars such as the Crab or Vela pulsars gives a Galactic population of about 10^8 old neutron stars (10^{-3} of all stars in the Galaxy). If, however, most supernovae leave behind either much more slowly spinning neutron stars ($P \gtrsim 0.3$ sec) whose spin-down energy loss could not power Crablike nebulae, or neutron stars whose magnetic fields grow to Crab-pulsar-like strength only after tens of thousands of years, then the total Galactic population of neutron stars formed in supernova explosions could approach 5×10^8.

It is now fairly clear what a young Crab-type remnant neutron star does, but very much less clear how it does it.

The Crab pulsar appears to inject about 10^{38} electrons (e^-) and/or positrons (e^+) per second into the surrounding nebula with typical energies of order 10^{12} eV. A luminosity of 10^{38} particles sec^{-1} is about 10^4 times greater than would be expected to come from the Crab pulsar unless there is very copious e^+–e^- production in its immediate neighborhood. Kennel (1984) and colleagues have shown that an excellent fit to the spectral and spatial features of the nebula's optical and X-ray radiation is indeed obtained from an unseparated e^+–e^- wind of this magnitude and the shock that results from it. There is also evidence that similar relativistic winds come from other young pulsars: X-ray synchrotron nebula are observed around pulsars as long as the neutron star spin-down power ($-I\Omega\dot{\Omega}$, where I is the theoretical neutron star moment of inertia, Ω its spin rate, and $\dot{\Omega}$ its deceleration) exceeds about 10^{34} erg sec^{-1} (Helfand 1982; Cheng and Helfand 1938). (Typical mature pulsars have $I\Omega\dot{\Omega} \sim 10^{32}$ erg sec^{-1}.)

Important clues to the mechanisms by which the Crab pulsar generates the particle input to its superrelativistic wind come from observations of its rotating beams of radiation. These carry away less than 10^{-2} of its spin-down energy loss. They consist of a pair of beams, probably fan shaped, 140° apart in phase, with an observed spectrum containing microwave, IR, optical, soft X-ray, MeV, GeV, and probably 10^{12} eV photons. Coincidences in timing and in beam phase separation seem to imply that all of these photons, with a spectrum extending over 18 orders of magnitude in frequency, come from the same physical location around the pulsar. There are many models (really semiqualitative suggestions),

TABLE 2. Crab Pulsar Versus Electric Power Generator[a]

Property	Crab Neutron Star in Vacuum	Electric Utility Power Generator
P (sec)	3×10^{-2}	2×10^{-2}
R (cm)	10^6	10^2
$\mathbf{B}$ (G)	10^{12}	10^4
$\mathbf{E}$ (V cm^{-1})	10^{12}	10
ΔV (Vs)	10^{18}	10^3

[a] The pulsar electric field and potential drop are estimated for an empty magnetosphere.

but certainly no consensus on how the pulsar accomplishes this. Part of the theoretical problem is the richness of relativistic quantum as well as classical electrodynamic and plasma phenomena that may be relevant. Even at 10^2 stellar radii from the Crab pulsar (near the "light cylinder" radius, $c\Omega^{-1}$, where corotating plasma around the neutron star must spin at the speed of light) the steady power flux away from the star reaches 10^{15} W cm^{-2}—equivalent to that of the total electric power generated on the earth flowing through a hair-thin wire or an entire nuclear war every second in one square meter.

There is, so far, no terrestrial laboratory analogue for a young, spinning, magnetized neutron star: A spinning object will generally break apart when parts of it move with respect to other parts at speeds faster than the sound velocity (v_s) within the object. For strong terrestrial matter $v_s \sim 10^5$ cm sec^{-1}. The parameters, electric fields, and potential drops from a 60-Hz electric power generator are compared in Table 2 to those outside of a Crablike neutron star spinning in a vacuum.

In the presence of such huge electric fields the region outside of the spinning neutron star will become almost completely filled out to the light cylinder with a corotating plasma (the "magnetosphere") flowing to short out components of **E** parallel to **B**. Some of this plasma can be pulled from the stellar surface (or, in one model, from a Keplerian disk around the star). An e^-–e^+ plasma will be created whenever **E** and ΔV are sufficiently large to support the production of γ-rays of high enough energy to materialize as pairs. Such γ-rays can create pairs in three ways:

1. Conversion of photons on strong magnetic fields (Sturrock 1970).

$$\gamma + \mathbf{B} \rightarrow e^+ + e^- + \mathbf{B}$$

This is important for γ-rays in the MeV–GeV range as they begin to cross the strong magnetic field ($\sim 10^{12}$ G) near the neutron star. (For special problems with stronger fields see Shabach and Usar 1982.)

2. Conversion by collisions with X-rays (γ_x).

$$\gamma + \gamma_x \rightarrow e^- + e^+$$

for multi-GeV γ-rays.

3. Conversion by collisions with soft (optical or IR) photons (γ_ν).

$$\gamma + \gamma_\nu \rightarrow e^- + e^+$$

for 10^{12} eV γ-rays.

The needed energetic γ-rays can be radiated (often by the e^-–e^+ particles they themselves create) by three mechanisms:

(*a*) Curvature radiation from accelerated e^- or e^+ constrained to move along curved magnetic field lines.

(*b*) Inverse Compton scattering of extreme relativistic e^- or e^+ on soft photon fluxes that give large local photon number densities.

(*c*) Synchrotron radiation from extreme relativistic electrons spiraling along magnetic field lines.

Because of plasma flow away from the star, large electric fields and potential drops may build up along certain magnetic field lines both near the light cylinder (outer magnetosphere) and much closer to the stellar surface (inner magnetosphere) until pair creation (or Keplerian disk sources) supplies net local charge deficiencies.

A necessary condition for copious outer magnetosphere e^-–e^+ production (by processes 2 and 3) is that electric fields there, in the absence of pair production, are able to accelerate electrons to enough energy to create the needed energetic photons despite radiation reaction or soft photon inverse Compton drag. An apppoximate cutoff for copious outer magnetosphere pair production occurs when even the complete absence of charged plasma on those magnetic field lines that penetrate the light cylinder (it is on these "open" field lines that outflowing plasma must be continually replenished) does not give a potential drop along **B** exceeding about 10^{14} V (Cheng et al. 1984). For a surface magnetic field $B(\equiv 10^{12} B_{12} G)$ this limit corresponds to

$$\frac{B_{12}}{P^2} \sim 20 \tag{1}$$

Pair production in the much stronger magnetic field near the neutron star will no longer be sutained if ΔV along **B** falls below around 10^{12} V (Sturrock 1970; Ruderman and Sutherland 1975), corresponding to the limit

$$\frac{B_{12}}{P^2} \sim 0.2 \tag{2}$$

These limits are simply related to spin-down energy loss rates in conventional pulsar spin-down models by

$$I\Omega\dot{\Omega} = \dot{E} \sim R^6 c^{-3} B^2 \Omega^4 \sim \left(\frac{B_{12}}{P^2}\right)^2 10^{32} \text{ erg sec}^{-1} \tag{3}$$

Copious outer magnetosphere pair production would then not be expected if $\dot{E} < 4 \times 10^{34}$ erg sec^{-1} in agreement with the observation mentioned above, that pulsars with $\dot{E} > 10^{34}$ erg sec^{-1} have synchrotron X-ray halos attributed to a strong outflow from them of extreme relativistic e^- and/or e^+. When $\dot{E} < 4 \cdot 10^{30}$ erg sec^{-1}, even inner magnetosphere pair production is expected to be extinguished. For most theoretical models this would also imply that the spinning neutron stars would no longer be radiopulsars. There is, at present, no consensus that the extraordinarily rich and broad electromagnetic radiation spectra from the young Vela and Crab pulsars is associated with outer magnetosphere pair generation, to say nothing of the precise source location or the specific radiation mechanisms.

As pulsars spin down their radiation, spectra become much more limited, as indicated in Table 3.

The "measured" surface B [from Eq. (3)] is shown for a local population of pulsars in Fig. 5. Because of selection effects and uncertain beaming corrections, it is not possible to read with confidence evolutionary tendencies or pulsar birth rates from these data. (Many, perhaps

TABLE 3. Properties of Spinning Neutron Stars

Pulsar	Age (yr)	Number in Galaxy	Strong Radiation Beams
Crab	930	$\lesssim 10$	RF, optical, X-ray, γ-ray
Vela	10^4	$\lesssim 10^2$	RF, γ-ray
Typical pulsar	10^6	10^4	RF
"Dead" pulsar	10^{10}	10^8	

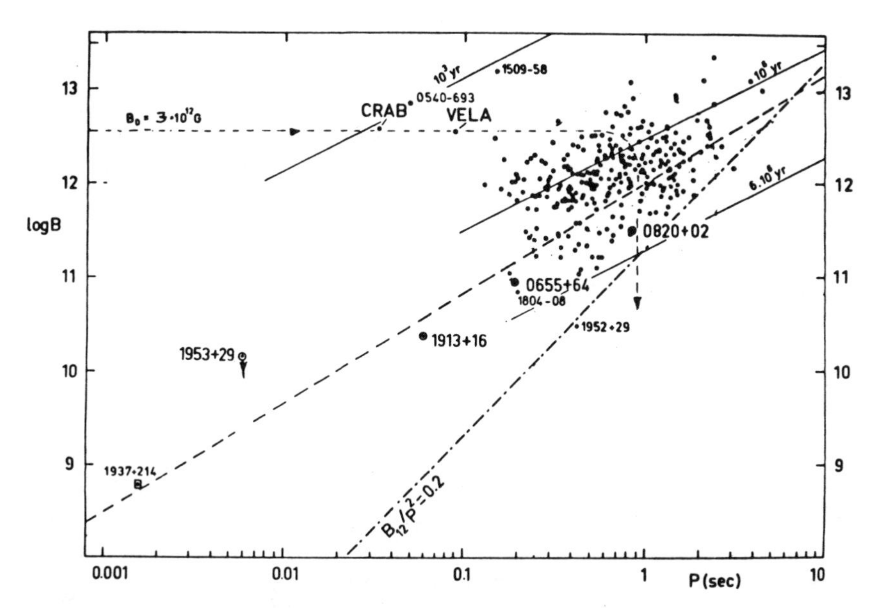

FIGURE 5. The surface (dipole) magnetic fields of pulsars (B) from Eq. (3) versus their periods (P) adapted from van den Heuvel (1984) and Rhadhakrishnan and Srinivasan (1981). An equivalent representation is in Alpar et al. (1982). The four binary pulsars are indicated by circled dots; the isolated millisecond pulsar is shown with a square. The binary 1953 + 29 has only an upper bound for $\dot{P}$ and B (Boriakoff 1984). Spin-down ages $P\dot{P}^{-1}$ in (BP) space are solid diagonals labeled 10^3, 10^6, and 6×10^6 years. The dashed trajectory with $B_0 = 3 \times 10^{12}$ G indicates the model evolution for P and $\dot{P}$ related by Eq. (3) when the neutron star magnetic field decays exponentially on a scale 3×10^6 years. The dash–dot line is the theoretical radiopulsar turnoff of Eq. (2). The dashed diagonal is the theoretical birth curve for pulsars spun up by Eddington limit accretion from a companion according to Eq. (4).

most, may be born spinning much more slowly than the Crab pulsar.) A common interpretation of the distribution of $B \propto (P\dot{P})^{1/2}$ versus P is that the surface magnetic fields of typical neutron stars decay to a much smaller value on a time scale of several million years. This would quickly carry them across the $\Delta V \sim 10^{12}$ V threshold of Eq. (2), after which they would no longer be observable as radiopulsars. Corroborative support comes from the observation that most radiopulsars are moving rapidly ($\sim$180 km sec^{-1}) away from the Galactic plane where they were born, but are still sufficiently close to the plane that their true age since birth must be less than 10^7 years. There is no consensus on how (or even when) pulsar magnetic fields are formed or on why (or even whether) they decay (Flowers and Ruderman 1977; Blandford et al. 1983). However, estimates for the lifetime of currrents in the crystalline crust of a neutron star do suggest that magnetic fields that owe their existence or stability to such currents may decay after several million years. Other field relaxation processes have also been suggested (Chanmugan 1984). Thus observations seem to indicate and current theory is at least comfortable with a population in our Galaxy of $1–5 \times 10^8$ old ($10^7–10^{10}$ years) low-magnetic-field neutron stars.

III. OLD NEUTRON STARS

Several developments in the past few years have served to emphasize that many of these old neutron stars are not only not inert but at times are even much livelier than when they were young.

It has been known since the early days of X-ray astronomy that there exists a population of neutron stars that accrete matter from nearby normal companion stars. These are observed by the thermal kiloelectron-volt X-rays from accretion or from explosive bursts from the sudden fusion of infalling hydrogen- and helium-rich matter. There are probably between 35 and 300 such X-ray binaries in the Galaxy. For most there are indications that the binary is very old and that the neutron star in it has a much lower magnetic field than that of most radiopulsars (too small to affect the infall of matter onto the neutron star or to modulate thermal X-rays from its rotating surface).

Recently, two new classes of apparently old neutron stars have become foci of interest: γ-ray bursters and resurrected pulsars.

A. Resurrection of Pulsars[3]

Figure 5 includes four radiopulsars that are members of binaries. Two of them, PSR 1953+29 and 1913+16 (the Hulse–Taylor neutron star binary), are not in the parameter space defined by other pulsars, and one, PSR 0655+64, lies at its boundary. All these pulsars have the anomalously low surface magnetic fields [as defined by Eq. (3)] seemingly characteristic of "old" neutron stars but the very short periods otherwise found only in young pulsars. They are, indeed, all probably old neutron stars. However, when their companions were younger, there was a transfer of stellar matter from the companions to Keplerian accretion disks around the neutron stars. This surrounding disk could have spun up a neutron star until the neutron star spin velocity and that of its corotating mag-

[3] For reviews and references see Reynolds (1984).

netosphere matched that of the inner edge of the accretion disk (Rhadhakrishnan and Srinivasan 1981; Damashek et al. 1982). While magnetospheric corotation for *isolated* magnetized neutron stars can extend to the light cylinder, for ones *accreting* from a nearby companion the magnetosphere terminates at the inner edge of the accretion disk, where disk inflow pressure balances the neutron's magnetic pressure. Thus the inner disk radius and the ultimate spin-up velocity of the neutron star depend on the accretion rate and the neutron star's magnetic field (Ghosh and Lamb 1974; Davidson and Ostriker 1973; van den Heuvel 1977).

While accretion continued, such binaries would have been observable as X-ray sources. Only after accretion stopped would they be expected to become radiopulsars if (B,P) for the spun-up neutron star again lies above the $\Delta V \sim 10^{12}$V (pair-production) turn-off line. The resurrected pulsar period is

$$P \sim B_8^{6/7} \left(\frac{\dot{M}}{10^{-8} M_\odot \text{ yr}^{-1}} \right)^{-3/7} \text{sec,} \tag{4}$$

where $\dot{M}$ is the mass accretion rate during spin-up. The maximum value for $\dot{M}$ is $\sim 10^{-8}\ M_\odot$ year, the "Eddington limit" at which gravitational pull is balanced by X-ray radiation pressure. This value is approached in many observed accreting X-ray sources. It is used for the otherwise parameter-free theoretical spun-up radiopulsar birth line in the (B,P) space of Fig. 5. The three binary pulsars with measured $\dot{P}$ lie near this birth line, and the deviations are toward that side of the zero-age birth line where reduced $\dot{M}$ and/or postrebirth spin-down would put them. A fourth binary pulsar, PSR 1953+29 (Boriakoff 1984), has only a measured upper bound for $\dot{P}$ and B. The most interesting of the very-short-period pulsars is PSR 1937−214, the first millisecond pulsar (Backer et al. 1982). It too lies on the binary accretion spin-up birthline, which encourages the hypothesis that it was also spun-up by accretion from a companion (Alpar et al.).[4] (A total accreted mass of about 0.1 $M_\odot$ would be needed.) Unlike the other putative spun-up pulsars, PSR 1937−214 does not have a companion. Three kinds of explanations[5] have had numbers of proselytizers:

[4] Like PSR 1953+29 it also lies almost exactly in the Galactic plane, suggesting that both may have been formed not in canonical supernova explosions, but by accretion onto a white dwarf until the dwarf mass exceeded the Chandrasekhar limit. The resulting implosion and mass loss may then have been sufficiently gentle to keep the intact binary from acquiring a large recoil velocity.

[5] For a review see, for example, van den Heuvel (1984) and Ruderman and Shaham (1983).

1. *Parthenogenesis*. There never was another star involved in the birth of the millisecond pulsar; it was formed from the collapse of a single low-magnetic-field high-spin stellar core.
2. *Transfiguration*. The Hulse–Taylor binary pulsar consists of two neutron stars in an orbit that is contracting because of gravitational radiation. In about 4×10^8 years they will coalesce and may leave behind a single hot, high-spin-angular-momentum, low-magnetic-field neutron star (or a black hole).
3. *Resurrection and sacrifice*. The neutron star is spun-up into the radiopulsar regime by an accretion disk that is well fed by mass loss from a companion. The companion is then tidally disrupted by the gravitational field of the neutron star.

When the companion is degenerate (or fully convective) its radius (R_2) generally increases as its mass (M_2) decreases (cf. Fig. 6). Avoiding catastrophic tidal disruption by a heavier neutron star is then not always easy to accomplish if the companion is too close. Tidal disruption seems always to occur to a companion white dwarf with $M_2 > 0.7\ M_\odot$ around a cold $1.4\ M_\odot$ neutron star whenever the binary separation is close enough to allow *any* mass to be pulled from the companion to the neutron star (van den Heuvel 1984). It may also happen to a very light degenerate (H_e) companion when its mass loss to the accreting neutron star brings $M_2 < 4 \times 10^{-3}\ M_\odot$, especially if the tidal disruption time of the secondary (several minutes) is less than the expansion time of the neutron star's accretion disk as matter is rapidly fed into it (Ruderman and Shaham 1983). Catastrophic tidal instability of a light degenerate companion may also be triggered when the binary is perturbed by a passing star (Webbink et al. 1984). Even if only gravitational radiation is removing angular momentum from the binary, this critical stage will be reached in less than the age of the Galaxy.

One suggested scenario (Ruderman and Shaham 1984) for the spin-up of the now isolated millisecond pulsar is that angular momentum was transferred to an old low-magnitude-field neutron star by accretion from a low mass companion as is presently observed in so-called Galactic bulge X-ray binaries. The companion may have begun its life as a light main sequence star burning hydrogen in a central core. When its mass decreased to of order $10^{-1}\ M_\odot$, its core burning would cease, leaving a cooling dwarf star with a very small largely helium core, much denser than the rest of the star. The heavy helium core would not be convectively mixed as the companion cooled to degeneracy. The R_2 versus M_2 behavior of such a star is sketched in Fig. 6. As mass is lost to the neutron star

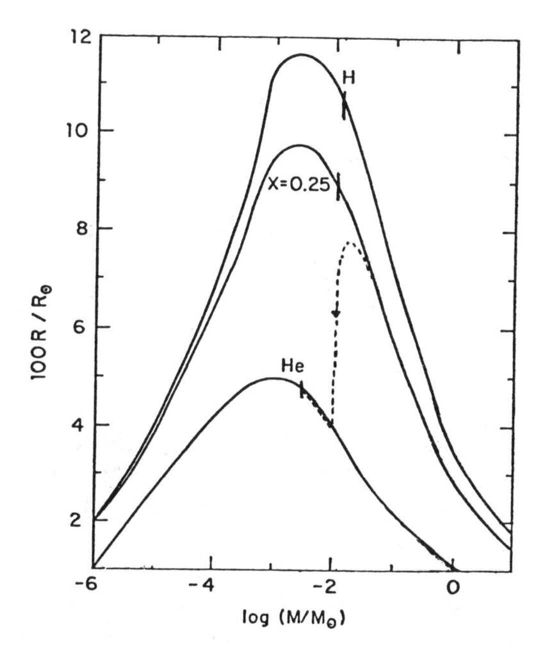

FIGURE 6. Radius (R) versus mass (M) for cold degenerate stars. The upper solid curve is for pure hydrogen; the lower is for pure helium and the $x = 0.25$ is for a cosmological mixture of 25% He and 75% H by mass. Solid-state corrections of Zapolsky and Salpeter (1969) are included. The dashed trajectory indicates the evolution of a mainly H white dwarf with a 10^{-2} $M_\odot$ He core. The vertical bars indicate where the stars will be after 2×10^{10} years when they have 1 $M_\odot$ neutron star companions to which they are continually brought closer (always filling their Roche lobes) by the loss of angular momentum to gravitational radiation.

from the outer hydrogen layers of the inhomogeneous degenerate dwarf, R_2 first grows, then contracts as the He core is approached, and then expands again after the core is revealed. For stable mass transfer the binary separation $\hat{a} \sim R_2(M_{NS}/M_2)^{1/3}$ has a similar behavior. If the He core boundary is idealized as a zero thickness shell, dR_2/dM_2 becomes infinite as the shell is reached. This would mean that $\hat{a}$ shrinks at that stage without any mass transfer; there would then be no accretion disk around the neutron star. The "sudden" resumption of R_2 expansion when the He core is reached, in the absence of an accretion disk to transfer angular momentum to the companion and rapidly expand $\hat{a}$, would then certainly tidally disrupt the so-modeled companion. Another scenario for removing a companion star (van den Heuvel 1984) exploits the tidal instability of a heavy 0.7 $M_\odot$ dwarf companion when it fills its Roche lobe. In either case tidal disruption is invoked after the companion gets "too close for comfort."[6]

A very different possible mechanism for destroying a very light weakly bound companion of the millisecond pulsar is based on the stimulated evaporation of the companion by an intense (approximately 10^{36} erg $\sec^{-1}$) wind of $e^{\pm}$, MeV γ-rays, etc., which carries away the pulsar's enormous but diminishing rotational energy.

B. γ-Ray Bursts[7]

Cosmic bursts of soft γ-rays (typically 10^{-5} erg cm^{-2} of 10 keV–10 MeV X-rays over several seconds) are detected at the earth at a rate of about 1 every 64 hours. They are isotropically distributed and have a number versus intensity distribution that now appears consistent with a homogeneous distribution. Thus the sources may be so close that the disk structure of our Galaxy is irrelevant, or part of a halo population, or extragalactic. The short time scale for intensity variations—typically 10^2 ms but less than 0.3 ms in one case—argues that the spatial size of the radiating region is far less than canonical stellar radii. Recent observations of substantial (nonthermal) 10–40 MeV components in some bursts is rather compelling evidence for the local source hypothesis. An extragalactic source would need an intrinsic X-ray luminosity over 10^4 times larger than a local one to give the same observed bursts. However, the X-ray photon density near a compact extragalactic source would then be so large that the short mean free path for e^{-}–e^{+} production by 10–40 MeV γ-rays would not permit those γ-rays to escape from the source.

Therefore, modelers are comfortable with the existence of 10–40 MeV components to γ-ray bursts from compact sources only when the characteristic distance to them is not greater than a few hundred light-years. A further indication of the source population comes from observational bounds on the repetition rate for bursts from any one source. Observations suggest that the repetition rate does not exceed about 1 per year but that some, perhaps all, burst sources do repeat. [At least one source (an anomalous one called the "March 5 event") and possibly a second (the "40

[6] In this confusion of the biblical story, in the beginning a companion transferred original spin to the neutron star. By this passover it resurrected the dead neutron star into a live radiopulsar, sacrificed itself, and became wholely a ghost.

[7] For reviews and references, see Liang and Petrosian (1985), Hurley (1983), Woosley (1984a,b), Lamb (1984), Strong et al. (1975), and Ruderman (1975).

longitude source'') have been observed to repeat in γ-rays. Other evidence for repetition comes from historical records of optical transients presumed coincident with γ-ray bursts (e.g., Schaefer 1981).] The maximum average repetition rate gives a needed population of γ-ray burst sources of about a hundred within, say, 300 light-years. This number is comparable to the expected number of all ''dead'' old neutron stars in that volume. The needed compactness of the source size and the relatively large (compared with ordinary stellar events) burst energy ($\sim 10^{37}$ ergs) are additional circumstantial evidence pointing to part of the local old neutron star population as the source of γ-ray bursts. Although isolated old neutron stars may occasionally have structural ''glitches,'' these would not be sufficiently frequent or large or release energy rapidly enough to be the γ-ray burst sources. Speculation and modeling have been directed mainly toward old neutron stars in binaries, accreting so slowly from underluminous companions that neither are yet detectable as steady sources. Burst energy would come either from accretion disk instabilities or from subsurface nuclear fusion explosions.

The release of 10^{38} ergs of γ-rays would require an accreted mass of 10^{18} g if all the energy came from gravitational binding (accretion disk instability), 10^{19} g if hydrogen fusion was the source, and 10^{20} g for explosive helium burning. For a nominal 3-year burst repetition interval the explosive fusion models would then imply thermal X-ray sources during the accretion in the range 10^{31}–10^{32} erg sec^{-1}, low enough that they might escape detection as steady X-ray sources. If, as expected, the accretion is through the intermediary of a Keplerian accretion disk around the neutron star, the neutron star's spin will ultimately approach the period P given by Eq. (4). Since $(10^{18}–10^{20})g$ $(3\text{ yr})^{-1} \sim (10^{-16}–10^{-14})$ $M_\odot$ yr^{-1}, $P \sim 3^{\pm 1} B_8^{6/7}$ sec. Thus the old neutron stars, which are the most attractive candidates for γ-ray burst sources, should be rotating with a period that depends on their magnetic fields in a manner similar to that for resurrected pulsars. There are some indications of periodicity in γ-ray burst observations. The 1979 ''March 5 event'' certainty has an 8-sec periodicity (e.g., Barat et al. 1979), but it is a very atypical repeating burstor. Wood et al. (1981) discuss a 4.2-sec period for the October 29, 1977, burst. Barat et al. (1984) consider the plausibility of a 5.7-sec period in the January 13, 1979, burst data. Evans et al. (1980) discuss others in the several second range. If these periods are those of the accretion disk spin-stabilized neutron star that is presumed to be the burst source, they would indicate $B \sim 10^9$ G. This is quite reasonable for an old neutron star, since it is similar to that of those which, with enormously stronger accretion rates, seem to have been spun up to be millisecond pulsars. Moreover, such a weak magnetic field would easily allow 20–40 MeV γ-rays to escape from the neutron star surface without pair creation, much more difficult to accomplish if B had the 10^{12} G which seems typical for younger neutron stars (at least those that are radiopulsars). However, if the magnetic field is so weak it may be harder to see how the rotation of the star would so strongly control the window for observing high-energy radiation ($\hbar eB/mc \sim 10$ eV for $B \sim 10^9$ G). Further, Woosley and others (1984b) (Petrosian and Liang, 1985) have martialed arguments that the qualitative features of γ-ray bursts from thermonuclear explosions are very difficult to interpret unless $B \sim 10^{12}$ G. Finally, a notch in the (time-averaged) burst photon energy spectrum has often been interpreted as evidence for cyclotron absorption (or emission) with B a few times 10^{12} G. Thus, there are also indications that this family of old neutron stars has strong magnetic fields. Much remains to be understood.

IV. EPILOGUE

The following are a list of questions put to the 1984 Aspen Center for Physics workshop "New Directions in Pulsar Physics" by the workshop organizers.[8] Generally, a consensus on answers was not reached, and it is a somewhat idiosyncratic outline of neutron star research remaining to be done.

1. *Magnetic Fields*

Why do radiopulsars have similar magnetic fields at similar "ages"?

Why do millisecond pulsars have such weak fields? Why do Galactic bulge X-ray source neutron stars have such weak fields?

Is B primordial or generated during the life of the neutron star or both?

Do core magnetic fields decay?

2. *Radioemission from Pulsar Magnetospheres*

Are e^+–e^- pairs necessary?

Does pulsar turnoff (on) coincide with pair production turnoff (on)?

Where are pairs made?

Why doesn't backflow heat pulsars so that they are observed as X-ray sources?

Do all radiopulsar magnetospheres also emit γ-ray beams?

[8] C. Kennel, D. Pines, M. Ruderman, and J. Shaham.

Why is only the Crab a strong optical pulsar?
Why are the Vela and Crab pulsars such strong γ-ray sources?
What is the origin of the pulsar radioemission?
How is radiopulse fine structure related to background plasma instabilities and/or large-amplitude wave instability?
What is the origin of the occasional giant radiopulses from the Crab pulsar?

3. *Beyond the Light Cylinder*

What comes out?
How anisotropic is the wind?
Are ions emitted? e^+–e^- pairs?
Is the plasma charge separated or an MHD wind?
How large is acceleration beyond the light cylinder?
Can isolated pulsars produce relativistic jets?
When are synchrotron nebulae formed?

4. *Basic Pulsar Plasma Physics*

How do superrelativistic shocks accelerate particles?
What is the plasma physics of the strong-magnetic-field–strong-radiative-capture regime?
What is the appropriate description of large-amplitude waves and particle acceleration?

5. *Birth and Death of Pulsars*

Are supernovae abundant enough to give the needed birth rate?
Can pulsars be formed "quietly"?
Are pulsar corpses still spinning down?
How might they be detected?
Do dipoles align with the spin axis near pulsar turnoff?
How much of torque decay is from field decay and how much is from alignment?

6. *Neutron Star Structure and Mass*

What are the current handles on stellar radius, internal structure, and mass?
Is there evidence for a range of neutron star masses?
Are masses and red shifts known well enough to pin down the neutron star equation of state?

7. *Neutron Star Cooling*

What role is played by magnetic fields?
What is the current status of the observational constraints?
Is it physically reasonable to argue that the Crab pulsar has an

internal temperature some 25 times that of the Vela pulsar?
Should there be a spin-modulated thermal X-ray pulse from Vela?

8. *Timing Noise, Glitches, and Postglitch Behavior*

Does pulsar timing noise scale with $\dot{P}$? with $\dot{P}/P$?
What kind of noise has thus far been observed?
What are the likely physical origins of timing noise?
Why are the Vela pulsar glitches so different from those observed for the Crab pulsar?
Can one fit all observed postglitch behavior of all pulsars with the pinned vorticity model?

9. *Millisecond Pulsars*

Was the millisecond pulsar 1937 + 214 formed in an isolated event or in a binary?
If it was formed in a binary, what happened to the companion?
Why are the two most rapidly spinning pulsars almost exactly in the Galactic plane?
How old is the millisecond pulsar?
Does it have a synchrotron nebula?

10. *γ-ray Bursts*

Are neutron stars the sources of γ-ray bursts?
Are the stars isolated or in binaries?
Are bursts a result of accretion instabilities or pyconuclear explosions?
Are the stellar magnetic fields large ($B \gtrsim 10^{12}$ G) or small ($B \lesssim 10^{10}$ G)?

11. *Peculiar Objects that May Be Neutron Stars*

Describe the following:
(a) Cyg X-3.
(b) Geminga.
(c) March 5 event.
(d) SS 433.
(e) Nemesis (the proposed companion star to the sun).

12. *Superenergetic γ-rays from Neutron Stars*

Is Cyg X-3 an accreting neutron star and a source of 10^{15} eV γ-rays? How are the γ-rays produced so efficiently?
What is the source of claimed 10^{12} γ-rays from the Crab and Vela pulsars?
What other pulsars should be sources of pulsed γ-rays of energy $\gtrsim 10^{12}$ eV?

13. *Environmental Impacts*
Do pulsars depress the stock market, increase crop yields, and play beautiful music?

ACKNOWLEDGMENTS

I wish to thank Professor Sidney Drell and the Theoretical Group at SLAC for their kind hospitality and the Aspen Physics Center, where this work was begun. The work was supported in part by the Department of Energy, contract DE-AC03-76SF00515, and by the National Science Foundation.

REFERENCES

Alpar, M., Cheng, A., Ruderman, M., and Shaham, J., 1982, *Nature,* **300,** 728.

Baado, W., and Minkowski, R., 1942, *Ap. J.,* **96,** 188, 199.

Backer, D., Kulkarni, S., Heiles, C., David, M., and Gross, W., 1982, *Nature,* **300,** 615.

Barat, C., Hurley, K., Niel, M., Vedrenne, G., Cline, T., Desai, U., Schaeffer, B., Teegarden, B., Evans, W., Fenimore, E., Klebasadel, R., Laros, J., Estulin, I., Zenchenke, V., Kuznetsov, A., Kurt, V., Ilovaisky, S., and Motch, C., 1984, *Ap. J. Lett.,* **286,** L5.

Barat, C., Chambor, G., Hurley, K., Niel, M., Vedrenne, G., Estulin, I. V., Kurt, V. G., and Zenchenko, V. M., 1979, *Astron. Astrophys. Lett.,* **79,** L24.

Blandford, R., Appelgate, J., and Hernquist, L., 1983, *Mon. Nat. Roy. Astron. Soc.,* **204,** 1025.

Boriakoff, V., 1984, in Reynolds, S., *Proceedings of NRAO Conference on Millisecond Pulsars,* Green Bank, 6–8 June 1984.

Chanmugam, G., 1984, in Reynolds, S., *Proceedings of NRAO Conference on Millisecond Pulsars,* Green Bank, 6–8 June 1984.

Cheng, A., and Helfand, D., 1983, *Ap. J.,* **271,** 271.

Cheng, K.-S., Ho, C., and Ruderman, M., 1984, to be published; see also Ho, C., "A Model of Pulsed Radiation from the Crab Pulsar" and Cheng, K.-S., "The Radiation Mechanisms of the Vela Pulsar," Columbia University Department of Physics Ph.D. Theses, 1984.

Clark, D., and Stephenson, R., 1977, *The Historical Supernovae,* Pergamon, Oxford.

Damashek, M., Backus, P., Taylor, J., and Burkhardt, R., 1982, *Ap. J.,* **253,** L57.

Davidson, K., and Ostriker, J., 1973, *Ap. J.,* **179,** 585.

Evans, W., Klebesadel, R., Laros, J., and Terrell, J., 1980, *Nature,* **286,** 784.

Flowers, E., and Ruderman, M., 1977, *Ap. J.,* **215,** 302.

Gnosh, P., and Lamb, F., 1979, *Ap. J.,* **234,** 296.

Gold, T., 1968, *Nature,* **218,** 731.

Gold, T., 1969, *Nature,* **221,** 25.

Helfand, D., 1984, report to the Aspen Physics Center Workshop, "New Directions in Pulsar Physics," August 13–31, 1984.

Helfand, D., 1982, *Supernova Remnants and Their X-ray Emission,* IAU Symposium 101.

Hewish, A., Bell, S., Pilkington, J., Scott, P., and Collins, R., 1968, *Nature,* **217,** 709.

Hoyle, F., Narlikar, J., and Wheller, J., 1964, *Nature,* **203,** 914.

Hurley, K., 1983, *Adv. Space Res.,* **3,** 163, and Proceedings of AIP Conference "Positron and Electron Pairs in Astrophysics," M. Burns, A. Harding, and R. Ramaty (eds.), to be published.

Kardashev, N., 1965, *Soviet Astron. A. J.,* **8,** 643.

Kennel, C., 1984, *Ap. J.* (in press).

Lamb, D., 1984, Proceedings of Eleventh Texas Symposium on Relativistic Astrophysics, Evans, D. (ed.), Ann. NY Acad. Sci., **422,** 237.

Liang, E., and Petrosian, V. (eds.), 1985, Proceedings of Stanford Workshop "γ-Ray and Neutron Star Phenomena," July 31–August 2, 1984, AIP, to be published.

Manchester, R., and Taylor, J., 1977 *Pulsars,* W. H. Freeman, San Francisco.

Michel, R. C., 1982, *Rev. Mod. Phys.,* **54,** 1.

Mitton, S., 1978, *The Crab Nebula,* Scribners, New York.

Pacini, F., 1967, *Nature,* **216,** 567.

Radhakrishnan, V., and Srinivasan, G., 1981, Proceedings of the 2nd Asian-Pacific Regional IAU Meeting, B. Hidayat (ed.) (in press).

Reynolds, S. (ed.), 1984, *Proceedings of NRAO Conference on Millisecond Pulsars,* Green Bank, 6–8 June 1984.

Richards, D., and Comella, J., 1969, *Nature* **222,** 551.

Ruderman, M., and Sutherland, P., 1985, *Ap. J.,* **196,** 51.

Ruderman, M., 1975, Proceedings of Seventh Texas Symposium on Relativistic Astrophysics, P. Bergmann, E. Fenyves, and L. Motz (eds.), *Ann. NY Acad. Sci.,* **262,** 164.

Ruderman, M., and Shaham, J., 1983a, *Nature,* **304,** 425.

Ruderman, M., and Shaham, J., 1983b, *Comments Astrophys.,* **10,** 15.

Ruderman, M., and Shaham, J., 1984, *Ap. J.* (in press).

Schaefer, B., 1981, *Nature,* **294,** 722.

Shabad, A., and Usov, V., 1982, *Nature,* **295,** 215.

Shklovski, I., 1968, *Supernovae,* Wiley, New York.

Staelin, D., and Peifenstein, E., 1968, *Science,* **162,** 1481.

Strong, I., Klebesade, R., Evans, N., 1975, Proceedings of Seventh Texas Symposium on Relativistic Astrophysics, P. Bergmann, E. Fenyves, and L. Motz (eds.), *Ann. NY Acad. Sci.,* **252,** 145.

van den Heuvel, E., 1977, *Ann. New York Acad. Sci.,* **302,** 14.

van den Heuvel, E., 1984, "Models for the Formation of Binary and Millisecond Radio Pulsars," in Reynolds, S., 1984, *Proceedings aof NRAO Conference on Millisecond Pulsars,* Green Bank, 6–8 June 1984.

Webbink, R., Rappaport, S., and Savanije, G., 1984, *Ap. J.* (in press).

Woltjer, L., 1964, *Ap. J.,* **140,** 1309.

Wood, K., Byram, E., Chubb, T., Friedman, H., Meekins, J., Share, G., and Yentis, D., 1981, *Ap. J.,* **247,** 263.

Woosley, S. (ed.), Proceedings of AIP Santa Cruz Summer Workshop "High Energy Transients," July 1983. See especially preprint, Ref. 51.

Woosley, S., 1984b, "The Theory of γ-ray Bursts", preprint for above proceedings.

Zapolsky, H., and Salpeter, E., 1969, *Ap. J.,* **158,** 809.

3.

Supernova Theory

HANS A. BETHE

Hans A. Bethe is the John Wendell Anderson Professor of Physics Emeritus at Cornell University. He has been one of the leading contributors of this century to our understanding of quantum mechanics and its applications to a wide variety of phenomena. He served as the director of the Theoretical Physics Division at Los Alamos during the Manhattan Project, and since that time has devoted considerable effort to the issues of nuclear arms and nuclear energy. He is the recipient of numerous prizes and awards, including the Nobel Prize in 1967 for his work on thermonuclear reactions in the sun.

I. OBSERVATIONS

A. Neutron Star or Black Hole?

According to the standard theory, a massive star ($M > 8M_{\odot}$) at the end of its evolution will undergo gravitational collapse of its center, followed by an outgoing shock that will eject the major part of the star's mass. The central star remaining becomes a neutron star.

Recent observations by Helfand and others (Helfand 1984; Becker 1982) have thrown doubt on the last conclusion. Helfand has observed compact X-ray sources, using the Einstein orbiting observatory. Compact sources exist that are associated with supernova remnants (SNR), such as the Crab, that can be seen optically. Helfand finds that all compact X-ray sources in SNRs are also associated with an observable pulsar. But only about 20% of the SNRs have a compact X-ray source and a pulsar. Assuming that 50% of the supernovae (SN) are Type I (which should not leave a central, residual star), Helfand concludes that only 40% of the SN of Type II leave a neutron star behind. The remaining SN II must then be assumed to leave a black hole. (The statistics unfortunately are rather poor.)

A particularly convincing case is Cas A. This SN, which occurred about 1667 (it was not observed at the time), has a large mass, estimated at about $20M_{\odot}$, and is therefore certainly Type II. But it has neither a compact X-ray source nor a pulsar. (It does have X-ray emission from the shock wave that the SN sent out into the surrounding interstellar gas.)

The close association between X-ray sources and pulsars also affects the statistics. Lyne (1982) made a careful statistical study of the rate of birth of pulsars and of SNs. He concluded that the birth rates are about the same, within statistical error. But he assumed, in accord with other astrophysicists, that only 20% of the pulsars are observable from the Earth, while the beams from the others go in other directions. The association of X-rays with pulsars indicates that we see (essentially) all pulsars. Assuming this, Lyne's statistics show that the birth rate of pulsars is only about 20% that of SNs.

Our later discussion of the theory will show that dying stars of 8 to 10 or 12 $M_{\odot}$ indeed turn into neutron stars, while heavier ones probably do not.

II. INFALL

A. Pre-SN Configuration

Many authors have computed the evolution of massive stars to the pre-SN stage. Figure 1 shows the density distribution according to one such

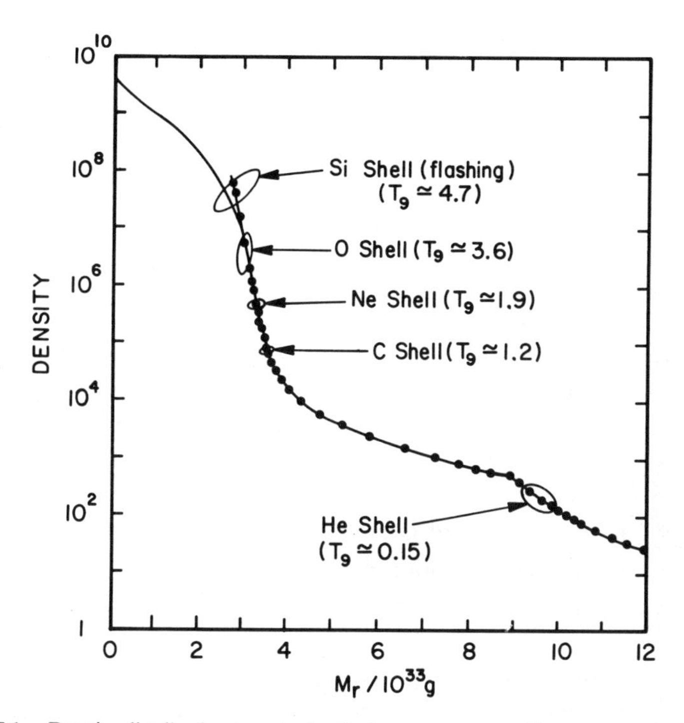

FIGURE 1. Density distribution in a star just before supernova collapse versus the enclosed mass. Shells of nuclear reactions are indicated, including their temperatures. (From Arnett 1977.)

calculation, by Arnett (1977). Other pre-SN results will be discussed in Sections III and IV. They agree substantially on the density distribution.

The density has a maximum at the center of the star and falls off fairly steeply toward the outside. In Fig. 1 are marked the locations where certain nuclei undergo reactions, H, He, C, Ne, O, and Si. In the core, the Si reactions have been completed, and the matter has been converted into Fe and similar nuclei. These can no longer release energy by nuclear reactions and are therefore inert.

The core is then a degenerate gas of electrons (of course, with nuclei interspersed). At low temperature, such a gas is stable as long as its mass is less than the Chandrasekhar limit,

$$M_{\mathrm{Ch}}^{0} = 5.82 Y_e^2 M_\odot \tag{1}$$

where Y_e is the electron fraction (i.e., the number of electrons per nucleon). In the course of a few days, as the Si reactions proceed, the Fe core grows to M_{Ch}. Once this is reached, or more accurately the limit of Eq. (2) below, the core becomes unstable, gravity exceeds the force from electron pressure, and the core collapses.

B. Entropy

The collapse is essentially adiabatic; that is, each material element preserves the entropy it has originally [deviations will be discussed in Section E.]. We generally consider the entropy per nucleon in units of the Boltzmann constant. In the presupernova configuration calculated by Arnett (Fig. 1), the entropy of the material near the center is as follows:

0.30	in the translational motion of nuclei
+ 0.48	in electrons
+ 0.06	in dissociation of nuclei (i.e., α-particles and neutrons)
+ 0.09	in internal excitation of nuclei
= 0.93	total

The total entropy is quite small, less than one unit.

The entropy in electrons, S_e, is especially important because it increases the stable Chandrasekhar mass to

$$M_{\mathrm{Ch}} = M_{\mathrm{Ch}}^{0}\left[1 + \left(\frac{S_e}{\pi Y_e}\right)^2\right] \tag{2}$$

C. Start of Collapse

As the core of the star contracts, its temperature increases by adiabatic compression. The temperature finally becomes high enough for nuclei to dissociate. As Fowler and Hoyle (1964) pointed out, this brings about a really rapid collapse, which takes about a second. In some models, electron capture (Section E) also contributes appreciably.

Either α-particles or neutrons may be released by the nucleus. The amount of dissociation, as a function of temperature and density, can be calculated from the Saha equation. Dissociation takes energy, and the only reservoir of energy is in the electron entropy. Thus, as the temperature increases, the electrons effectively are cooled. According to (2), this decreases the Chandrasekhar mass, there is more imbalance between electron pressure and gravitation, and the collapse becomes rapid. Velocities of the infalling material equal to one-half the free-fall velocity are typical.

D. Equation of State

To describe the infall of the stellar core, it is essential to have a good equation of state (EOS). The EOS of the electrons is of course elementary, the pressure being given by

$$\frac{p}{\rho} = 2.8(\rho_{10} Y_e)^{1/3} \frac{\text{MeV}}{\text{nucleon}} \tag{3}$$

with a small correction for the (now small) entropy in electrons. Equation (3) gives the main contribution to the pressure. Here ρ_{10} is the density in units of 10^{10}g cm^{-3}.

The first satisfactory EOS was derived by Lamb et al. (1978), L^2PR. The material consists of two phases, compact nuclei and a dilute nucleon gas. The latter consists mostly of neutrons, the protons being in the nuclei. The pressure in the two phases is equal, as are the chemical potentials of both neutrons and protons. The nucleons are assumed to interact by a Skyrme force. This gives the correct density (0.16 nucleons per fermi^3) and binding energy (16 MeV per nucleon) for equilibrium nuclear matter. Then statistical mechanics is used to determine all physical quantities as functions of the overall density and the temperature.

Perhaps the most important result is that most of the matter is in complex nuclei as long as the entropy S is less than about 2.7 (per nucleon, in units of k_B); above $S = 2.7$, most of the matter is in free nucleons (in that case, of course, there are many free protons as well as neutrons). When the material density ρ is greater than about $\rho_0 = 0.16$ fm^{-3}, the nuclei form a uniform mass of nuclear matter. A defect of the L^2PR equation of state is that the compression modulus of the nuclear matter is too high, $K \simeq 350$ MeV instead of the (somewhat uncertain) observed value of $K = 220$ MeV. This is due to the use of Skyrme forces; they generally have this property. Probably this is also the cause of some lack of smoothness of the adiabatic index,

$$\Gamma = \left(\frac{\partial \log p}{\partial \log \rho}\right)_S \tag{4}$$

at $\rho < \rho_0$. At $\frac{1}{2}\rho_0 < \rho < \rho_0$, the configuration gets inverted; that is, there is now a continuum of nuclear matter, with bubbles of dilute nucleon gas interspersed.

After several other attempts, Cooperstein (1984) has constructed an EOS which is even more satisfactory than L^2PR. As far as possible, it is semiempirical. The observed compression modulus K is used; the density of states of excited nuclei is expressed in terms of an effective mass m^* of nucleons, and m^* is considered a function of the density ρ, which goes to the limit $0.7m$ for large ρ, where m is the bare nucleon rest mass. This limit is given by nuclear matter theory. Results are checked against the Hartree–Fock calculations by Bonche and Vautherin (1981) at $S = 1$. The transition from nuclei to bubbles is found to be very smooth, while the disappearance of bubbles near ρ_0 gives a mild phase transition. It is essential that the nuclei at $\rho < \rho_0$, as well as the nuclear matter in the bubble phase, have densities $\rho_n < \rho_0$; this serves to decrease the surface energy. The result for Γ, Eq. (4), is very smooth (see Fig. 2), a great advantage for numerical computation of the dynamics of collapse of the star.

E. Electron Capture

With increasing density, the chemical potential (Fermi energy) of electrons increases; it is approximately

$$\mu_e = 11.1(\rho_{10} Y_e)^{1/3} \text{ MeV} \tag{5}$$

As μ_e gets large, it becomes energetically favorable to capture electrons in nuclei. The elementary reaction is

$$p + e^- \rightarrow n + \nu$$

Capture by free protons is relatively fast, but the concentration of protons, X_p, is generally small. (X_p is the number of free protons divided by the total number of nucleons, bound and free.) It is convenient to compare X_p with X_n, the concentration of neutrons,

$$X_p = X_n \exp\left(\frac{-\hat{\mu}}{T}\right) \tag{6}$$

where $\hat{\mu} = \mu_a - \mu_p$ is the difference between the chemical potentials of neutron and proton. Approximately,

$$\hat{\mu} = 190(0.45 - Y_e) \text{ MeV} \tag{7}$$

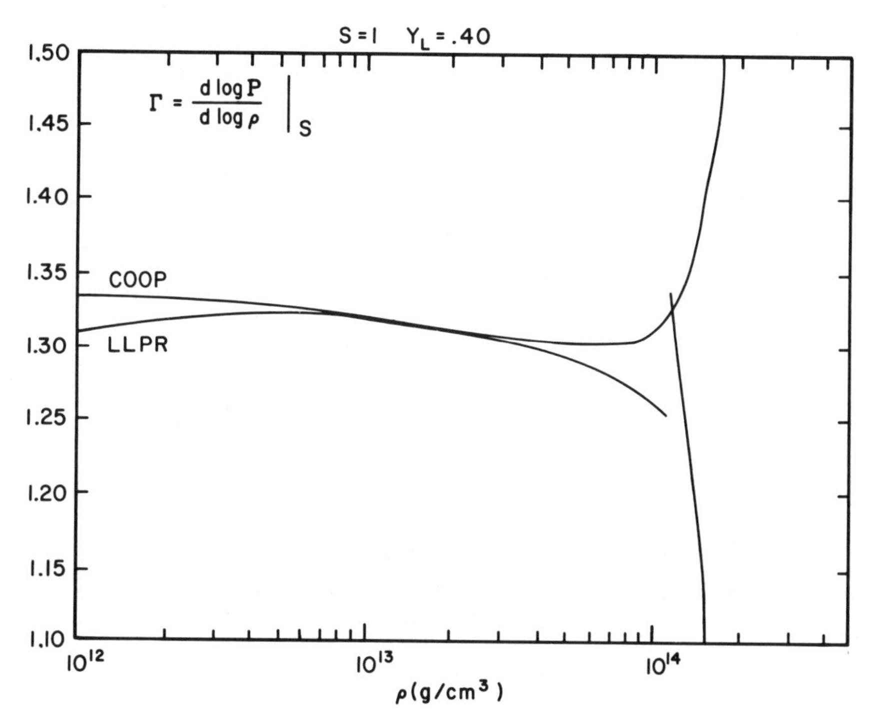

FIGURE 2. The equation of state of Cooperstein (1984) compared with that of Lamb, Lattimer, Pethick, and Ravenhall (1978). Shown is the adiabatic index Γ. In Cooperstein's case, this is very smooth and close to $\frac{4}{3}$, up to a density over 10^{14}. The LLPR equation is similar but has a break when nuclei turn into bubbles, and Γ becomes very small at $\rho \simeq 1.5 \times 10^{14}$. (From Cooperstein 1984.)

and $T \simeq 1$ MeV. The neutron concentration X_n is a few percent. Hence, X_p becomes extremely small once $Y_e < 0.4$, and so therefore does the electron capture by protons. Typically, the center of the star already starts out with an initial value of Y_e given by $Y_{ei} = 0.42$, and there is therefore rather little electron capture by protons. Farther out in the star, the initial Y_{ei} is higher.

Bethe et al. (BBAL 1979) argued that electron capture by complex nuclei should be stronger than that by free protons, because the protons in the $f_{7/2}$ shell of a nucleus like Fe can become neutrons in the $f_{5/2}$ shell by an allowed Gamow–Teller transition. However, Fuller (1982) pointed out that the $f_{5/2}$ shell is full once there are 38 neutrons, which occurs very soon; thereafter, only forbidden transitions can occur, and these tend to be less probable than capture by free protons.

The most complete treatment of this problem has been given by Cooperstein and Wambach (1984). They confirm Fuller's result that electron capture by nuclei is likely to be less important than that by free protons. They find that at a density $\rho = 10^{11}\ \mathrm{g\ cm^{-3}}$, the quantity Y_e has decreased to about 0.41. In more detail, they show that most of the captures in complex nuclei occur due to thermal excitation: Either a proton is in an excited shell or a neutron in a normally full shell is missing. This means that the nucleus, in the β-transition, is apt to gain much less energy than it would at zero temperature, and hence there is only a modest increase of entropy (or even a decrease) associated with electron capture by nuclei. In capture by protons, the entropy actually decreases. According to BBAL, Eq. (B.7), the entropy change is, per capture:

$$\delta S \simeq \frac{\frac{1}{6}\mu_e - \hat{\mu}}{T} \tag{8}$$

At $\rho = 10^{11}\ \mathrm{g\ cm^{-3}}$, $Y_e = 0.41$, we have

$$\mu_e = 17.6\ \mathrm{MeV},\ \hat{\mu} = 7.6\ \mathrm{MeV}$$

therefore, if $T = 1$ MeV,

$$\delta S = -4.7 \tag{8a}$$

per electron capture.

F. Neutrino Trapping

Electron capture would continue indefinitely and would lead to very small values of Y_e if neutrinos could always escape freely. However, because of weak neutral currents, neutrinos are scattered by nucleons; consequently, at high material density they are trapped in the star. The mean free path of neutrinos has been calculated by Lamb and Pethick (1976) and is

$$\lambda = \frac{1.0 \times 10^{10}}{\rho_{10}\epsilon_\nu^2}\left(\frac{N^2}{6A}X_H + 1 - X_H\right)^{-1}\ \mathrm{cm} \tag{9}$$

where ϵ_ν is the neutrino energy in MeV, N is the (average) number of neutrons and A is the mass number of the heavy nuclei present, and

X_H is the mass fraction of the material that is in heavy nuclei. The term $N^2/6A$ represents the coherent scattering by the nucleons in one heavy nucleus; the Weinberg angle has been taken to be $\sin^2 \theta_W = 0.25$, which simplifies the expression.

Trapping occurs when λ is about $R/10$, where R is the distance from the center. In typical collapse calculations,

$$\rho R^3 = C \times 10^{31} \text{ g} \tag{10}$$

where C is usually about 3. The neutrino energy is about

$$\epsilon_\nu \simeq \tfrac{5}{6}\mu_e \tag{11}$$

if the electrons are captured by protons. X_H is near 1. Then the condition for trapping is fulfilled when

$$\rho = \rho_{tr} \simeq 3 \times 10^{11} \text{ g cm}^{-3}; \tag{12}$$

ϵ_ν is then about 15 MeV. This ρ_{tr} has been used when estimating the resulting electron concentration as $Y_e = 0.41$. For higher initial entropy the final Y_e would be lower (e.g., $Y_e \simeq 0.38$ for $S = 1.1$).

After trapping, electron capture continues, but the total lepton concentration remains fixed,

$$Y_L \equiv Y_e + Y_\nu \simeq 0.41 \tag{13}$$

The energy difference,

$$\Delta\mu = \mu_e - \mu_\nu - \hat{\mu} \tag{14}$$

is made into thermal energy when an electron is captured, so that the entropy increases by $\Delta\mu/T$. Ultimately, $\Delta\mu$ goes to 0, and equilibrium is established. At this point, approximately, $Y_\nu = 0.09$, $Y_e = 0.32$.

G. Energy and Entropy

Before collapse, the entropy S is usually about one unit (i.e., k_B per nucleon, cf. Section B). The only change is due to electron capture. Before trapping, if capture is by protons, S decreases, and probably even capture by nuclei does not lead to an increase of S [Sec. (e)]. After trapping of neutrinos, entropy increases by $\Delta\mu/T$ per capture [see Eq. (14)]. The net result is probably that S does not change appreciably. This is important for the energy available to the shock [Sec. 3(b)].

The energy of the stellar core before collapse is close to zero because its radius is large. A small amount of energy is transferred to the core from the mantle by PdV work. The main change of energy is due to the emission of neutrinos. The average energy per neutrino is about $0.4\mu_e$ at the time of trapping, which amounts to about 8 MeV. Thus, the energy loss in neutrinos is about

$$\int dm\, [Y_{ei}(m) - 0.41]\ 8\ \text{MeV} \tag{15}$$

where the integral is over the nucleon number in the core. Depending on the initial Y_e, this is about 0.5 foe, where 1 foe $\equiv 10^{51}$ ergs (*F*ifty-*O*ne *E*rgs).

The observed energy in a supernova is of the order of 1 or a few foe; the total gravitational energy released in the collapse is about 100 foe (see Section V.C).

H. Dynamics

The dynamics of the infall has been computed more than a dozen times. In the Newtonian approximation it is governed by the equation of motion

$$\ddot{r} = -\frac{1}{\rho}\frac{\partial p}{\partial r} - \frac{GM(r)}{r^2} \tag{16}$$

where $M(r)$ is the mass included in the sphere of radius r, and p is obtained from density ρ and temperature T (or entropy S) by means of the equation of state. The result of one dynamic calculation, by Arnett (1977) is given in Fig. 3 which shows the infall velocity u $= -\dot{r}$ as a function of position r, about 2 milliseconds (msec) before complete collapse. The figure also shows the local sound velocity,

$$a = \left(\frac{\gamma p}{\rho}\right)^{1/2}. \tag{17}$$

Clearly, there are two regions: For $r < 40$ km, u is proportional to r. This means this core contracts homologously, the density distribution remaining similar to itself at all stages. In the core, the infall velocity u

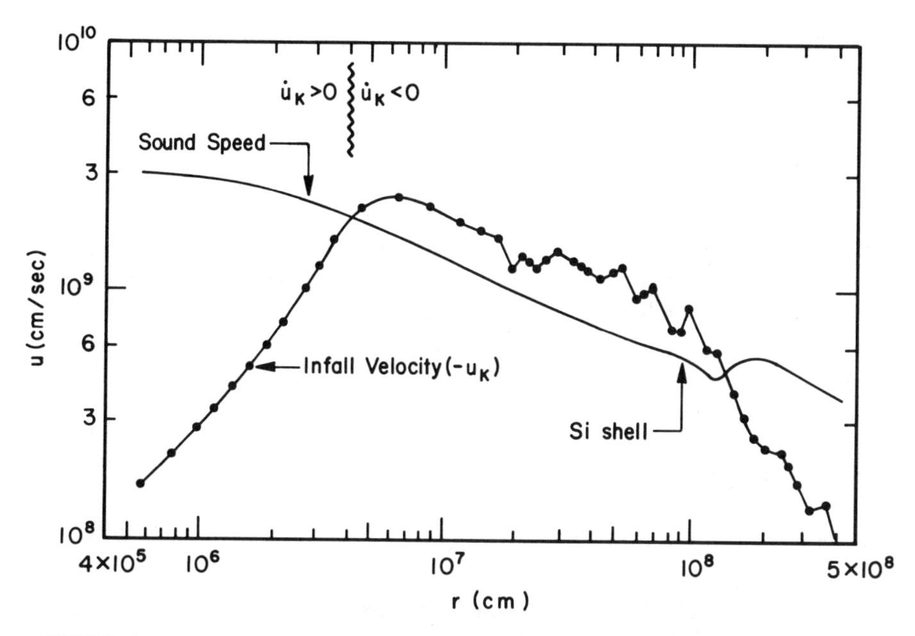

FIGURE 3. A late stage in the collapse. The infall velocity is about proportional to the distance r from the center, for $r < 4 \times 10^6$ cm; this is the homologous core. For larger r the infall velocity is greater than the local sound velocity. (From Arnett 1977.)

is smaller than the local sound velocity, thus the parts of the core can communicate with each other, which makes the homologous contraction possible. For $r > 40$ km, however, $u > a$, the material moves supersonically. The value of u in this region is

$$u \simeq \alpha \left(\frac{GM}{r} \right)^{1/2} \tag{18}$$

where $(GM/r)^{1/2}$ is the free-fall velocity. The gas pressure reduces this by the factor α, which is about $\frac{1}{2}$ in actual computations. At $r \simeq 40$ km, the curves of u and a cross: This is the sonic point. Pressure waves coming from the inside cannot cross this point since they move outward with velocity a relative to the material while the material moves inward with velocity u. Slightly outside the sonic point, the infall velocity u has a maximum.

Goldreich and Weber (1980) have shown analytically that the core should collapse homogeneously. An even more complete analytical solution has been given by Yahil and Lattimer (1982), who showed that all physical quantitites are functions of

$$x = \frac{r^{3/2}}{t} \tag{19}$$

where t is the time before collapse. The exponent $\frac{3}{2}$ is correct only for $\gamma = \frac{4}{3}$; it is different for other γ. This analytical solution is very useful to get approximate values of many physical quantities during collapse, but is not accurate enough to predict success or failure of the subsequent shock.

III. THE SHOCK

A. Start of the Shock

When the density near the center becomes greater than the normal nuclear density, $\rho_0 = 10^{14}$ g cm^{-3}, the nuclear matter becomes almost incompressible. This means that the infall velocity becomes very nearly zero, first at the center, and soon also at larger r. This sudden change of material velocity then generates a pressure wave which moves out with sound velocity. At this pressure wave the density will increase by a fraction.

$$\frac{\Delta\rho}{\rho} = \frac{u}{a} \tag{21}$$

As we know from Fig. 3, the ratio u/a is small near the center of the star but increases with r. Therefore, near the center we get a mild pressure wave which leaves the entropy essentially constant. Thus the core of the star will remain essentially unshocked.

As the pressure wave proceeds to larger r, u/a will increase. This will cause an increase of the entropy which, according to the general theory of shock waves, is

$$\Delta S = \tfrac{1}{12}\gamma(\gamma + 1)\left(\frac{u}{a}\right)^3 \tag{22}$$

This equation shows that the nature of the pressure wave changes when $u/a \approx 1$. As we discussed before, $u/a = 1$ is the sonic point. Once we go beyond the sonic point, the pressure wave becomes a shock which has a substantial discontinuity in density, entropy and velocity.

This shows that the sonic point has an important function beyond the fact that sound waves from the interior cannot move out beyond that point. The sonic point essentially marks the surface of the unshocked core.

In the shock wave the entropy S increases and soon reaches 6–10 units. At $S \approx 2.7$, the equation of state shows that nuclei dissociate into nucleons. This dissociation costs about 9 MeV energy per nucleon, but it clearly increases the pressure. The sudden increase in pressure makes the material in the shocked region reverse its direction of motion and drives it out of the star. The shock therefore is the basis of the supernova phenomenon. The infall compresses the core of the star beyond the equilibrium density ρ_0, the core then bounces back and expels the outside material with high velocity. The energy for this expulsion comes of course from the gravitational energy set free in the collapse.

All numerical computations agree qualitatively with this description. Quantitatively, however, the shocks have very different strengths according to the detailed assumptions made.

B. Sample Calculations of Shocks

Among the many computations of supernova shocks, we select that by Cooperstein, Bethe, and Brown (1984). In these calculations an equation of state of the nuclear material was chosen which we consider particularly reasonable. The initial, presupernova conditions were chosen to be similar to the evolution calculations by Weaver, Woosley, and Fuller (1983), with some modifications that facilitate the shock propagation. Two calculations were made in which the mass of the Fe core equals 1.25 and $1.35 M_\odot$, respectively. The smaller mass led to a successful shock; for the larger mass the shock was unsuccessful.

Figures 4 and 5 give a sequence of snapshots for the two shocks. The abscissa in each case is the enclosed mass in solar masses, $M(r)$. The ordinate is the total energy per unit mass, epm, given by

$$\text{epm} = -\frac{GM(r)}{r} + \epsilon_i \tag{23}$$

expressible in MeV/nucleon using,

$$1\ \text{MeV/nucleon} = 0.96 \times 10^{18}\ \text{erg g}^{-1} \tag{24}$$

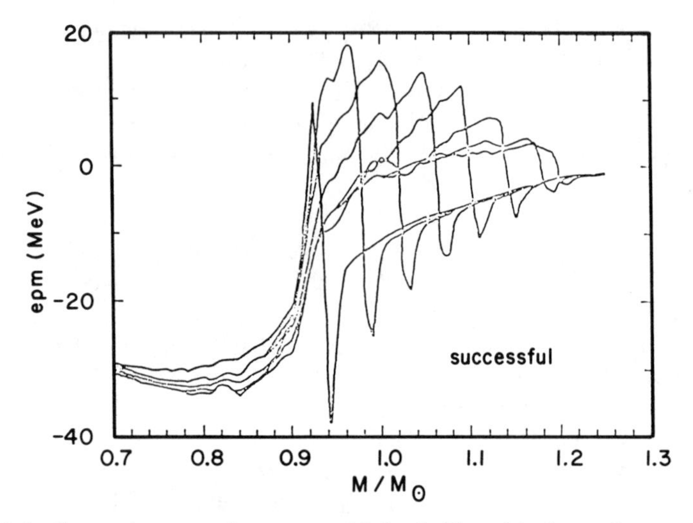

FIGURE 4. Successive stages in a successful shock. Plotted is the total energy per unit mass (epm), in megaelectron volts per nucleon. Positive epm means that the material could escape from the gravitational attraction of the star if no other forces were acting on it. The sharp drop in epm is the shock front. The earliest time is represented by the curve farthest to the left. (From Cooperstein 1984.)

and ϵ_i is the internal energy, including dissociation energy, per unit mass. The various curves correspond to successive times in the shock propagation, the left-most curve being the earliest time.

It will be noticed that in the successful shock, epm remains positive at all times. At the latest time, the shock has gone almost to the surface of the Fe core. Once it goes beyond that surface, it meets Si and soon afterwards O^{16}; once O^{16} nuclei are hit by the shock, they are sufficiently heated to undergo nuclear reactions in which extra energy is released. This energy is added to the shock energy; therefore, when the shock has penetrated into the Si region, it will almost surely be successful.

Beyond the mass region in which epm > 0, the epm falls abruptly to a negative value. This is the region outside the shock in which the negative gravitational energy dominates. If we wish to find the total energy in the shock, it is reasonable to integrate the epm from the point where it becomes positive to $M = \infty$. The region outside the shock has to be included in this integration because as the shock moves on, it sweeps up these regions of negative energy.

In the unsuccessful shock, once the shock has moved beyond $M/M_\odot = 1.1$, epm becomes negative. In this case, therefore, the shock gets stuck at about $M/M_\odot = 1.1$. Any further infalling material will simply

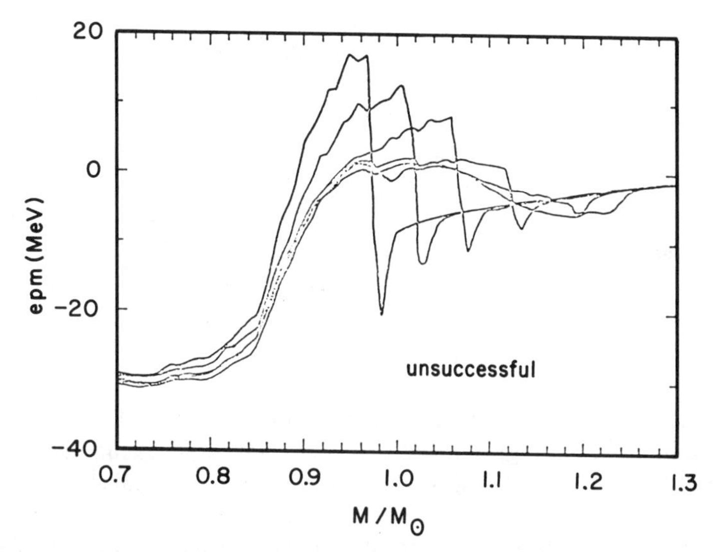

FIGURE 5. Same as Figure 4, but for an unsuccessful shock, with an Fe core mass of $1.35M_\odot$. In the later curves (right-hand side), epm is negative *inside* the shock. ((rom Cooperstein 1984.)

accrete to the existing core. If nothing else happens, the entire mass of the original star will accrete to the central core. This would thereby get a very large mass, and become a black hole. However, in Section V we show that even in this case we are likely to obtain a supernova.

As already mentioned, many calculations have been made of the supernova shock. The results vary substantially, because of different initial conditions, because of different equations of state and because of different treatment of the neutrino diffusion. Nearly all of them follow the same qualitative behavior and for most of them the shock gets stuck at some value of M. In the following, we shall try to analyze the causes of success or failure.

C. The Net Ram Pressure

A very useful basis of discussion is the Virial theorem,

$$4\pi pR^3 - E(R) = \int_0^R \left(3\frac{p}{\rho} - \epsilon_i\right) dm - \int r\ddot{r}\, dm \qquad (25)$$

Here ϵ_i is the internal energy per unit mass, and $E(R)$ the total energy included in the sphere R. The integral goes over elements of mass; r is the position and $\ddot{r}$ the acceleration of mass element m. The theorem is correct for any upper limit R, or $M(R)$. The theorem is particularly simple if the matter is stationary, that is, for $\ddot{r} = 0$.

The theorem simplifies further if we assume that there are two inner regions of the star in which the adiabatic index γ is constant. In particular, we assume that the central core has matter of density $\rho > \rho_0$ so that its equation of state is that of dense nuclear matter. We assume there

$$p \sim \rho^{\gamma_N}, \ \gamma_N = 3 \tag{26}$$

This dense center is assumed to extend out to $r = R_1$. It is surrounded by a medium-dense region in which the pressure is essentially given by relativistic leptons, predominantly electrons. Here we take $\gamma = \frac{4}{3}$.

Because each region has a constant γ, it is a polytrope. Regardless of the values of γ, we have the polytrope equations which are similar to the Virial theorem,

$$4\pi p(R_1)R_1^3 = V_N + 3W_N \tag{27}$$

$$4\pi p_2 R_2^3 - 4\pi p_1 R_1^3 = V_e + 3W_e \tag{28}$$

$$W = \int (p/\rho)\, dM \tag{29}$$

$$V = -G \int (M/r)\, dM \tag{30}$$

where the subscript e (for electron) refers to the outer polytrope. M is always the enclosed mass, V is seen to be the gravitational potential energy, and W is related to the internal energy. Because the inner region is nearly incompressible, so that it has nearly constant density, we have very accurately

$$V_N = -\frac{3}{5}\frac{GM_N^2}{R_1} \tag{31}$$

For a polytrope one can show

$$W + 2\frac{\gamma - 1}{\gamma} V = M\left(\frac{p}{\rho} - G\frac{M}{r}\right) \tag{32}$$

Using this, and using the fact that the outer polytrope has $\gamma_e = \frac{4}{3}$ one can show

$$4\pi p_2 R_2^3 - E(R_2) = -V_N \frac{3\gamma_N - 4}{3\gamma_N} - E_x \qquad (33)$$

Here we have inserted the quantity E_x which represents the deviation of the outer polytrope from $\gamma = \frac{4}{3}$, that is, the fact that the internal energy is actually

$$\epsilon_i = 3\frac{p}{\rho} + \epsilon_x \qquad (34)$$

where ϵ_x is mainly due to excitation energy of the nuclei, and

$$E_x = \int_{R_1}^{R_2} \epsilon_x \, dm \qquad (35)$$

Unfortunately, so far it is not possible to calculate the mass in the dense center M_N from first principles; it has to be obtained from numerical computation. The computations by Cooperstein give

$$M_N = 0.27M_\odot \qquad (36)$$

Using this, we get

$$NR \equiv 4\pi p_2 R_2^3 - E_2 = 8.0\text{foe} - E_x \qquad (37)$$

The pressure p_2 serves to drive the shock outward. The pR^3 term also serves to oppose the ram pressure due to infalling material. Because of this, we have called the quantity on the left-hand side of (37) the "Net Ram Pressure." As can be seen from the derivation, the main contribution to Net Ram comes from the dense core of the star in which

$$3\frac{p}{\rho} - \epsilon_i \gg 0 \qquad (38)$$

The significance of the quantity in (38) is best seen from the Virial theorem, (25).

The shock will be successful if p_2 is large. The 8 foe in Eq. (37) are clearly very helpful. The energy in nuclear excitation, E_x, is harmful. It can easily be shown that

$$E_x \approx 2.6\ S_x^2 \tag{39}$$

where S_x is the average entropy in nuclear excitation in the core of the star outside the very dense center.

As the shock wave moves out, the Net Ram decreases because the integrand in the first integral of (25) is generally negative; this will be discussed in the next section.

D. Dissocation

As the shock moves out, the nuclei are dissociated. As we previously mentioned, nuclear matter is mainly in the form of complex nuclei for $S < 3$, but is dissociated into nucleons when S is higher. We have stated previously that S is commonly between 6 and 10 in the shock, while it is of the order of 1 initially. Thus, as the shock enters new material, its nuclei will be dissociated.

Full dissociation of Fe into nucleons requires an energy of 8.7 MeV per nucleon. The shock-liberated nucleons will exert pressure; since they are nonrelativistic, and nondegenerate

$$\frac{p_N}{\rho} = \tfrac{2}{3}(\epsilon_{iN} - D), \tag{40}$$

where the subscript N denotes the contribution from nucleons and D denotes the dissociation energy. Hence, the contribution of nucleons to the integrand in (25) is

$$3\frac{p_N}{\rho} - \epsilon_{iN} = (\epsilon_{iN} - D) - D = \tfrac{3}{2}T - D, \tag{41}$$

where $\frac{3}{2}T$ is the thermal energy per nucleon in units of k_B. The contribution of electrons is clearly zero. Equation (41) shows that the nucleon contribution is positive if the temperature $T > 6$ MeV but is negative if the reverse is true. In the very inner part of the shock, where the entropy increases from 2.7 to its final value, T is commonly about 6 MeV so that this inner part contributes very little to the Virial theorem (25). Farther out, however, once the entropy has settled down to its final value in the shock, the integrand (41) is generally negative. Thus, the Net Ram gets eroded by dissociation.

It is more convenient and more accurate to discuss the effect of dissociation by using the Net Ram, rather than by explicitly considering the shock energy and subtracting the dissociation energy. Using the Net Ram, one deals with smaller quantities, so that the difference in (41) is more accurate. The fact remains, however, that (41) is negative in most of the shocked region. This continues until the temperature has decreased to about 1.4 MeV; at this point, dissociation is no longer into nucleons, but stops at α-particles (and neutrons). In this case, the dissociation energy goes down to $D = 2.22$ MeV/nucleon. The p/ρ also decreases, to one-fourth of its value for nucleons, but the main effect is that the negative integrand in (41) is much diminished. Dissociation from this point on is much less damaging to the shock.

We may distinguish five regions in the star, as follows:

1. The dense core ($\rho > \rho_0$). This gives a large positive contribution to the Net Ram.
2. The unshocked core ($\rho < \rho_0$, $S < 2.7$). In this region we have nuclei, not nucleons. The pressure comes mainly from electrons. Losses of Net Ram occur (a) from excitation of nuclei and (b) from the fact that the nuclear contribution to pressure is negative, mainly due to the Coulomb energy.
3. The inner shock ($2.7 < S < 8$). Here we have nucleons. The extra pressure from the nucleons is balanced by the dissociation energy so that the contribution to (41) is nearly 0.
4. The main shock ($S \approx 8$, $\ddot{r} = 0$). There is a loss because $T < 6$ MeV. In this region, the gravitational potential energy and the internal energy nearly compensate.
5. The shock front. Here $\ddot{r} < 0$ because behind the shock the material velocity, which is strongly positive at the shock front itself, settles down to about zero. This region gives a positive contribution to the Net Ram in (25). It is somewhat difficult to get a reliable calculation of this contribution. (In the successful calculation by Cooperstein, the contributions are about as given in Table 1.)

The quantity we really want to know is pR^3. For this we need to know also the enclosed energy. As previously stated, the star energy starts from zero but decreases because neutrinos are lost. This will be discussed in the next section, with the result $E = -1.8$. This gives, then,

$$4\pi pR^3 = 0.4 \tag{42}$$

This quantity is barely positive, so that the shock is just barely successful.

TABLE 1. $4\pi pR^3 - E \equiv$ "Net Ram Pressure" (in foe)

1. From dense core	+8.0	
2. (a) Excitation	−2.8	
(b) Nuclear pressure	−1.3	3.9
3. Inner shock	0	
4. Main shock (dissociation)	−1.7	2.2
At R_B: enclosed energy (neutrino losses)		−1.8
Ram pressure $4\pi pR^3$		0.4

E. Electron Capture and Neutrino Loss

It was shown in Section II.F that neutrinos are trapped once the density becomes greater than about 10^{11} g cm^{-3}. The shock, however, moves out to regions of decreasing density. Once the shock has progressed to a place where the unshocked density is about 10^{11} g cm^{-3}, the neutrinos caught in the shock can escape. The criterion is that escape occurs outside radius R_ν defined by

$$\int_{R_\nu}^{\infty} \frac{dr}{\lambda_\nu} \simeq \frac{2}{3}, \tag{43}$$

where λ_ν is the mean free path for neutrinos. It happens that this criterion is fulfilled just about when the density outside the shock is 10^{11} g cm^{-3}. Generally, we shall call the point in the star where (43) is satisfied the neutrinosphere. Once the shock has progressed beyond the neutrinosphere, neutrinos caught inside R_ν can be emitted. Of course these neutrinos must first make their way through the dense material inside R_ν to the neutrinosphere, (see Bethe 1982).

Once the shock is beyond R_ν, neutrinos that are produced behind the shock and outside R_ν can escape without appreciable hindrance. The shock, as we have discussed, dissociates complex nuclei into nucleons. The electrons that are present in this material can now easily be captured by the protons. Therefore, the region outside R_ν will be a strong source of neutrinos. The electron capture, and hence the source of neutrinos, is proportional to the material density. It is negligibly small for $\rho < 10^{10}$ g cm^{-3} but strong between this density and that at the neutrinosphere. This process of electron capture and neutrino escape from the region outside R_ν has been studied particularly by Burrows and Mazurek (1982). The effect of this depends chiefly on the speed of the shock. At a shock velocity of 10^9 cm/sec, the energy loss is of the order $\frac{1}{2}$ foe. For higher shock velocity, the loss is smaller, but for lower shock velocity it becomes very large. Thus, if the shock is apt to fail, this condition is aggravated by the electron capture and subsequent neutrino emission in this region of relatively low density.

F. Criterion for Success of the Shock

As has been discussed in Section B, the shock is likely to be successful if the difference

$$\Delta M = M_{\text{init}} - M_{\text{fin}} \tag{44}$$

is small. Here M_{init} is the mass of the initial degenerate core before collapse; M_{fin} is the mass of the unshocked core. The difference (44) is the amount of material that the shock has to traverse. In doing so, it loses energy or Net Ram pressure because of dissociation of the material. It is therefore desirable to have a small initial core mass and a large final one. This criterion was already recognized by Arnett and Hillebrandt in 1981.

M_{fin} is essentially the mass of a Chandrasekhar core for the electron fraction Y_e which remains after collapse. As we discussed in Section II.E, with the best estimates of electron capture, the final lepton concentration Y_L is about 0.41. Taking into account that at the end of collapse the fraction of neutrinos $Y_\nu = 0.09$, one obtains

$$M_{\text{fin}} \approx 0.8 M_\odot \tag{45}$$

M_{init} has to be deduced from calculations of the evolution of the star before it becomes a supernova. This is a long and complicated calculation. Results vary according to the assumptions made. Thus, Weaver, Zimmerman, and Woosley (1978) found $M_{\text{init}} \approx 1.5 M_\odot$. On the basis of the calculations of Cooperstein, and of others, it is very unlikely that such an initial mass will lead to a successful shock. Taking into account some additional nuclear reactions in the evolution calculation, Weaver, Woosley, and Fuller (1983) found $M_{\text{init}} \approx 1.35 M_\odot$. The distribution of entropy in this calculation is still such that Cooperstein's calculations would not give a successful shock. However, this M_{init} is marginal, and improvements in the computation may easily make the shock successful. Such improvements are presently under way.

In their more recent calculation of evolution, Woosley and Weaver find $M_{\text{init}} \approx 1.6 M_\odot$ or possibly more, for a star of total mass $26 M_\odot$. It seems clear that for such a star the initial shock cannot be successful. However,

Woosley points out that stars of total mass $15M_\odot$ may yield a presupernova core mass $M_{init} = 1.4M_\odot$. The question therefore remains open up to what mass the initial shock can be successful.

IV. SMALL STARS

A. Evolution of Stars of 8–11 $M_\odot$

Nomoto (1984) has investigated the evolution of stars of $8\text{–}11M_\odot$. Stars of $M < 8M_\odot$ are believed to eject a large part of their material and then to become white dwarfs [for a review, see Trimble (1982)]. Stars of $M > 8M_\odot$ burn the carbon in their core, so that the predominant elements in the core are O, Ne, and Mg.

Nomoto shows that if $M < 10\ M_\odot$, the Ne in the core does not ignite. Ignition of Ne requires a temperature $T > 1.2 \times 10^9$ K. The point of highest temperature is located not at the center, but fairly far out in the star. Calculations of evolution show that the ignition temperature can be reached only if the point of maximum temperature lies at $M(r) > 1.37\ M_\odot$. Stars whose He core has a mass of less than $2.5M_\odot$ do not satisfy this condition, whereas $M_{He} = 2.6M_\odot$ does satisfy it. Therefore, the $2.6M_\odot$ star will ignite Ne, but the 2.4 or $2.2M_\odot$ star will not.

Ne is the element in which the nuclear reaction can be ignited most easily; O and Mg require higher temperatures but will undergo nuclear reactions once Ne is burning. The masses are given purposely in terms of the mass of the He core because the surrounding H envelope can be partially expelled by radiation pressure during the evolution. Nomoto assumes that the original mass of the star while it is in the main sequence is four times the mass of the He core and therefore concludes that the maximum main sequence mass that will avoid Ne ignition is $10M_\odot$. More recent work by Woosley et al. (private communication) indicates that the ratio of main sequence mass to He core mass is likely to be 4.5 rather than 4, which would raise the upper limit of this group of stars to $11M_\odot$.

If Ne is not ignited, the core of the star will gradually contract, without getting additional energy from nuclear reactions. After a while, the electrons in the core will become degenerate; for $M_{He} = 2.4M_\odot$ this happens about 100 years before the star becomes a supernova. We then have a degenerate core containing O, Ne, and Mg.

The exact masses are not very reliable, so the upper limit of mass of this type of star may be 10, 11, or even $12M_\odot$. Since the abundance of stars falls rapidly with increasing mass, the range from 8 to $12M_\odot$ contains about half of all the stars with $M > 8M_\odot$.

As we shall see in the next section, it is essential that the core of these stars contain nuclei that still can release some energy upon further synthesis. Perhaps there is still another class of stars, where the degenerate core contains Si; this would then extend the mass range to somewhat higher masses.

B. Collapse and Shock

Stars in this intermediate mass range become supernovae by a rather intricate mechanism that has been proposed and investigated by Hillebrandt et al. (1983). The sequence of events, while complicated, is quite logical.

1. First, electrons are captured by Mg and Ne. This goes relatively slowly because rather high energy is required to change a proton in these nuclei into a neutron. As the electrons are captured, the core contracts because the electron pressure is diminished. The material therefore is heated adiabatically, especially near the center.

2. The temperature gets high enough for the nuclear reaction O + O to be ignited. This happens at a central density of 2.5×10^{10} g cm^{-3}. The reaction sets free some energy, but this energy is small compared with the internal energy of the material. Therefore, no detonation wave will occur, but merely slow burning, deflagration. The deflagration front moves fairly slowly outward in the material, but since the material is generally moving inward, the front stays at about a constant radius, 170 km. The deflagration reduces the infall velocity of the material to about 70% and reduces its density because of the higher temperature. Both of these effects will be important for the shock (item 6 below).

3. The material after deflagration is matter in nuclear statistical equilibrium (NSE). It is at a temperature 1.3 MeV. This material then contains a fair fraction of free protons, which greatly accelerates the electron capture. This phase begins when the central density is about 2.7×10^{11}. The entropy is about 1 k/nucleon, and the electron concentration has gone down to $Y_e = 0.36$ at the center. Because of the faster electron capture, the contraction now proceeds rather rapidly, and this phase lasts only for about 20 milliseconds (msec).

4. The Chandrasekhar mass becomes less than the mass of the O + Ne + Mg core. At this point, the collapse becomes very rapid; it takes about 4 msec. As in the general theory of supernova collapse, the center reaches nuclear density, $\rho = 3 \times 10^{14}$ g cm^{-3}. The core mass at this point is about $1.1\ M_\odot$.

5. A bounce occurs, as in the general theory. The shock forms at 0.63 $M_\odot$ while in the usual theory discussed in Section III.B it formed at $0.8M_\odot$ The difference is due to the smaller electron concentration; in our present case, it is $Y_L = 0.35$ rather than 0.4.

6. A shock is formed and moves outward. Now the previous deflagration of O is a great advantage because the material ahead of the shock has lower density and smaller infall velocity than it would have had with deflagration. The best quantity to consider is the ram pressure, ρu^2, of the infalling material. This ram pressure is four to ten times smaller because of the deflagration. Therefore, the shock does not lose strength as it goes outward; its velocity remains at about 5×10^9 cm sec^{-1}, very high for any shock.

7. The model has the further characteristic that the density decreases drastically at the surface of the core. It drops from about 10^8 to 10^{-4} g cm^{-3} at the surface of the core, going to the He mantle. The shock, coming through this enormous decrease of density, is greatly accelerated.

At an enclosed mass of about $1.2M_\odot$ the material behind the shock reaches escape velocity and is therefore likely to be ejected. The matter up to $M = 1.2M_\odot$ remains behind; it is safely below the stability limit for a neutron star.

This mechanism therefore leads to a supernova phenomenon and leaves a neutron star as a remnant. It may be the predominant way by which neutron stars are formed in the supernova process.

C. Discussion

The details of the supernova theory outlined in Section B may be subject to some change, in particular the numbers for times, densities, and masses. However, the mass of the residual star is likely to be fairly reliable. It is determined by the Chandrasekhar mass for the particular distribution of Y_e in the star. Since electron capture initiates the entire process, it is logical that the final Y_L and Y_e must be small. Hence, the Chandrasekhar mass also will be small.

The resulting mass of the neutron star is rather low. Most observed neutron star masses are about $1.4M_\odot$, and these are gravitational masses, whereas the $1.2M_\odot$ derived in Section B is the rest mass of the baryons in the star. The baryon mass for the observed neutron stars is about 10% greater than the gravitational mass, and is thus $1.55M_\odot$.

Hillebrandt et al. mention the possibility that some of the mass initially escaping may fall back because of rarefaction. We do not believe this is likely: Because of the very sharp decrease of density, outgoing material is accelerating rather than slowing down.

The theory of Section B is likely to explain such neutron stars as that in the Crab Nebula. The measurements of masses have all been done in binaries, and it is not clear that the Crab pulsar also has a mass of $1.55M_\odot$. The small mass of the residual star is a slight defect of the Hillebrandt theory.

Another important feature of supernovae is the synthesis of nuclei. It is generally believed that most of the nuclei we see in the world were originally produced in heavy stars and ejected in supernovae. The nuclei synthesized in the supernova described in Section B have a rather anomalous distribution. In the core, the synthesis is likely to go all the way to Fe or Ni because the shock is very strong and the temperature correspondingly high. There are very few nuclei of intermediate mass. This seems to be compatible with the spectrum of the Crab Nebula. In the mantle of the star we have He, which cannot be processed by the shock because the density is too low. The envelope of H may have been lost to a large extent before the supernova event. It is therefore a confirmation of the model that the ratio He/H in the Crab is abnormally high.

The rather meager evidence from the Crab Nebula is thus compatible with the model of Section B. However, it is clear that other types of supernovae are needed to explain the existing abundance of elements in the sun and other stars. Thus, it is possible that the medium mass stars discussed in this chapter may be the only ones that lead to neutron stars, but there must also be supernovae produced by other, presumably heavier stars.

V. LARGE STARS

A. Presupernova Evolution

The standard paper on the evolution of massive stars before they become supernovas is due to Weaver, Zimmerman, and Woosley (1978). For stars of initial mass of 15 and $25M_\odot$, they find the following results: The Fe core, with degenerate electrons, has a mass of about $1.6M_\odot$. The electron concentration Y_e at the center is 0.43–0.44. The central temperature is about 5×10^9 K, and the central density is 5×10^9 or 3×10^9 g cm^{-3}, for the 15- and 25-$M_\odot$ stars, respectively. Thus, the core of the smaller star is somewhat more degenerate.

As has been pointed out in Section III, such a star could not explode by means of the shock wave generated in the first bounce.

In 1983, Weaver, Woosley, and Fuller (WWF) obtained a considerably more promising presupernova configuration. The main change in assumptions was that they included additional nuclei in the chain of nuclear reactions, namely, isotopes more neutron-rich than the dominant, α-particle-type nuclei like Mg^{24}. The more neutron-rich nuclei have larger cross sections for both nuclear reactions and electron capture. With these assumptions, WWF find Fe cores of 1.35–$1.4M_{\odot}$. These would be small enough that there would be hope for a successful first shock as outlined in Section III.

In spring 1984, however, the situation changed again. Recent experiments on the reaction $C^{12} + He^4 \rightarrow O^{16} + \gamma$ have given a larger cross section than earlier ones. This means that in the stellar shell containing C and O, the abundance of C is smaller. As a consequence, for the star of $25M_{\odot}$, C is exhausted before the SN collapse. This makes the core evolve as if it were a core of a larger star, and gives a mass of $1.8M_{\odot}$ or more for a Fe core, leading almost surely to failure of the shock described in Section 3.

Woosley (private communication) argues that this result is specific for the $25M_{\odot}$ star. In a preliminary result for $15M_{\odot}$, he finds a core of $1.35M_{\odot}$, just on the border between success and failure. On the other hand, stars of still larger initial mass will have large Fe cores (e.g., for initial mass $100M_{\odot}$ Woosley obtains a core of at least $1.8M_{\odot}$, surely not leading to a successful first shock).

It is therefore likely that at least very many of the stars above $12M_{\odot}$, and possibly all of them, will fail to explode by means of the shock due to the first rebound. We must therefore consider a mechanism that could produce a supernova explosion after the initial shock has gone into accretion. Two such mechanisms will be discussed in the next two sections.

B. Rotating Stars

Bodenheimer and Woosley (1983) have discussed the influence of rotation on the dynamics of a massive star. Rotation is very common in stars; for example, massive stars in the main sequence (O stars) have commonly a rotational velocity of about 200 km sec^{-1} at their surface. How the angular velocity is distributed inside the star is essentially unknown, and BW have to make appropriate assumptions.

They assume that the centrifugal force has a constant ratio to the gravitational attraction, independent of r. It is convenient to put this in the form

$$\frac{T_{\text{rot}}}{W_{\text{grav}}} = \frac{r^2\omega^2}{GM(r)/r} = \beta = \text{const.} \tag{46}$$

Equation (46) implies a much faster rotation near the center than farther out, but this may not be very important for the calculation. In their main calculation they assume $\beta = 0.12$. This value of β means that the angular momentum of the inner $8M_\odot$ is about equal to that of a main sequence O star.

The collapse of the inner core is treated approximately only. It is assumed to be similar, but not quite the same as in the absence of rotation. An accretion shock is formed. The behavior of the core is replaced by a boundary condition at 1500 km. This simplification may strongly affect the subsequent dynamics of the outer material. The material outside 1500 km is treated by two-dimensional hydrodynamics.

After 2 sec, the centrifugal force has stopped the infall in the equatorial plane. Along the rotation axis, material continues to stream in throughout the calculation. In the equatorial plane, however, the centrifugal force now combines with nuclear energy generation. The material in the core at this time is somewhat more than $3M_\odot$. Near this enclosed mass, the presupernova star consists mostly of O^{16}. Since the temperature is now quite high, the O nuclei react with each other and release considerable energy. The energy generated by this, together with the centrifugal force, produces outflow of material near the equatorial plane at about 4.5 sec after collapse of the center. The outflow reaches velocities up to about 7000 km sec^{-1}.

After 20 sec, an energy of about 2 foe has been produced by the O + O reaction. However, most of this energy remains locked up in the accreting central core. Only 0.12 foe is kinetic energy of the material flowing out in the equatorial plane. The total mass of this material is $0.5M_\odot$, most of it having been processed by the O + O reaction. These "O ashes" have ample momentum to propel the He and H in the mantle and envelope of the star.

By calculating eight additional models, BW found out that both the rotation and the burning are essential to cause the material to flow out in the equatorial plane. The dissociation of nuclei that would take place at higher temperature turns out to be unimportant. The ratio β of centrifugal force to gravity can be diminished to 0.06 without stopping the outflow.

There is a lot of nucleosynthesis in this model. The temperature remains about 3–4 × 10^9 K during the entire burning process. The composition of the resulting material is then similar to that found by Woosley, Arnett, and Clayton (1973): All the elements from Si^{28} to Cr^{50} are produced in nearly the proportions found in the sun. Material that has gone to higher temperature than 4 × 10^9 is unlikely to be ejected but will be part of the residual central core; thus, the ejecta will not contain appreciable elements in the Fe group.

Assuming there have been 10^9 supernovae of this type, each ejecting $0.5M_\odot$ of Si group elements, this would provide 5 × 10^8 $M_\odot$ of these elements in the Galaxy, which is the requisite amount for a Galactic mass of 4 × 10^{11} (Cameron 1982). At the present rate of supernovae of this type, only about 10^8 would be expected during the life of the Galaxy, but it is likely that massive stars, and hence supernovae of this type, were more frequent in early epochs.

Observations show copious O and ashes from O burning in some supernova remnants. This is especially true in the fast-moving knots in Cas A, which move with a velocity of 3000–9000 km sec^{-1}. On this model, we would have to assume that these knots represent material from the O burning shell that has not lost much momentum in going through the mantel and envelope; the speed then is in accord with the predicted maximum of 7000 km sec^{-1}.

Equatorial expulsion should lead to a toroidal geometry. Such geometry has been observed in several supernova remnants (e.g., in N-1320 in the Large Magellanic Cloud, which also has knots rich in O). In one example, the torus has been actually measured in X-ray imagery. Cas A shows a differential Doppler shift, also indicating an anisotropic remnant.

The core mass at 20 sec after bounce is $3.8M_\odot$. This is clearly far too high for a neutron star, so the residual star should become a black hole. In order to do so, it must get rid of a large angular momentum. This could lead to substantial gravitational radiation, but BW suggest that other instabilities may also be important.

C. Revival by Neutrinos

An alternative way to get a supernova shock, which works especially for stars that do not have substantial rotation, was suggested by James Wilson (1982) and further analyzed by Bethe and Wilson (1984). The main feature is that some of the neutrinos emitted by the core of the star will be absorbed by the infalling material farther out and will heat that material sufficiently so that its internal pressure leads to the expulsion of the mantle and envelope of the star.

(i) Neutrino Emission. The core of the star contains a lot of excess energy. In particular, in the inner core (inner $0.8M_\odot$) there are still many electrons giving pressure. Given time, these electrons will be captured by the nuclei, and the resulting neutrinos will slowly diffuse out. Once they get to the neutrinosphere, which is approximately at $1.1M_\odot$, they can escape permanently.

The rate of this "neutronization" of the inner core has been calculated by Burrows, Mazurek, and Lattimer (1982). They find that about half the leptons are emitted in half a second.

As the leptons are lost, the pressure in the inner core decreases substantially. Therefore, the inner core will contract, and the outer core, between $0.8M_\odot$ and the neutrinosphere, will follow. In this process, a large amount of gravitational energy is released. This energy is emitted in the form of neutrinos.

To estimate the energy in these neutrinos, we note that the final neutron star (if this is formed) has a binding energy of about 10% of its mass. If we further estimate the mass of that star as $1.4M_\odot$, its binding energy is

$$B = 2.5 \times 10^{53} \text{ erg} \tag{48}$$

We shall assume now that a loss of half the leptons also means a loss of half the energy; this then happens in half a second. Essentially all this energy comes out in neutrinos, about half of it in electron neutrinos and antineutrinos (the rest is in μ and τ neutrinos). Then the luminosity in ν_e and $\bar{\nu}_e$ in the first half second is

$$\text{L}_\nu \simeq 1.3 \times 10^{53} \text{ erg sec}^{-1} \tag{49}$$

This energy will be essentially blackbody emission from the neutrinosphere. The emission per unit area is

$$J_\nu = \left(\frac{c}{2}\right) 1.20 \times 10^{26} T^4 \text{ erg cm}^2\text{sec}^{-1} \tag{50}$$

counting both ν_e and $\bar{\nu}_e$. We have used $c/2$ rather than $c/4$ because the neutrinos go predominantly outward at the neutrinosphere. The total emission from that sphere, of radius R_ν, is then

$$\text{L}_\nu = 2.3 \times 10^{49} R_{6\nu}^2 T_\nu^4 \tag{51}$$

where R_6 is the radius in units of 10^6 cm. Comparing (49) and (51), we find

$$R_{6\nu}T_\nu^2 \simeq 75 \tag{52}$$

The neutrinosphere may be defined by the condition

$$\int \frac{dr}{\lambda} = \frac{2\epsilon^2}{\lambda_0} \int \rho \, dr = \frac{2}{3} \tag{53}$$

Here λ is the mean free path of neutrinos, ρ the material density, and ϵ their root mean square energy. With the neutrino spectrum at the neutrinosphere (see Section 6c),

$$\epsilon^2 = 6.5T^2 \tag{54}$$

λ_0 is a standard mean free path which, in accord with Eq. (9), we take to be

$$\lambda_0 = 10^{20} \tag{55}$$

The factor 2 in (53) takes into account approximately the ratio of total neutrino cross section to elastic cross section. The $\frac{2}{3}$ on the right-hand side of (53) takes into account approximately the fact that the neutrinos emerge at some angle with the radius.

Assuming now that the density near the neutrinosphere is given by

$$\rho = \frac{10^{31}C}{r^3} \tag{56}$$

with C a constant, which is substantially fulfilled in computer calculations, we find

$$\int_{R_\nu}^{\infty} \rho \, dr = \tfrac{1}{2}\rho(R_\nu)R_\nu \tag{57}$$

Inserting all this in (53), we obtain

$$\rho(R_\nu)R_\nu T^2 = 1.0 \times 10^{19} \tag{58}$$

and using (52) we get

$$\rho(R_\nu) = 1.3 \times 10^{11} \text{ g cm}^{-3} \tag{59}$$

It is remarkable that from such general and simple consideration, one can deduce the density at the neutrinosphere. This again agrees well with the result of computations.

Now use, again on the basis of computations,

$$C = 1\text{–}3 \tag{60}$$

Then we find

$$R_{6\nu} = 4.2C^{1/3} \tag{61}$$

$$T_\nu = 4.2C^{-1/6} \text{ MeV} \tag{62}$$

Thus the neutrinosphere is at a temperature of about 4 MeV.

(ii) Neutrino Absorption. The net energy gain from the absorption of neutrinos by matter far outside the neutrinosphere is

$$\frac{dE}{dt} = K(T_\nu)\left[\frac{L_\nu}{4\pi R_m^2} - \left(\frac{T_m}{T_\nu}\right)^2 acT_m^4\right] \tag{63}$$

where K is the opacity, L_ν the luminosity in neutrinos, R_m the radius at which the neutrinos are absorbed, T_m the temperature at that radius. The second term in (63) represents the neutrino energy emitted by the matter under consideration. There are several refinements of this second term.

In order to have positive heating of the matter, it can be shown from (63) that we must have

$$\frac{T_m}{T_\nu} < \left(\frac{R_\nu}{2R_m}\right)^{1/3} \tag{64}$$

Typically, $R_m \approx 5R_\nu$; therefore, (64) requires $T_m \lesssim 0.45T_\nu$. With $T_\nu \approx 4$ MeV, the matter temperature must be appreciably less than 2 MeV.

It is important to note that (63) is independent of the density of the matter. The heating rate dE/dt means the energy gain per unit mass of the matter.

(iii) Requirement for Outward Acceleration. For matter to be accelerated outward it is necessary that the force due to the pressure gradient be greater than that due to gravity, that is,

$$\frac{1}{\rho}\left|\frac{\partial p}{\partial r}\right| > \frac{GM(r)}{r^2} \tag{65}$$

The included mass, $M(r)$, is about 1.65 $M_\odot$ in Wilson's calculation. This means that

$$GM \simeq 22 \times 10^{25} \text{ cm}^3 \text{ sec}^{-2} \tag{66}$$

To satisfy (65), the ratio p/ρ must be large and the pressure must be a steep function of r. The ratio p/ρ can be estimated from the energy absorbed, and it turns out that

$$\frac{rp}{\rho} \simeq (3.5 \pm 1) \times 10^{25} \text{ cm}^3 \text{ sec}^{-2} \tag{67}$$

nearly independent of time and of r in the relevant region. To fulfill (65), it is then necessary that the pressure fall very steeply with r.

In the actual computation this comes about by the ongoing shock actually receding for a period of time. After this has happened, the pressure distribution behind the shock is very steep. It is enough to satisfy (65).

It is not yet known whether this result will be obtained for an arbitrary stalled shock or whether specific conditions have to be satisfied to make it true. My feeling is that it is a general feature.

Once (65) is fulfilled for some region of the matter, acceleration goes very fast. The reason is that the gravitational acceleration [i.e., the rhs of (65)] is very large, of order 10^{11} cm sec^{-2}. Thus, even if the difference between lhs and rhs is relatively small, it takes much less than 0.1 sec to attain an outward velocity of the order of 10^9 cm/sec. The time required for heating the material, on the other hand, is of the order of several tenths of a second. Thus, once (65) is fulfilled, some material in the star will rapidly move outward, while other material farther in still moves inward due to gravity.

Figure 6 gives the motion of various mass points in the computation by Wilson. Between $t = 0.2$ and 0.5 sec, some mass points remain suspended at $r \simeq 200$ km, but then suddenly fall rapidly toward the accreting central core. At about $t = 0.5$ sec, however, newly shocked material is suddenly turned around and propelled outward. This happens just after the shock (dashed line) has receded substantially, all in accord with the theoretical description we have just given. There is a clear separation between the material that moves outward, $M > 1.665 M_\odot$, and that moving inward.

(iv) Quasi-vacuum. With some material moving outward and other material inward with velocities of the order 10^9 cm sec^{-1}, there takes place

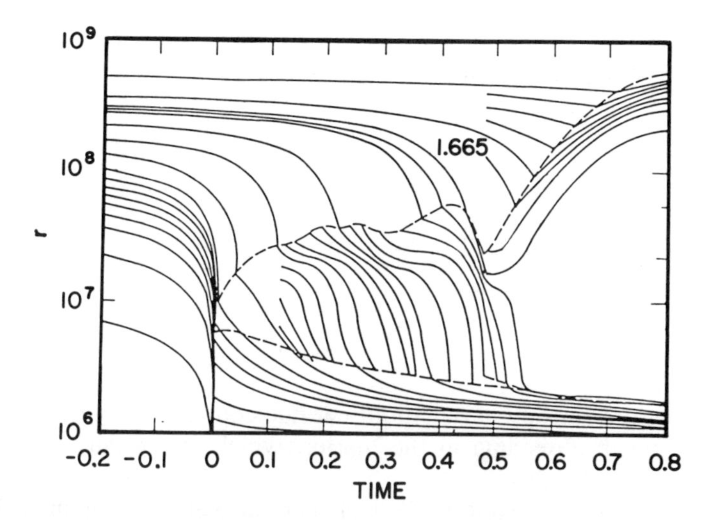

FIGURE 6. Motion of various mass points when the shock gets revived by neutrino absorption. Time is in seconds, $t = 0$ is collapse; r is in centimeters. The curve relating to enclosed mass of 1.665 $M_\odot$ turns around and moves outward after $t = 0.48$ sec. The inner dashed line is the neutrino sphere, the outer one is the shock. Note that the shock moves inward between about 0.4 and 0.5 sec. The empty space on the right-hand side of the figure is the "vacuum" described in the text; it is filled with electromagnetic radiation.

an obvious separation of the matter in the star, a bifurcation. In between the two parts of matter, the density is very small; we have essentially a vacuum.

Obviously, there must be pressure in the vacuum; otherwise the acceleration of the outgoing material could not take place. The pressure is due to electromagnetic radiation and to pairs of positive and negative electrons that are closely coupled to the radiation.

The density in this quasi-vacuum decreases rapidly with time. According to the continuity equation, and neglecting unimportant terms, we have

$$\frac{d\ln\rho}{dt} = -\frac{\partial u}{\partial r} \tag{68}$$

In the computations by Wilson, the rhs is about 30 sec^{-1}. The density decreases indeed at the rate indicated by the continuity equation.

The role of the electromagnetic radiation in the quasi-vacuum is very important. It provides the necessary pressure and does so at low temperature. As we showed in (64), T_m must be less than 2 MeV. The ratio of radiation energy (including electron pairs) to material energy is typically

$$\frac{w_r}{w_m} = 25\text{–}200 \tag{69}$$

Accordingly, the entropy in radiation is very high; we have

$$S_r = \frac{2w_r}{w_m} \tag{70}$$

where S_r is the entropy in radiation, per nucleon. This may be compared with the entropy residing in the matter, which is typically 8 ± 2 per nucleon.

We believe that the bifurcation of material velocity obtained in this calculation is typical for bifurcation in other cases. If the direct shock described in Section III is successful, it also must ultimately lead to bifurcation, with some material being propelled outward whereas other material falls back on the core. This bifurcation probably happens only when the shock has proceeded outward to a point where radiation pressure dominates over material pressure. In the direct shock of Section III, this probably happens at $r > 2000$ km. Once this shock radius has been attained, bifurcation can take place with the pressure in the intervening region being provided by radiation. While the material density in this region may be smaller than farther out, we do not need to be afraid of the "vacuum" sucking in the material that has reached escape velocity. That vacuum has a high pressure and is easily able to support the outgoing material.

(v) Final energy. In the case computed by J. Wilson, at the end of the computation, the energy of the material outside the bifurcation was 0.40 foe. This is somewhat lower than usual supernova energies, but not by much.

The mass of the remaining core in Wilson's calculation is $1.66M_\odot$. It is likely that this can lead to a neutron star; in this case, the gravitational mass of that star should be about 10% lower, or $1.5M_\odot$. This is in the neighborhood of the masses of observed neutron stars, namely about $1.4M_\odot$. It is also below the likely upper limit for neutron star mass, which is about $1.8M_\odot$. However, the Wilson calculation was done with the presupernova condition as obtained by Weaver, Woosley, and Fuller (i.e., an Fe core of about $1.4M_\odot$). If that mass were to increase appreciably, the mass of the residual star in Wilson's theory would become greater than the upper limit for a neutron star.

(vi) Nucleosynthesis. Only the material outside the bifurcation will be ejected. Most of the time, that material is at a low density, of order 10^6 g cm^{-3}. In these conditions very little nucleosynthesis can take place. So it is likely that the ejected material would contain mostly O, C, and a few neighboring elements. It is unlikely that much material in the Si group can be produced.

VI. CONCLUSIONS

A. Variability of Supernovae

It will be clear from these discussions that supernovae of "Type II" in reality fall into many different classes.

Stars of original mass between 8 and $10M_{\odot}$ seem well understood (Section IV.D). Their core consists of O, Ne, and Mg, so that when they collapse, considerable energy is produced by nuclear reactions. This facilitates the later explosion. They release an energy of about 2 foe, and they leave a neutron star behind. This class, which we shall call Type II-A, may have the Crab as a prototype. If we assume that the number of stars of mass greater than M is proportional to M^{-2} (which is reasonably fulfilled by the existing star population), about 35% of all stars above $8M_{\odot}$ fall into this category.

The absolute masses are somewhat uncertain; they could easily be 10 or 20% higher. Moreover, there is the possibility that slightly heavier stars will have a Si core and may behave somewhat similarly. This has not been investigated. Conservatively, we might consider this class of stars to extend up to $12M_{\odot}$.

Stars of $M > 12M_{\odot}$ are less clear as yet. They all have Fe cores. Those of moderate total mass may have a successful first shock, as described in Section III. Whether this is so depends on future revised calculations of the mass of the Fe core in the presupernova state. If this turns out to be about $1.35M_{\odot}$, the chance for a successful first shock is good. But success also depends on further refinement of the calculations of this first shock, as described in Section III. If there are stars with an Fe core and a successful first shock, they will also leave a residual neutron star and will release about $\frac{1}{2}$ foe of energy. We shall call these Type II-B.

Certainly stars of $25M_{\odot}$ and higher would have a large Fe core, and therefore will not have a successful first shock. If the limit is indeed $25M_{\odot}$

and if the population distribution is as M^{-2}, only about 10% of the supernovas will fall into this category. However, it is quite possible that all stars of $M > 12M_{\odot}$ have this property, then perhaps as many as 50% of all supernovae of Type II may have this characteristic.

Given that the first shock fails, there are two possibilities: If the star has substantial rotation, it will expel much of its material in the equatorial plane, as described in Section V.B. These stars almost certainly leave a black hole rather than a neutron star at the center. The supernova remnant will be highly anisotropic, as has been observed in a few instances. We shall call these supernovae Type II-C. The energy release could be quite variable for this type but may again be of the order $\frac{1}{2}$ foe. However, this type probably will have fast-moving knots consisting of the material originally expelled in the equatorial plane.

If a large star does not have much rotation and does not have a successful first shock, it probably makes a supernova by delayed neutrino heating, as described in Section V.C. The energy release is again of the order $\frac{1}{2}$ foe, but the expulsion is likely to be fairly symmetrical. We may call these Type II-D.

Summarizing, it is likely that only the smaller stars, Type II-A, have an energy release well in excess of 1 foe. For the heavier stars, the present models give only about $\frac{1}{2}$ foe.

The residual star is a neutron star in Type II-A, and in II-B if that type exists. It is a black hole in II-C and probably in most cases of Type II-D.

Helfand's observations, discussed in Section I.A indicate that only a rather small fraction, maybe 30–40%, of supernovae leave neutron stars behind. This would be explained most easily if Type II-B does not exist at all. Then essentially only Type II-A would lead to neutron stars, and this would be in accord with Helfand's estimate.

B. Nucleosynthesis

The most important elements that should be ejected from SN II are in the medium-weight elements from Mg to Ca. The supernovae of Type II-C (i.e., those that expel their material in the equatorial plane) are likely to produce and eject these elements copiously, together with O and C. However, because of the complicated dynamics of these anisotropic explosions, accurate predictions for Type C may be difficult.

Type II-B, having a successful first shock, is likely to give similar and satisfactory results for the production of nuclei in the Si group. However, as we pointed out in Section VI.A, Type II-B may not exist at all.

Type II-A apparently gives almost exclusively Fe + Ni, and He from the envelope. This type of supernova is therefore not very successful for supplying a galaxy with the medium-weight nuclei.

Similar arguments probably apply also to Type II-D, the stars that explode by delayed neutrino heating.

C. Neutrino Emission

The total energy in neutrinos is likely to be very large, of the order of 100 foe, for any supernova of Type II. This should make them fairly easy to observe, once the apparatus is set up and a supernova occurs in our galaxy. It is uncertain, however, how much we shall be able to learn from these observations.

In every Type II supernova, there will be two short pulses of neutrinos, followed by a much longer pulse. The short pulses consist of the neutrinos emitted during the infall, and then those emitted when the shock reaches the neutrinosphere. Each of them should last a few milliseconds; the second is superposed on the first, and is probably stronger. If the times of these pulses can be measured with some accuracy, we could learn something about the dynamics of the supernova.

The long pulse lasts about a second and is due to the diffusion of neutrinos out of the unshocked core and to the cooling of the entire core. It involves of the order 100 foe.

The neutrinos emitted in the infall mostly have an energy comparable to the Fermi energy of the electrons being captured, that is, about 10 MeV. The other neutrinos have a modified thermal spectrum, as will be discussed. The temperature of the neutrinosphere when the shock first reaches it may be of order 5 MeV, while in the pulse arising from the slow cooling it will be about 4 MeV. These numbers are for Type II-B, -C, and -D. Type II-A has not yet been discussed in detail. It probably does not make much difference whether the central star becomes a neutron star or a black hole.

The spectrum of the neutrinos is not a simple Boltzmann spectrum. This is because the lower-energy neutrinos have a longer mean free path than the higher-energy ones. The high-energy neutrinos therefore come from a region farther out in the star that is likely to be cooler than the neutrinosphere. Therefore, the spectrum falls faster at high energies than exponentially.

We have calculated the spectrum with the assumption that the density is proportional to $1/r^3$, and the temperature proportional to $1/r$. Then the spectrum is

$$f(\epsilon)\,d\epsilon = \epsilon^5\,d\epsilon\,\exp - \left(\frac{\epsilon}{T_\nu}\right)^{4/3} \tag{71}$$

This spectrum agrees quite well with the numerical calculations by Jim Wilson.

The strong falloff of the spectrum at high energies is unfortunate because higher-energy neutrinos are easier to detect than lower-energy ones. This is partly compensated for by the fact that at least in the longer pulse (of about a second) there are about as many antineutrinos as neutrinos. The first short pulse, from the infall, consists essentially only of neutrinos; the second short pulse (when the shock reaches the neutrinosphere) is also predominantly neutrinos.

ADDENDUM

In the year since this chapter was written, there have been many changes in our understanding of the supernova mechanism.

The mechanism for small stars, discussed in Section II.C, apparently does not work. This was shown by Burrows and Lattimer and independently by Wilson.

The standard mechanism of the prompt shock, discussed in Section II.B, got a new lease on life from the work of Baron, Coopersmith and Kahana. They noted that the use of general relativity can help the shock propagation, and this help is enhanced if nuclear matter is very compressible. G. E. Brown has given good arguments that this is indeed the case. His arguments are not generally accepted, but I believe now that Type II-B exists.

The calculations by Woosley and Weaver (1986) of the presupernova evolution have given rather definite results. Stars at least up to $15M_\odot$ have Fe cores of about $1.35M_\odot$; stars of $M_\odot \gtrsim 20M_\odot$ have cores of $\gtrsim 1.7M_\odot$. The latter are therefore clearly Type II-D, and leave a black hole behind. The former are probably Type II-B and lead to neutron stars. It is not clear yet at which star mass the limit lies, nor whether it is a sharp limit. But these calculations are at least in qualitative agreement with the observations of Helfand reported in Sec. I that only a fraction of supernovae of Type II leave neutron stars behind.

REFERENCES

Arnett, W. D., 1977, *Ap. J.*, **218,** 815.

Becker, R. H., Helfand, D. J., and Szymkowiak, A. E., 1982, *Ap. J.*, **255,** 557.

Bethe, H. A., Brown, G. E., Applegate, J., and Lattimer, J. M., 1979, *Nucl. Phys.*, **A324,** 487.

Bethe, H. A., 1982, in *Supernovae,* M. J. Rees and R. J. Stoneham (eds.) Reidel, Dordrecht, p. 405.

Bethe, H. A., and Wilson, J., 1985, *Ap. J.*, **295,** 14.

Bodenheimer, P., and Woosley, S. E. 1983, *Ap. J.*, **269,** 281.

Bonche, P., and Vautherin, D., 1981, *Nucl. Phys.*, **A372,** 496.

Burrows, A., and Mazurek, T. J., 1982, *Ap. J.*, **259,** 330.

Burrows, A., Mazurek, T. J., and Lattimer, J. M., 1981, *Ap. J.*, **251,** 325.

Cameron, A. G. W., 1982, in *Essays in Nuclear Astrophysics,* C. Barnes et al. (eds.) Cambridge University Press, Cambridge.

Cooperstein, J., 1985, *Nucl. Phys.*, **A438,** 722.

Cooperstein, J., and Wambach, J., 1984, *Nucl. Phys.*, **A420,** 591.

Cooperstein, J., Bethe, H. A., and Brown, G. E., 1984, *Nucl. Phys.*, **A429,** 527.

Fowler, W., and Hoyle, F., 1964, *Ap. J. Suppl.*, **9,** 201.

Fuller, G., 1982, *Ap. J.*, **252,** 741.

Goldreich, P., and Weber, S., 1980, *Ap. J.*, **238,** 991.

Helfand, D. J., and Becker, R. H., 1984, *Nature,* **307,** 215.

Hillebrandt, W., Nomoto, K., and Wolff, R. G., 1984, *Astr. Ap.*, **133,** 175.

Lamb, D. Q., and Pethick, C. J., 1976, *Ap. J.*, **209,** L77.

Lamb, D. Q., Lattimer, J. M., Pethick, C. J., and Ravenhall, D. G., 1978, *Phys. Rev. Lett.*, **41,** 1623.

Lyne, A. G., 1982, in *Supernovae,* M. J. Rees and R. J. Stoneham (eds.), Reidel, Dordrecht, p. 405.

Nomoto, K., 1984, *Ap. J.*, **277,** 791.

Trimble, V., 1982, *Rev. Mod. Phys.*, **54,** 1183.

Weaver, T. A., Zimmerman, G., and Woosley, S. E., 1978, *Ap. J.*, **225,** 1021.

Woosley, S. E., and Weaver, T. A., 1986, *Ann. Rev. Astron. Astrophys.*, in press.

Wilson, J. 1982, in *Proc. Univ. of Illinois Meeting on Numerical Hydrodynamics.*

Woosley, S. E., Arnett, W. D., and Clayton, D. D., 1972, *Ap. J.*, **175,** 731.

Woosley, S. E., Arnett, W. D., and Clayton, D. D., 1973, *Ap. J. Suppl.*, **26,** 231.

Yahil, A., and Lattimer, J. M., 1982, in *Supernovae,* M. Rees and R. J. Stoneham (eds.), Reidel, Dordrecht, p. 53.

4.

Solar Neutrinos

JOHN N. BAHCALL

John N. Bahcall is Professor of Theoretical Physics at the Institute for Advanced Study in Princeton, New Jersey. Under his guidance, the Institute has built up one of the most outstanding centers for astrophysical research in the world. He has contributed much to our understanding of high-energy astrophysics, including the solar neutrino problem, X-ray astronomy and quasars, and galactic dynamics.

I. INTRODUCTION

There is a serious discrepancy between the standard theory of how the sun shines and the most direct observational test of this theory, assuming conventional physics applies.

I will try to give you a feeling for this subject without deluging you with details. The topics I will cover are, in order: an outline of the theory of stellar evolution, a description of the main sources of solar neutrinos, a summary of the results of the ^{37}Cl experiment, an analysis of the uncertainties in the calculated rates, a list of the principal new solar neutrino experiments that have been proposed, and a somewhat more detailed description of the proposed ^{71}Ga experiment, which is the consensus choice for the next major experiment. You can find more detailed information by consulting technical reviews that also list the many important references (see, e.g., Bahcall and Davis 1966, 1976, 1982; Salpeter 1968; Cleveland, Davis, and Rowley 1981; Fowler 1982; Zatsepin 1982; Haxton 1984; Bahcall 1978, Paper I; Bahcall et al. 1982, Paper II; Bahcall et al. 1984). In what follows I shall quote—without explicit reference—numerical results on neutrino cross sections from Paper I and numerical results on the standard solar model from Paper II.

Any neutrino experiment measures event rates, the product of flux times neutrino interaction cross sections. I will *not* discuss here that half of the subject that involves the calculation of the neutrino cross sections (the physics for each of the nuclei of interest is described in Paper I and in Bahcall 1981). The reason is that I am sure the calculations are correct to the stated accuracy. In fact, part of my personal definition of a "good" solar neutrino experiment is that I can calculate the capture rate to an accuracy of 10% or better for the target of interest (including the effects of nuclear excited states and all the relevant atomic physics effects). On this point, you will just have to trust me (or read the technical papers).

One may well ask, "Why devote as much effort in trying to understand a backyard problem like the sun's thermonuclear furnace when there are so many exciting and exotic discoveries occurring in astronomy?" Most natural scientists believe that we understand the process by which the sun's heat is produced—that is, in thermonuclear reactions that fuse light elements into heavier ones, thus converting mass into energy. However, no one has found an easy way to test the extent of our understanding because the sun's energy source is deep in the interior, where it is hidden by an enormous mass of cooler material. Hence conventional astronomical instruments can only record the photons emitted by the outermost layers of the sun (and other stars). The theory of solar energy generation is sufficiently important to the general understanding of stellar evolution that one would like to find a more definitive test.

There is a way to directly and quantitatively test the theory of nuclear energy generation in stars like the sun. Neutrinos are the only particles produced by thermonuclear reactions that have the ability to penetrate from the center of the sun to the surface and escape into space. Thus neutrinos offer us a unique possibility of "looking" into the solar interior.

An experiment designed to capture neutrinos produced by solar thermonuclear reactions is a crucial test of the theory of stellar evolution. The theory of stellar evolution by thermonuclear burning is widely used in interpreting many kinds of astronomical information and is a necessary link in establishing such basic data as the ages of the stars and the abundances of the elements. The parameters of the sun (its age, mass, luminosity, and chemical composition) are better known than those of any other star, and it is in the simplest and best understood stage of stellar evolution, the quiescent main sequence stage. Thus an experiment designed to capture neutrinos produced by solar thermonuclear reactions is a crucial one for the theory of stellar evolution. I hoped originally that the application of a new observing technique would provide added insight and detailed information. It is for all of these reasons (a unique opportunity to see inside a star, a well-posed prediction of a widely used theory, and the hope for new insights) that so much effort has been devoted by astrophysicists to the solar neutrino problem.

Physicists have also devoted a great deal of effort to understanding the solar neutrino problem. The main reason is that solar neutrino experiments offer the opportunity of testing the stability of the neutrino over proper times that are inaccessible in the laboratory. In laboratory experiments, one can use distances that are at most a few km and that involve high-energy neutrinos (tens or hundreds of GeV), or observe extremely high-energy neutrinos that pass through the earth's diameter. By contrast, solar neutrino experiments involve a source to detector distance of 10^{13} cm, with a characteristic energy of MeV. Thus elements of the neutrino mass matrix as small as 10^{-6} eV can be investigated with the aid of solar neutrino experiments, while it seems that the best one can do with terrestrially based experiments is about $10^{-1.5}$ eV. One of the major goals of modern theoretical physics is the unification of all of the known interactions in a Grand Unified Theory. A common characteristic of many existing models of unified theories is that neutrino mass matrix elements of order 10^{+1}–10^{-6} eV are expected.

Many solutions to the solar neutrino problem, modifying either the physics or the astronomy (and in some cases both), have been proposed. Even if one grants that the source of the discrepancy is astronomical, there is no general agreement as to what aspect of the theory is most likely to be incorrect. As indicated earlier, many of the proposed solutions of the solar neutrino problem have broad implications for conventional astronomy and cosmology. Some of them would change the theoretical ages of old stars or the inferred primordial element abundances. On the other hand, some theories of the weak interactions have been proposed in which neutrinos may disappear by mixing or decay in transit from the sun to the earth, but for which there are no terrestrially measurable consequences. It is conceivable that one of these theories of the weak interactions is correct and that the standard solar model is not in conflict with observations.

The solar neutrino problem may be caused by an inadequancy in our understanding of how nuclear fusion occurs in the interior of the sun or by an incompleteness in our description of how neutrinos propagate. The immediate task of solar neutrino research is to decide which of these alternative explanations—involving neutrino *production* or neutrino *propagation*—is correct. In different terms, one may say that the current challenge for solar neutrino scientists is to show whether the physicists or the astronomers are wrong.

II. STELLAR EVOLUTION

Table 1 lists everything that I think you need to know about stellar evolution. There are many details used in stellar evolution calculations, but you don't have to know them in order to understand the main results of solar neutrino experiments. Table 1 summarizes the principles that are required for constructing solar models, the ideas that are tested by solar neutrino experiments.

The first principle is hydrostatic equilibrium, which in practice is used together with the special assumption of spherical symmetry (established to the accuracy required here by optical observations of the solar surface).

TABLE 1. Three-Minute Course in Stellar Evolution Principles

Hydrostatic equilibrium: spherical sun
Nuclear energy source
Energy transport by radiation and convection
Uniform primordial composition = surface composition
Evolution (age = 5×10^9 years)
Bottom line: only ^{37}Cl experiment inconsistent with standard theory

[In mathematical terms, hydrostatic equilibrium in a spherical sun implies that the radial derivative of the pressure equals minus the density times the radial derivative of the gravitational potential—$dP/dr = -p\ (d\psi/dr)$, r: distance from the center of the Sun; P: pressure, p: density; ψ: gravitational potential.] The second principle is that the energy source for sunlight is nuclear; the rates of the nuclear reactions depend on the density (p) and the temperature (T), and the composition (X_i). The practical part of this principle is that the rate at which the nuclear reactions produce energy when integrated over the whole sun is equal to the observed solar luminosity *today*. The "today" is an essential part of this principle.

The third principle is that the energy is transported from the deep interior to the surface via steady-state radiation and convection [and not by transient instabilities or waves]. In practice, for most (but not quite all) of the models, the great bulk of the energy is transported by radiation. The key quantities are the gradient of the temperature (dT/dr) and the opacity of the solar matter.

The characteristic densities in the solar interior are 150—100 gm cm^{-3} and the central temperature is about 15×10^6 K. The primordial helium abundance by mass always turns out to be $Y = 0.25 \pm 0.01$, for all the parameters that seem plausible for the standard solar model.

The assumption that the initial composition was uniform and is equal to the presently observed surface composition is closely related to the question of which radiative opacity should be used. It is easy to show that the surface composition of the sun has not changed much because of nuclear reactions since the sun was formed. It is not quite so obvious that nothing much has been added to the solar surface since the sun was born. However, the generalization of this assumption—primordial surface compositions for main sequence stars—is widely used in interpreting stellar and galactic observations and is the basis for making the standard calculations. One of the principal observational arguments supporting this assumption is the uniformity of surface compositions that are observed for stars throughout the galaxy.

The final principle is that the sun evolves because it burns its own (limited) nuclear fuel. We believe that the sun has been shining for something like 5 billion years so far. One mocks up this evolution by computing several quasistatic models that march along in time.

The bottom line of this brief course in stellar evolution is that within our store of observational information about stars, only the ^{37}Cl experiment of Ray Davis and his colleagues (especially B. Cleveland, K. Rowley, J. Evans, and D. Harmer) is inconsistent with the standard theory of stellar evolution. It is the only place where we do not see a way out of observational difficulties unless we modify something among the basic assumptions.

III. NUCLEAR FUSION IN THE SUN

I shall now outline the conventional wisdom regarding nuclear fusion as the energy source for main sequence stars like the sun. It is assumed that the sun shines because of fusion reactions similar to those envisioned for terrestrial fusion reactors. The basic solar process is the fusion of four protons to form an alpha particle, two positions (e^+), and two neutrinos (ν), that is, $4p \rightarrow \alpha + 2e^+ + 2\nu_e$. Almost all (98.5%) of the energy generation in the present-day sun comes from the *p–p* chain, with only about 1.5% from the CNO cycle. The principal reactions are shown in Table 2 with a column indicating in what percentage of the solar terminations of the proton–proton chain each reaction occurs. The rate for the initiating proton–proton (*pp*) reaction, number 1 in Table 2, is largely determined by the total luminosity of the sun. Unfortunately, these neutrinos are below the threshold, which is 0.81 Mev, for the ^{37}Cl experiment. Several of the proposed new experiments, especially the ^{71}Ga and ^{115}In experiments, will be primarily sensitive to neutrinos for the p–p reaction.

TABLE 2. The Proton–Proton Chain in the Sun

Number	Reaction	Solar *p–p* Terminations (%)	Maximum Neutrino Energy (MeV)
1	$p + p \rightarrow {}^2H + e + \nu$	99.75	0.420
	or		
2	$p + e^- + p \rightarrow {}^2H + \nu$	0.25	1.44 (monoenergetic)
3	${}^2H + p \rightarrow {}^3He + \gamma$	100	
4	${}^3He + {}^3He \rightarrow {}^4He + 2p$	88	
	or		
5	${}^3He + {}^4He \rightarrow {}^7Be + \gamma$	12	
	with		
6	${}^7Be + e^- \rightarrow {}^7Li + \nu$		0.861 (90%); 0.383 (10%) (both monoenergetic)
7	${}^7Li + p \rightarrow 2\,{}^4He$		
	or		
8	${}^7Be + p \rightarrow {}^8B + \gamma$	0.02	
9	${}^8B \rightarrow {}^8Be^* + e^+ + \nu$		14.06
10	${}^8Be^* \rightarrow 2\,{}^4He$		

The *pep* reaction (number 2), which is the same as the familiar *pp* reaction except for having the electron in the initial state, is detectable in the ^{37}Cl experiment. The ratio of *pep* to *pp* neutrinos is approximately independent of which model (see below) one uses for the solar properties. Two other reactions in Table 2 are of special interest. The capture of electrons by ^{7}Be (reaction 6) produces detectable neutrinos in the ^{37}Cl experiment. The ^{8}B beta decay, reaction 9, was expected to be the main source of neutrinos for the ^{37}Cl experiment because of their relatively high energy (14 Mev), although it is a rare reaction in the sun (see Table 2). There are also some less important reactions from the carbon–nitrogen–oxygen (CNO) cycle that give rise to the neutrino-producing isotopes, ^{13}N and ^{15}O, but we shall not discuss them in detail, since the CNO cycle is believed to play a rather small role in the energy-production budget of the sun.

A *minimum* event rate for solar neutrino experiments can be calculated that is consistent with the hypothesis that the sun is currently producing nuclear energy by light element fusion at the same rate at which photons escape its surface and the additional hypothesis that nothing happens to the neutrinos on the way to the earth from the sun. This minimum rate would be obtained if the only nuclear reactions that occurred in the solar interior were numbers 1–4 of Table 2 (ending with the ^{3}He–^{3}He reaction). In this case, the *p–p* neutrino flux is 6.5×10^{10} cm^{-2} sec^{-1} and the pep flux is 1.6×10^{8} cm^{-2} sec^{-1}.

IV. THE ^{37}Cl SOLAR NEUTRINO EXPERIMENT

The chlorine solar neutrino detector is based on the neutrino capture reaction (Davis 1964, 1969, 1978; Cleveland, Davis, and Rowley 1984; Pontecorvo 1946; Alvarez 1949):

$$\nu + {}^{37}\mathrm{Cl} \underset{\text{decay}}{\overset{\text{capture}}{\rightleftarrows}} {}^{37}\mathrm{Ar} + e^{-} \tag{1}$$

which is the inverse of the electron capture decay of ^{37}Ar. The radioactive decay occurs with a half-life of 35 days. This reaction was chosen for the first solar neutrino experiment because of its unique combination of physical and chemical characteristics, which were favorable for building a large-scale solar neutrino detector. Neutrino capture to form ^{37}Ar in the ground state has a relatively low-energy threshold (0.81 Mev) and a favorable cross section, nuclear properties that are important for observing neutrinos from ^{7}Be, ^{13}N, and ^{15}O decay and the pep reaction.

The ^{37}Cl reaction is favorable from a chemical point of view. Chlorine is abundant and inexpensive enough that one can afford the many hundreds of tons needed to observe solar neutrinos. The most suitable chemical compound is perchloroethylene, C_2Cl_4, a pure liquid, which is manufactured on a large scale for cleaning clothes. The reaction product, ^{37}Ar, is a noble gas. The neutrino capture process produces an ^{37}Ar atom with sufficient recoil energy to break free of the parent perchlorothylene molecule and penetrate the surrounding liquid, where it reaches thermal equilibrium.

The ^{37}Cl detector was built by Ray Davis deep underground to avoid the production of ^{37}Ar in the detector by cosmic rays. This was done with the cooperation of the Homestake Gold Mining Company (Lead, South Dakota), which excavated a large cavity in its mine (~1500 m below the surface) to house the experiment. The final detector system consists of an ~400,000-liter tank of perchlorothylene, a pair of pumps to circulate helium through the liquid, and a small building to house the extraction equipment.

A set of 59 experimental runs carried out in the ^{37}Cl experiment over the last 14 years shows that the ^{37}Ar production rate in the tank is 0.47 ± 0.04 ^{37}Ar atoms per day in the 615 tons of C_2Cl_4 (Cleveland, Davis, and Rowley 1984; Davis 1978). Even though the tank is nearly a mile underground, a small amount of ^{37}Ar is produced by cosmic rays. An evaluation of data obtained by exposing 7500 liters of C_2Cl_4 at various depths underground suggests that the cosmic ray production rate in the detector may be 0.08 ± 0.03 ^{37}Ar atoms per day.

It is possible that all the capture rate that has been observed is due to background effects. This possibility must be investigated by performing further background experiments with a more sensitive ^{37}K detector and by evaluating the ^{8}B neutrino flux with independent solar neutrino detectors (see Section VII).

If the small background rate cited previously is used, then a positive signal corresponding to:

$$2.1 \pm 0.3 \text{ SNU}(1 - \sigma \text{ error}) \tag{2}$$

is inferred (see Davis 1978; Cleveland, Davis, and Rowley 1984). Here I use the convenient notation 1 SNU $= 10^{-36}$ captures per target particle per second (where SNU stands for Solar Neutrino Unit and is pronounced SNEW, the characteristic product of solar neutrino flux times cross section).

Table 3 shows the predicted neutrino fluxes and capture rates for a standard solar model (Bahcall, et al. 1984; Paper II). Most of the capture

TABLE 3. Predicted Capture Rates for a Standard Solar Model[a]

Neutrino Source	Flux at Earth (10^{10} cm^{-2}sec^{-1})	Capture Rate (SNUs)	
		^{37}Cl	^{71}Ga
p–p	6.1	0	70
^{8}B	0.0004	4.3	1.2
pep	0.015	0.2	2.5
^{7}Be	0.40	1	27
^{13}N	0.05	0.08	2.6
^{15}O	0.04	0.2	3.5
Total		6 SNU	107 SNU

[a] The ^{71}Ga rate is about 10% higher when transitions to excited states of ^{71}Ge caused by ^{8}B solar neutrinos are included.

rate for the chlorine experiment is expected to come from the rare ^{8}B neutrinos, which are produced only about once in every 10,000 terminations of the solar proton–proton chain.

V. UNCERTAINTIES IN THE PREDICTIONS

How serious is the discrepancy between theory and observation? Can the difference be explained by errors in some of the input parameters?

In order to answer these important questions regarding uncertainties, I calculated (see Paper II)—together with Walter Heubner, Steve Lubow, Peter Parker, and Roger Ulrich—the partial derivative of the expected flux from each neutrino source with respect to every important input parameter. These calculations required the construction of hundreds of "standard" solar models that were iterated to obtain—with the changed parameters—final solar models with the observed luminosity and surface composition. I also did a few test cases of very large changes in input parameters to verify that the effect of conceivably big differences in input parameters could be estimated well with the aid of the partial derivatives.

The uncertainties in the experimental parameters were evaluated using the original data (which was reanalyzed in some cases). The uncertainties in the theoretical quantities (e.g., the opacity) were estimated by using the range in published evaluations by competent workers.

There are many different input parameters, so the estimation of the final error is not easy. I defined an overall "effective 3-σ" limit by eval-

TABLE 4. Uncertainties in SNUs from Individual Parameters

Parameter (or reaction)	^{37}Cl	^{71}Ga
p–p	0.6	2
$^{3}H3$–^{3}He	0.3	1.5
^{3}He–^{4}He	1.3	6
$^{7}Be(p,\gamma)^{8}B$	1.3	0.4
$L_{\odot}$	0.2	1
Heavy elements	0.8	3
Opacity	0.4	1
Neutrino cross sections	0.4	8
Total	2.2	11

uating the 3-σ range for each experimental quantity, taking the total published range of the theoretical estimates, and combining quadratically all the independent sources of uncertainty.

In practice, a number of judgments had to be made. In order to conform to the spirit of the "effective 3-σ" limit, I developed the following rule of thumb. An *effective 3-σ* error is to be so serious that if it is later discovered to have been made by a young investigator, it would result in that person not getting tenure. If a 3-σ error were committed by a senior scientist, it would mean that his colleagues were justified in making the offender a department chairman or dean.

Table 4 shows the recognized uncertainties in the chlorine experiment, computed using the apparatus and results of Paper II and the slightly revised reaction rates cited by Bahcall et al. (1984). The overall uncertainty in the predicted capture rate—computed with the aid of the rule of thumb stated above—is 2.2 SNU. Thus the discrepancy between theory and observation is indeed serious.

Parameter changes do not seem likely to resolve the conflict that has existed for the past 17 years. Indeed, one can reach this conclusion in an independent way by examining Fig. 1, which shows all the theoretical evaluations that I have been associated with since 1964 and the experimental determinations of the capture rate by Davis and his collaborators (the references to the individual papers are given in Bahcall and Davis 1982; see caption of Fig. 3, and in Bahcall et al. 1984). The calculated results haven't changed much in the past decade and a half (after the initial large uncertainties in some of the nuclear parameters were reduced

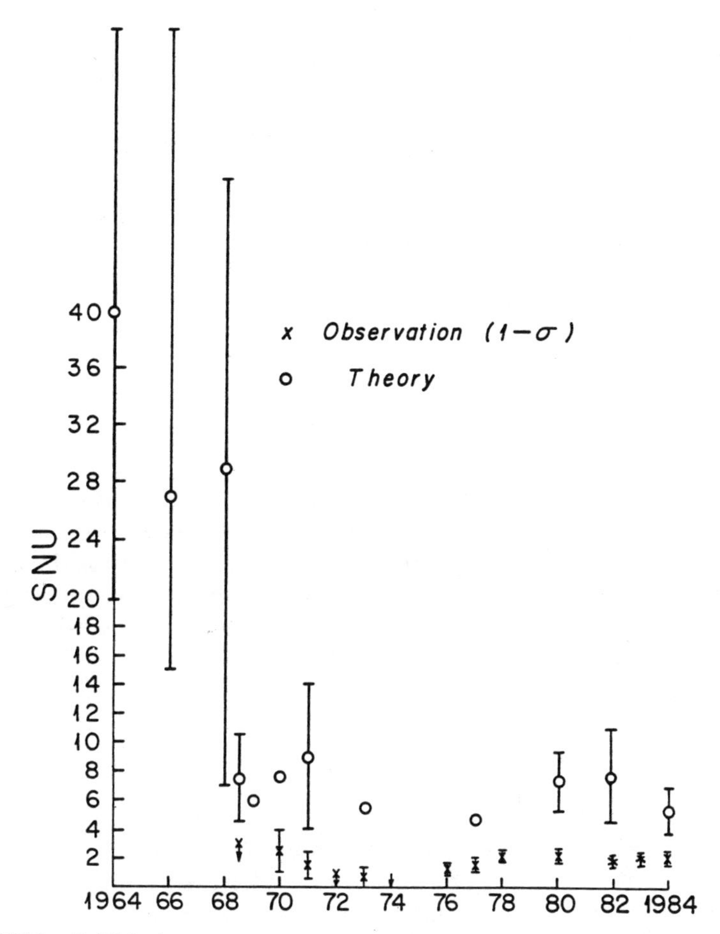

FIGURE 1. Published capture rates as a function of time. (For references, see Bahcall and Davis, 1982, and Bahcall et al. 1984.)

by laboratory experiments at low energies), despite the detailed investigation of many theoretical effects and the remeasurement of a number of the crucial nuclear cross sections. The inferred experimental capture rate has also remained unchanged for many years, once the small-number statistics were treated properly by Bruce Cleveland using the maximum likelihood method.

The gap between theory and observation has remained approximately the same since it first became apparent in 1968.

TABLE 5. Significance of Counting Rates in the ^{37}Cl Experiment [One Solar Neutrino Unit (SNU) = 10^{-36} captures per target per second]

Counting Rate (SNU)	Significance of Counting Rate
28	Expected if the CNO cycle produces the solar luminosity
6	Standard solar model
2 ± 1	Cocktail hour suggestions
0.3	Expected from the *pep* reaction, hence a test of the basic idea of nuclear fusion as the energy source for main sequence stars

VI. OBSERVATIONAL IMPLICATIONS

The ^{37}Cl experiment tests theoretical ideas at different levels of meaning, depending on the counting rate being discussed. The various counting rates and their significance are summarized in Table 5. It is obvious from a comparison of Table 5 with the experimental results given earlier that the value of 28 SNU's based on the CNO cycle is ruled out. More surprisingly, the best current models based on standard theory, which imply ~6 SNU's (Paper II; Bahcall et al. 1984) are also inconsistent with the observations. This disagreement between standard theory and observations has led to many speculative suggestions of what might be wrong (for reviews see Bahcall and Davis 1982; Rood 1978). A number of these ad hoc suggestions lead to expected counting rates that are consistent with the chlorine experiment, although none are generally accepted. (In fact, nearly all of the explanations *must* be wrong, since they are mostly mutually exclusive.) In one nine-year period, 1969–1977, Davis and I counted 19 independent ideas of what might be wrong (see Bahcall and Davis 1982). New ideas have been suggested at the rate of two or three per year since then.

I note that present and future versions of the ^{37}Cl experiment are not likely to reach a sensitivity as low as 0.2 SNU, the *minimum* capture rate (from reaction 2 of Table 2) that can be expected if the basic idea of nuclear fusion as the energy source for main sequence stars is correct (and electron neutrinos reach the earth from the sun with an undiminished flux).

VII. NEW EXPERIMENTS

Another experiment is required to settle the issue of whether our astronomy or our physics is at fault. Fortunately, one can make a testable dis-

TABLE 6. Solar Neutrino Experiments Being Actively Pursued by Different Groups[a]

Detector	Dominant Neutrino Source	Experimental Group
^{37}Cl	^{8}B	UP (USSR)
^{71}Ga	p–p	LASL; MPI; USSR
^{115}In	p–p	France; Oxford
^{7}Li	^{8}B, *pep*, ^{15}O	USSR (ORNL)
^{81}Br	^{7}Be	LASL-ORNL
γ_e-scattering	^{8}B	Proton decay detectors, Sydney
^{2}H	^{8}B	Canada-US
^{97}Mo, ^{98}Mo	^{8}B	LASL (Mo)

[a] UP = R. Davis Jr. at the University of Pennsylvania; Canada-US = currently forming Canadian–U.S. collaboration; France = several independent French groups; LASL = Los Alamos Scientific Laboratory under T. J. Bowles and R. G. H. Robertson; LASL (Mo) = Los Alamos Scientific Laboratory under G. A. Cowan; MPI = Max Planck Institute under T. Kirsten and W. Hampel; ORNL = Oak Ridge National Laboratory under G. S. Hurst; Oxford = Oxford University under N. Booth; Sydney = University of Sydney under L. Peak; USSR = Soviet scientists under G. Zatsepin.

tinction. The flux of low-energy neutrinos from the *pp* and *pep* is almost entirely independent of astronomical uncertainties and can be calculated from the observed solar luminosity, provided only that the basic physical ideas of nuclear fusion as the energy source for the sun and of stable neutrinos are correct. If these low-energy solar neutrinos are not detected in a future experiment, we will know that the present discrepancy between theory and observation is due at least in part to a departure from classical weak interaction theory, not just poorly understood astrophysics.

Table 6 lists the experiments that are currently being investigated by different groups around the world. Each of these experiments is difficult and will require many years of hard work. All of the experiments listed will provide valuable information about the solar interior or the propagation of neutrinos. We must have a variety of experiments in order to provide cross-checking and verification of observational results.

Over the past several years, a rare consensus has developed in which most astronomers, chemists, and physicists concerned with the question agree that the next step in resolving the solar neutrino problem should be a gallium experiment (see, for example, Bahcall et al. 1978 and Zatsepin 1982). This experiment has several advantages. The primary reason for the enthusiasm surrounding a gallium experiment is that a ^{71}Ga detector is mainly sensitive to the basic proton–proton neutrinos. Moreover, the chemistry for a GaCl solution is simple and has been demonstrated in a modular pilot experiment. The sensitivity of the gallium detector to the background is small and is self-monitoring (using ^{69}Ga, which is more sensitive to the background than is the neutrino-detector, ^{71}Ga). A recent analysis of the amount of gallium that is required for a solar neutrino experiment shows (Bahcall, Cleveland, Davis, and Rowley 1984) that 30 tons is sufficient for a very good experiment, and one can probably answer the most important questions with as little as 15 tons.

Table 3 (last column) shows the expected capture rate for a ^{71}Ga detector from different neutrino sources, calculated with the aid of a standard solar model. Note that the dominant source, almost 70%, comes from the basic *p–p* and *pep* reactions (numbers 1 and 2 in Table 2). Most of the remaining neutrino capture rate comes from the ^{7}Be reaction (number 6 of Table 2), which is intermediate in its sensitivity to solar interior details between the pp and the ^{8}B neutrinos. The *pp* flux is guaranteed to be present if nothing happens to the neutrinos on the way to the earth from the sun and if the sun is currently supplying by nuclear fusion the energy it is radiating in photons. Since essentially all the solar energy generation is expected to be produced by reactions dependent upon the basic *pp* reaction, and since the luminosity of the sun is observed accurately, most astrophysical uncertainties do not affect significantly the calculated *pp* flux. The theoretical rate of the *pp* reaction is tied closely to the observed solar constant. For the ^{7}Be neutrino flux, Table 2 indicates that approximately 12% of the solar luminosity is generated through this reaction branch. The flux of ^{7}Be neutrinos depends somewhat on the details of the solar interior, although not as sensitively as does the flux of rare ^{8}B neutrinos.

The *minimum* capture rate for the gallium experiment is 78 SNU if one assumes that the current solar energy production is equal to the surface (photon) luminosity and nothing happens to the neutrinos on the way to the earth. This minimum rate is well separated from the *maximum* rate that can be expected for the gallium experiment if the discrepancy between theory and observation for the chlorine experiment is attributed to the propagation of neutrinos (see Bahcall et al. 1984).

Table 4, last column, shows the most important known uncertainties in the prediction of the capture rate for a gallium solar neutrino experiment.

VIII. CONCLUSION

What this subject needs is a new experiment, or preferably, several new experiments. We are in the uncomfortable position of not knowing whether the discrepancy that has existed for a decade and a half is due to our inadequate understanding of how the sun shines or to an incomplete description of how neutrinos propagate. Neither pure theory nor laboratory experiments have resolved the problem. We must turn to nature for an answer.

ACKNOWLEDGMENT

This work was supported in part by NSF contract number PHY-8217352.

REFERENCES

Alvarez, L. W., 1949, *Univ. Calif. Rad. Lab. Rep.,* UCR-328.

Bahcall, J. N., 1978, *Rev. Mod. Phys.* **50,** 881, Paper I.

Bahcall, J. N., 1981, *Phys. Rev.,* **C24,** 2216.

Bahcall, J. N., Cleveland, B. T., Davis, R., Dostrovsky, I., Evans, J. C., Frati, W., Friedlander, G., Lande, K., Rowley, J. K., Stoenner, R. W., and Weneser, J., 1978, *Phys. Rev. Lett.* **40,** 1351.

Bahcall, J. N., Cleveland, B. T., Davis, R. and Rowley, J. K., 1984, *Phys. Rev. Lett.* (submitted).

Bahcall, J. N., and Davis, R., 1966, in *Stellar Evolution,* R. F. Stein and A. G. W. Cameron (eds.), Plenum Press, New York. p. 241.

Bahcall, J. N. and Davis, R., 1976, *Science,* **191,** 264.

Bahcall, J. N. and Davis, R., 1982, in *Essays in Nuclear Astrophysics,* C. A. Barnes, D. D. Clayton, and D. Schramm (eds.), Cambridge University Press, Cambridge, p. 243.

Bahcall, J. N., Huebner, W. F., Lubow, S. H., Parker, P. D., and Ulrich, R. K., 1982, *Rev. Mod. Phys.,* **54,** 767 Paper II.

Cleveland, B. T., Davis, R., and Rowley, J. K., 1981, *AIP Conference Proceedings No. 72, Weak Interactions as Probes of Unification,* G. B. Collins, L. N. Chang, and J. R. Ficenec (eds.), American Institute of Physics, New York. p. 322.

Cleveland, B. T., Davis, R., and Rowley, J. K., 1984, in *Proceedings of the Second International Symposium on Resonance Ionization Spectroscopy and Its Applications,* Adam Hilger Ltd., Bristol, England, Knoxville, Tennessee.

Davis, R., Jr., 1964, *Phys. Rev. Lett.,* **12,** 303.

Davis, R., Jr., 1969, *Proc. Int. Conf. Neutrino Physics and Astrophys (Moscow),* **2,** 99.

Davis, R., Jr., 1978, *Proc. of the Brookhaven Solar Neutrino Conference BNL 50879,* **1,** 1.

Fowler, William A., 1982. "The Case of the Missing Solar Neutrinos," *AIP Conference Proceedings No. 96, Science Underground,* M. M. Nieto, W. C. Haxton, C. M. Hoffman, E. W. Kolb, V. D. Sandberg, and J. W. Toevs (eds.), American Institute of Physics, New York. p. 80.

Haxton, W. C., 1984, talk presented at Conference on the Intersection Between Particle and Nuclear Physics, Steamboat Springs (to be published).

Pontecorvo, B., 1946, *Chalk River Lab. Rep.*, **PD-205.**

Rood, R. T., 1978, in *Proceedings of the Brookhaven Solar Neutrino Conference,* G. Friedlander (ed.), **1,** 175.

Salpeter, E. E., 1968, *Comments Nucl. Part. Phys.* **II,** 97.

Zatsepin, G., 1982, in *Neutrino '82,* Proceedings of International Conference on Neutrino Physics, Balatonfured, Hungary, p. 53.

RELATIVISTIC ASTROPHYSICS

5.

Black Holes: The Membrane Viewpoint

KIP S. THORNE

Kip S. Thorne is the William R. Kenan, Jr., Professor and Professor of Theoretical Physics at Caltech. He has been a pioneer in applying the tools and results of general relativity theory to explain observed astrophysical phenomena. In addition to his numerous research accomplishments, he is the coauthor of the influential textbook *Gravitation*.

I. INTRODUCTION

A. Overview of the Membrane Viewpoint

During the past 10 years theoretical physicists exploring general relativity's predictions about the properties of black holes have gradually been led to a new viewpoint. Previously, we regarded a black hole as a nearly spherical hole in space down which things can fall but out of which nothing can come; and we thought of the hole's edge (i.e., its "horizon") as an immaterial, vacuum boundary. Now we understand that the hole's horizon behaves as though it were a membrane with a variety of simple physical properties:

The horizon behaves as though endowed with a surface pressure P_H, which is simultaneously the source of the hole's gravitational attraction and the agent that holds the horizon out against its own attractive pull. The horizon rotates rigidly with an angular velocity Ω_H, and any object very near the horizon, infalling or not, is dragged inexorably into lock-step orbital motion with that angular velocity. The horizon is endowed with a surface temperature T_H, by virtue of which it emits thermal radiation ("Hawking radiation"); and it also has an entropy S_H, which participates in both the first and the second laws of thermodynamics. The horizon possesses a surface viscosity η_H; and when tides are raised on the horizon by the gravitational pulls of external bodies, this viscosity interacts with the horizon's time-varying deformation to produce a shear stress that both slows the hole's rotation and increases its entropy ("tidal dissipation").

The horizon can also be regarded as possessing a surface density of electric charge σ_H and a surface current $\mathcal{J}_H$. From this viewpoint, when positively charged particles fall into the hole from the external universe (inflowing current), their charge does not pass through the horizon into the hole's interior. Rather, it lands on the horizon and then moves about (surface current) until such a time as infalling negative charges (outflowing current) annihilate it. The surface charge and current satisfy the law of charge conservation, Gauss's law, Ampere's law, and Ohm's law—with a surface electrical resistivity $R_H = 377$ ohms. The surface current produces ohmic dissipation (entropy increase) in the horizon, and the surface charge and current interact with electromagnetic fields to produce a Lorentz force on the horizon.

These membranelike properties of the horizon are alien to the standard view of black holes as presented, for example, in Chapters 32–34 of Misner, Thorne, and Wheeler (1973) (cited henceforth as MTW). The standard viewpoint asserts that if any observer chooses to fall into a black hole, as he passes through the horizon he will see nothing special there: no charge, no current, no viscosity, no surface pressure. . . . That standard viewpoint is correct, of course. Nevertheless, it turns out that, as seen by an observer who remains always outside the hole, the horizon behaves as though it were endowed with these surface properties. In fact, one can rewrite the exact equations of general relativity for the hole's exterior in a form ("the membrane formalism") that embodies these properties in physically appealing ways. And since astrophysicists must always remain outside the horizons of black holes in order to publish their research, they can adopt the membrane formalism and viewpoint with impunity.

There is an advantage in adopting the membrane viewpoint: That viewpoint is accompanied by mental pictures and intuitive arguments that are powerful research tools when one wishes to study black holes immersed in a complicated, astrophysically realistic external universe—a universe with orbiting and accreting plasmas, electromagnetic fields, and gravitating bodies. For this reason I have adopted the membrane viewpoint in most of my own recent black hole research.

B. History of Research on the Membrane Viewpoint

The membrane viewpoint was motivated by the results of a number of black-hole calculations in the 1970s: Hawking's (1974, 1975) discovery that black holes should radiate thermally; the Bekenstein (1972a,b, 1973, 1974) and Hawking (1976) formulation of the laws of thermodynamics for black holes; the Hawking and Hartle (1972) discovery of the viscous spindown of a tidally deformed hole (see also Hartle 1973, 1974); the Hanni and Ruffini (1973) discovery that a black hole placed in an external electric field behaves as though its surface had acquired a polarized electric charge; and Znajek's (1976, 1978) discovery that when electric current is run through a black hole (with, e.g., positive charges flowing in at the poles and negative in at the equator) the horizon behaves as though it had an electric resistivity.

Motivated by these results, Damour (1978, 1979, 1982) rewrote the equations governing the evolution of a general black hole horizon in a form where one could identify explicitly terms that "looked like" electric conductivity, surface viscosity, surface pressure, surface temperature, entropy, and so on; and Znajek (1978) independently developed a nearly equivalent but less complete version of the equations.

The beautiful and elegant Damour–Znajek formalism made it obvious that the membrane viewpoint had sufficient richness and sufficient formal justification to become an important tool for astrophysical research. However, the Damour–Znajek formalism focused attention solely on the horizon, without paying much attention to the physics of the external universe in which the horizon lives. To make the Damour–Znajek formalism into a real tool for astrophysics required marrying it to an equally appealing description of the external universe. Such a description is provided by the "3+1" formulation of general relativity (Arnowitt, Deser, and Misner 1962; Smarr and York 1978; York 1979; Chapter 21 of MTW).

The 3+1 formalism chooses a preferred family of three-dimensional, spacelike hypersurfaces in spacetime (surfaces of "constant time") and treats them as though they were a single three-dimensional space that evolves as time passes ("decomposition of four-dimensional spacetime into three-dimensional space plus one-dimensional time"). The general relativistic physics of black holes, plasmas, and accretion disks takes place in this three-dimensional space; and the relativistic laws of physics that govern them, written in three-dimensional language, resemble the nonrelativistic laws to which astrophysicists are accustomed. Thus, the 3+1 formulation is well suited to carrying physicists' nonrelativistic intuition about plasmas, hydrodrynamics, and stellar dynamics into the arena of black holes and general relativity.

The marriage of the 3+1 formalism to the Damour–Znajek horizon formalism (Thorne and Macdonald, 1982; Price and Thorne 1986) gave rise to the membrane formalism for black holes. This article is a brief introduction to electromagnetic and thermodynamic aspects of that membrane formalism—an introduction with adequate detail to give the reader a general understanding of it, but not enough detail for a full working knowledge. The reader who wants a full working knowledge might best study the pedagogical monograph Black Holes: The Membrane Paradigm (Thorne, Price, and Macdonald, eds., 1986); cited henceforth as BHMP. Also of interest may be Macdonald and Thorne (1982) and Macdonald (1984) for applications of the membrane formalism to studies of black hole magnetospheres and of black holes as power sources in active galactic nuclei and quasars (subject of Section VI of this chapter), and Macdonald and Suen (1985) and Suen, Price, and Redmount (1986) for technical details of a variety of model problems that give insight into the membrane viewpoint.

C. Overview of this Chapter

Throughout this chapter we shall restrict attention to black holes that gravitationally dominate their own neighborhoods; that is, we shall insist that the total mass-energy in matter and nongravitational fields near a hole be negligible compared to the hole's own mass. Then regardless of how dynamical the matter and fields may be, they cannot induce the hole to vibrate significantly; and regardless of how they might be distributed, they cannot deform it significantly. Their interaction with the hole will involve an exchange of energy and angular momentum that may be very significant for the matter and nongravitational fields; but it will be significant for the hole only when integrated up over astrophysically long time scales. As a result, the black hole will be well approximated at any given time as a smoothly rotating, axisymmetric "Kerr" hole whose mass M and angular momentum J will evolve significantly only on time scales $>>\gg GM/c^3$. This restriction to "quiescent" black holes is not an essential feature of the membrane formalism, but it simplifies the formalism considerably.

The organization of this chapter is as follows: In Section II we shall present the $3+1$ split of spacetime into space plus time for the exterior of a quiescent black hole, and we shall study the resulting $3+1$ version of the laws of gravity and electromagnetism outside a black hole. In Section III we shall introduce the concept of the "stretched horizon" and shall study the membranelike laws of thermodynamics that it obeys. In Section IV we shall study the stretched horizon's electrodynamic properties. In Section V we shall present a series of electromagnetic model problems that illustrate the membrane formalism; among these are electric circuits in which a black hole plays, variously, the role of a resistor, the role of the rotor of an electric motor, and the role of a battery. In Section VI we shall sketch very briefly some astrophysical applications of the membrane formalism. Finally, in Section VII we shall list a number of features of the membrane formalism that are not treated in this chapter.

Although our presentation will allude to the formalism of general relativity in a number of places, it should be understandable to people who are not relativity aficionados. The general relativistic allusions are intended to help the aficionados understand the underpinnings of the membrane formalism. The nonaficionado should pass quickly through those allusions and focus on the main theme of the presentation: the three-dimensional description of black holes and its close resemblance to elementary three-dimensional physics.

To simplify our notation, we shall use "geometrized units" in which Newton's gravitational constant G and the speed of light c are set equal to unity. This means, in particular, that mass, energy, length, and time are all measured in the same units with conversion factors

$$1.0 = c = 2.998 \times 10^{10} \text{ cm sec}^{-1}, \qquad 1.0 = c^2 = 8.988 \times 10^{20} \text{ erg g}^{-1}$$

$$1.0 = \frac{G}{c^2} = 0.7425 \times 10^{-28} \text{ cm g}^{-1} \tag{1}$$

Note that in these units the mass of the sun is $M_\odot = 1.989 \times 10^{33}$ g $= 1.477 \times 10^5$ cm. The reader can restore the factors of G and c to any equation by requiring that the dimensions come out the same on both sides of the equality, in cgs units. Occasionally in our presentation the factors of G and c will be restored as a pedagogical device. For further discussion of geometrized units see, for example, pp. 35 and 36 of MTW.

II. THE 3+1 SPLIT OF SPACETIME

In this section we shall study the 3+1 description of physics outside a black hole. We shall begin (Section A) with the split of spacetime into absolute space and universal time, and the corresponding 3+1 split of the spacetime metric. Next (Section B) we shall introduce and study the gravitational analogs of the electric field and magnetic field, which live in absolute space. Finally (Section C), we shall study the 3+1 description of electromagnetism (the real thing, not the gravitational analog).

A. ZAMOs and the 3+1 Split of the Metric

In flat spacetime there are preferred families of observers: "inertial," or "Lorentz" observers; and corresponding to a given inertial family there is a preferred spacetime coordinate system x^α ($\alpha = 0, 1, 2, 3$) in which the family members are all at rest and the metric has the standard Lorentz form

$$ds^2 = -(dx^0)^2 + \delta_{ij}\, dx^i\, dx^j; \qquad \delta_{ij} \equiv (1 \text{ if } i = j;\ 0 \text{ if } i \neq j) \tag{2}$$

(Here and throughout we adopt Einstein's summation convention: When an index is repeated, once up and once down, it is to be summed; also we use Latin indices to denote components of three-dimensional, spatial vectors and tensors.) When one adopts the physical viewpoint of these inertial observers, one thereby splits spacetime into three-dimensional space with the three-dimensional Euclidean coordinates x^i and metric $ds^2 = \delta_{ij}\, dx^i\, dx^j$, plus one-dimensional time $t = x^0$. Correspondingly, one splits the four-dimensional electromagnetic field tensor $F_{\alpha\beta}$ into two parts:

the three-dimensional electric field $E_i = F_{i0}$ measured by the chosen observers and the three-dimensional magnetic field $(B_1, B_2, B_3) = (F_{23}, F_{31}, F_{12})$ that they measure.

In a generic curved spacetime there are no preferred families of observers—which is why relativists often insist on using a four-dimensional formalism with space and time unified into space plus time. However, in the special case of a quiescent black hole, curved spacetime does possess a preferred family of observers: the "zero-angular-momentum observers," or ZAMOs, of Bardeen (1973); and when one adopts the physical viewpoint of those ZAMOs, one thereby obtains the 3+1 split used in the membrane formalism. (In much of the membrane-formalism literature the ZAMOs are referred to as "fiducial observers," or "FIDOs.")

For relativity aficionados we exhibit explicitly the 3+1 split of the spacetime metric of a quiescent (Kerr) black hole. Written in a form that leads quickly to the 3+1 split, the metric is (cf. Chap. 33 of MTW or Section 11 and Chap. 6 of Chandrasekhar 1983)

$$ds^2 = -\alpha^2 \, dt^2 + g_{ij}(dx^i + \beta^i \, dt)(dx^j + \beta^j \, dt). \tag{3}$$

Here α, β^i, and g_{ij} are expressed in terms of the hole's mass M, angular momentum per unit mass $a \equiv J/M$, and "Boyer–Lindquist" spacetime coordinates $x^0 = t$, $x^1 = r$, $x^2 = \theta$, $x^3 = \varphi$ as

$$\alpha = \frac{\rho}{\Sigma}\sqrt{\Delta}, \qquad \beta^r = \beta^\theta = 0, \qquad \beta^\varphi = -\omega \tag{4a}$$

$$g_{rr} = \frac{\rho^2}{\Delta}, \qquad g_{\theta\theta} = \rho^2, \qquad g_{\varphi\varphi} = \tilde{\omega}^2, \qquad \text{other } g_{ij} = 0 \tag{4b}$$

where

$$\Delta \equiv r^2 + a^2 - 2Mr, \qquad \rho^2 = r^2 + a^2 \cos^2\theta \tag{4c}$$

$$\Sigma^2 \equiv (r^2 + a^2)^2 - a^2 \Delta \sin^2\theta, \ \omega = \frac{2aMr}{\Sigma^2}, \ \tilde{\omega} = \frac{\Sigma \sin\theta}{\rho} \tag{4d}$$

In the special case of a nonrotating hole, $a = 0$, these complicated formulas reduce to the standard "Schwarzschild" ones:

$$\alpha = \left(1 - \frac{2M}{r}\right)^{1/2}, \qquad \beta^i = 0,$$

$$g_{rr} = \left(1 - \frac{2M}{r}\right)^{-1}, \qquad g_{\theta\theta} = r^2, \qquad g_{\varphi\varphi} = r^2 \sin^2\theta$$

The explicit functional forms in Eqs. (4) and (5) are not of central interest for the 3+1 split. Focus attention instead on Eq. (3). It shows explicitly a split of the spacetime metric $g_{\mu\nu}$ into three parts: α, β^i, and g_{ij}. This is the analog for gravity of the electromagnetic split $F_{\mu\nu} \rightarrow (E_i, B_j)$.

Just as the inertial observers of flat spacetime see all events with the same time x^0 as physically simultaneous, so the ZAMOs outside a black hole see all events with the same time $t = x^0$ as physically simultaneous. For this reason we shall call t "universal time," and we shall mentally collapse all the hypersurfaces of constant t together into a single three-dimensional space, called "absolute space," in which physics occurs as universal time evolves.

Just as in flat spacetime one obtains the Euclidean metric $ds^2 = \delta_{ij}\, dx^i\, dx^j$ of three-dimensional space by setting $dx^0 = 0$ in the spacetime metric (2), so in a black hole spacetime one obtains the metric of absolute three-dimensional space by setting $dt = 0$ in the spacetime metric (3):

$$\begin{aligned} ds^2 = g_{ij}\, dx^i\, dx^j &= \left(\frac{\rho^2}{\Delta}\right) dr^2 + \rho^2\, d\theta^2 + \tilde{\omega}^2\, d\varphi^2 \quad \text{for rotating hole} \\ &= \left(1 - \frac{2M}{r}\right)^{-1} dr^2 + r^2\, d\theta^2 + r^2 \sin^2\theta\, d\varphi^2 \quad \text{for nonrotating hole} \end{aligned} \tag{6}$$

The space curvature inherent in this metric can be visualized by means of an "embedding diagram": Imagine extracting the equatorial plane, $\theta = \pi/2$, from the curved space of a nonrotating hole and embedding it with unchanged geometry in a flat Euclidean 3-space with cylindrical coordinates $\bar{r}$, $\bar{z}$, and $\bar{\varphi}$. The embedded surface will have an axisymmetric shape $\bar{z} = \bar{z}(\bar{r})$ computable by equating $\bar{\varphi} = \varphi$ and

$$\begin{aligned} ds^2 &= d\bar{z}^2 + d\bar{r}^2 + \bar{r}^2\, d\bar{\varphi}^2 \\ &= \left(1 - \frac{2M}{r}\right)^{-1} dr^2 + r^2\, d\varphi^2 \end{aligned} \tag{7}$$

The result is easily worked out to be (cf. pp. 612–615 of MTW)

$$\bar{z} = \sqrt{8M(\bar{r} - 2M)} \tag{8}$$

which is the paraboloid of revolution shown schematically in Fig. 1. For a rotating hole ($a \neq 0$) the embedding diagram is qualitatively the same.

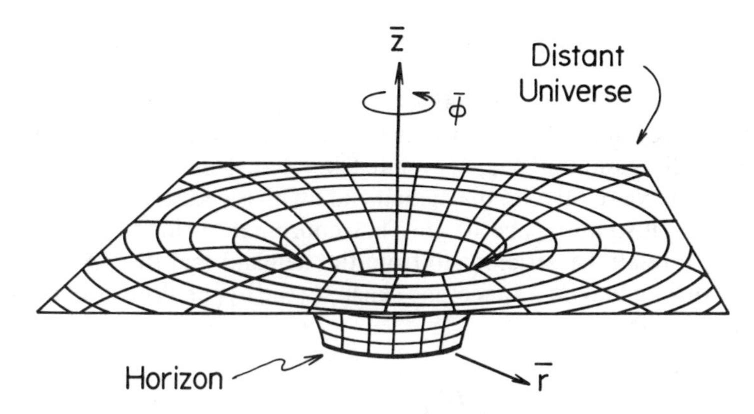

FIGURE 1. Embedding diagram for the curved space outside a nonrotating black hole. As discussed in the text, the intrinsic geometry of this trumpet-horn-like surface in Euclidean space is the same as the intrinsic geometry of the equatorial plane of the hole's curved absolute space. This figure is schematic, not exact.

Notice that this geometry has the property that, as we move inward from one equatorial circle surrounding the hole to another, the circumference decreases at a rate

$$\frac{d(\text{circumference})}{d(\text{distance})} = \frac{d(2\pi r)}{(1 - 2M/r)^{-1/2}\, dr} = 2\pi\left(1 - \frac{2M}{r}\right)^{1/2}. \tag{9}$$

At large radii $r \gg 2M$ this is the same rate, 2π, as in Euclidean space. But as we get closer and closer to the hole's horizon, $r = 2M$, the rate of decrease of circumference gets smaller and smaller until finally, at the horizon, the rate goes to zero and the geometry develops a "neck" (Fig. 1).

The rotation of the hole (angular momentum J; angular momentum per unit mass $a \equiv J/M$) "drags" absolute space into a tornadolike circulating motion, and correspondingly drags all physical objects in absolute space into circulating orbital motion; nothing physical can resist. The nearer one is in absolute space to the horizon, the stronger is this circulation; so if we let our spatial coordinates (r, θ, φ) succumb to it, our radial coordinate lines will quickly be sheared into spirals, and as time passes the spirals will get wound up more and more tightly by the hole, like thread being wound onto a spool. To avoid this, we insist that our coordinates (which, after all, are not physical objects) resist the tornadolike circulation and remain always and everywhere at rest relative to the distant stars. This, in fact, is the nature of the coordinates $(x^1, x^2, x^3) = (r, \theta, \varphi)$ that appear in Eqs. (3)–(6).

The ZAMOs, on the other hand, are physical observers and thus cannot resist the circulation. In fact, because they see all events in absolute space as physically simultaneous, they must be at rest in absolute space and thus must move relative to the distant stars with the same finite coordinate velocity as absolute space itself. That coordinate velocity can be computed by a relativity afficionado from the demand that the ZAMO world lines be orthogonal, in a spacetime sense, to the hypersurfaces of constant t. The result is

$$\left(\frac{dx^i}{dt}\right)_{\text{ZAMO}} = -\beta^i \tag{10}$$

where β^i are the metric coefficients of Eqs. (3) and (4a). Since $\beta^r = \beta^\theta = 0$, $\beta^\varphi = -\omega$, the ZAMOs orbit the hole with angular velocity relative to the distant stars $d\varphi/dt = \omega$. At large radii this ZAMO (and absolute space) angular velocity is $\omega \simeq 2J/r^3$. As r decreases, ω increases (stronger dragging at smaller radii), until at the horizon, that is, at radius

$$r_H = M + (M^2 - a^2)^{1/2} \tag{11}$$

ω becomes equal to the hole's angular velocity of rotation Ω_H:

$$\omega \to \Omega_H \equiv \frac{a}{2Mr_H} \qquad \text{as } r \to r_H \tag{12}$$

(All objects near the horizon, including ZAMOs and absolute space, get dragged into lockstep rotation with precisely this angular velocity.)

Because β^i describes the continual shifting of the star-fixed coordinates relative to the ZAMOs [Eq. (10)], it is sometimes called the "shift function."

In the membrane formalism a special role is played by physical quantities that the ZAMOs measure. Of special importance is the proper time τ measured by the clock that a ZAMO carries. From the spacetime metric (3) with $(dx^i/dt)_{\text{ZAMO}} = \beta^i$ we see that

$$\left(\frac{d\tau}{dt}\right)_{\text{ZAMO}} = \alpha \tag{13}$$

Thus, α measures the amount of proper time that elapses per unit universal time—which is why α is sometimes called the "lapse function."

Because α is unity far from the horizon ($\alpha \to 1$ as $r \to \infty$), universal time t is the same as ZAMO proper time there. Thus, $\alpha(r, \theta)$ can be thought of as the ratio of ZAMO clock ticking rates at location (r, θ) to ZAMO clock ticking rates at infinity—which is why α is sometimes called the "gravitational redshift function." As one moves inward from $r = \infty$ toward $r = r_H$, α decreases monotonically from unity to zero

$$\alpha \simeq 1 - \frac{M}{r} \qquad \text{at } r \gg r_H$$

$$\alpha \simeq \left[\frac{2(r_H{}^2 + a^2 \cos^2 \theta)(r_H - M)}{(2Mr_H)^2}\right]^{1/2} (r - r_H)^{1/2} \qquad \text{at } r \simeq r_H \quad (14)$$

Correspondingly, the gravitational redshift of clock ticking rates ($dt/d\tau = 1/\alpha$) becomes infinite at the horizon.

To recapitulate: When one adopts the physical viewpoint of the ZAMOs, four-dimensional spacetime splits up into three-dimensional absolute space plus universal time t; and the metric $g_{\mu\nu}$ of spacetime splits up into the 3-metric g_{ij} of absolute space, plus the shift function $\beta^i = -(dx^i/dt)_{\text{ZAMO}}$, plus the lapse function (gravitational redshift function) $\alpha = (d\tau/dt)_{\text{ZAMO}}$.

B. Gravitoelectric and Gravitomagnetic Fields

(i) Regions of Strong Gravity. Physical quantities measured by ZAMOs are described by scalars, 3-vectors, and 3-tensors that reside in absolute space. An important example is the "acceleration of gravity" $\mathbf{g}$ (abstract notation) or g_i (component notation) that a ZAMO measures using a standard earth-type gravimeter or accelerometer that he carries with himself. It turns out [e.g., Eq. (5) of Thorne and Macdonald 1982] that

$$\mathbf{g} = -\frac{1}{\alpha}\nabla\alpha, \qquad \text{that is, } g_i = -\frac{1}{\alpha}\frac{\partial\alpha}{\partial x^i} \qquad (15)$$

where ∇ is the gradient in absolute space. At large radii, where $\alpha \simeq 1 - M/r$, $\mathbf{g}$ has the Newtonian form $(-M/r^2)\mathbf{e}_{\hat{r}}$, with $\mathbf{e}_{\hat{r}}$ the unit radial vector;

nearer the hole it is stronger than Newtonian theory would predict; and very near the horizon it becomes divergently large:

$$\mathbf{g} \simeq -\left(\frac{M}{r^2}\right)\mathbf{e}_{\hat{r}} \qquad \text{at } r \gg r_H \tag{16a}$$

$$\mathbf{g} \simeq -\frac{1}{\alpha}\left(\frac{r_H - M}{2Mr_H}\right)\mathbf{e}_{\hat{r}} \qquad \text{as } r \to r_H \text{ and } \alpha \to 0 \tag{16b}$$

Since $m\mathbf{g}$ is the gravitational force exerted on a mass m at rest with respect to a ZAMO and $q\mathbf{E}$ is the electric force on a charge q, one can regard $\mathbf{g}$ as the gravitational analog of the electric field $\mathbf{E}$. For this reason $\mathbf{g}$ is sometimes called the *gravitoelectric field*. General relativity also predicts the existence of a gravitational analog $\mathbf{H}$ of the magnetic field $\mathbf{B}$. This gravitomagnetic field is given by

$$\mathbf{H} \equiv \frac{1}{\alpha}\nabla \times \boldsymbol{\beta} \tag{17}$$

where $\nabla\times$ is the curl in absolute space; and the lapse function $\boldsymbol{\beta}$ is sometimes called the *gravitomagnetic potential*. Whereas the $\mathbf{g}$ of a black hole has a monopolar form (analog of electric field of a point charge), $\mathbf{H}$ has a dipolar form (analog of magnetic field of a spinning charge distribution):

$$\mathbf{H} \simeq 2\left[\frac{\mathbf{J} - 3(\mathbf{J}\cdot\mathbf{e}_{\hat{r}})\mathbf{e}_{\hat{r}}}{r^3}\right] \qquad \text{at } r \gg r_H \tag{18a}$$

$$\mathbf{H} \simeq -\frac{1}{\alpha}\frac{4Jr_H}{(r_H{}^2 + a^2\cos^2\theta)^2}\cos\theta\,\mathbf{e}_{\hat{r}} \qquad \text{at } r \simeq r_H \tag{18b}$$

(Fig. 2). Note from (18a) that the gravitomagnetic dipole moment is the hole's angular momentum $\mathbf{J}$.

(ii) Regions of Weak Gravity. The magneticlike role of $\mathbf{H}$ shows up most clearly in the equation of motion for a low-velocity test particle of mass m moving along a trajectory $x^i(t)$ [$\mathbf{x}(t)$ in abstract notation] far from the hole, where gravity is weak:

$$m\frac{d^2\mathbf{x}}{dt^2} = m\left(\mathbf{g} + \frac{d\mathbf{x}}{dt} \times \mathbf{H}\right) \tag{19}$$

(Forward 1961; Braginsky, Caves, and Thorne 1977). This equation of

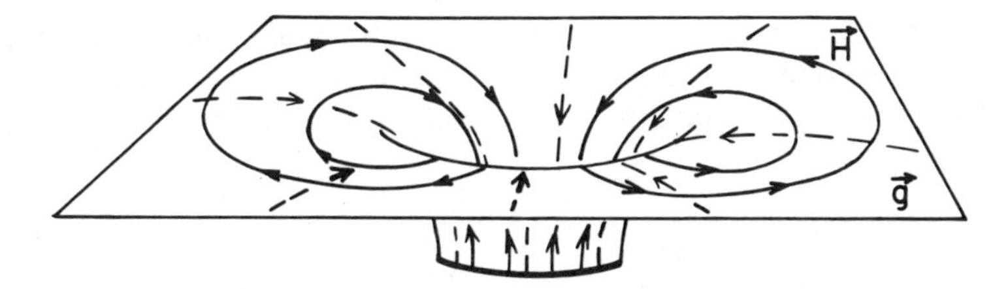

FIGURE 2. The gravitoelectric (ordinary gravitational acceleration) field **g** (dashed lines) and gravitomagnetic field **H** (solid lines) outside a black hole. These fields reside in the hole's absolute space, so they are shown drawn on an embedding diagram like that of Fig. 1. This figure is schematic, not exact.

motion [which is really the geodesic equation in the spacetime metric (3), converted to $3+1$ language] is identical to the Lorentz force law, with charge q replaced by mass m, electric field $\mathbf{E}$ replaced by gravitoelectric field $\mathbf{g}$, and magnetic field $\mathbf{B}$ replaced by gravitomagnetic field $\mathbf{H}$.

As an aside, aimed at reinforcing the electromagnetic analogy, consider a weakly gravitating, rotating body such as the earth or the sun, with all nonlinear gravitational effects neglected. Such a body can be described by the same $3+1$ split (3) as we have used for black holes; but now the Lorentz-force-like equation of motion (19) is valid inside and near the body as well as far away. Moreover, in this case we can compute the gravitoelectric and gravitomagnetic fields $\mathbf{g}$ [Eq. (15)] and $\mathbf{H}$ [Eq. (17)] from the linearized Einstein equations. It turns out (Forward 1961; Braginsky, Caves, and Thorne 1977) that for a time-independent body those Einstein equations are almost identical to Maxwell's equations:

$$\nabla \times \mathbf{H} = \nabla \times \nabla \times \boldsymbol{\beta} = -\nabla^2 \boldsymbol{\beta} = -16\pi G \rho_m \mathbf{v}, \qquad \nabla \times \mathbf{g} = 0 \tag{20a}$$

$$\nabla \cdot \mathbf{g} = -\nabla^2 \alpha = -4\pi G \rho_m, \qquad \nabla \cdot \mathbf{H} = 0 \tag{20b}$$

The only differences are minus signs (due to gravity being attractive rather than repulsive), a factor 4 in the $\nabla \times \mathbf{H}$ equation (presumably due to gravity being a spin-two field rather than spin-one), the presence of the gravitation constant G (which we usually set to unity), the replacement of charge density ρ_e by mass density ρ_m, and the replacement of current density $\mathbf{j}$ by the density of mass current $\rho_m \mathbf{v}$ (with $\mathbf{v}$ the velocity of the mass). By the standard manipulations of elecromagnetic theory one can verify that far outside the rotating, gravitating body $\mathbf{g}$ and $\mathbf{H}$ have the same monopolar and dipolar forms (16a), (18a) as those for a black hole, with the gravitoelectric monopole moment being the body's mass M and the gravitomagnetic dipole moment the body's angular momentum **J**. And from the linearized Einstein equations one can verify that the metric of absolute space far outside the body has the same form (4b) as for a black hole:

$$\mathbf{g} \simeq -\frac{M}{r^2}\mathbf{e}_{\hat{r}}, \qquad \mathbf{H} \simeq 2\left[\frac{\mathbf{J} - 3(\mathbf{J}\cdot\mathbf{e}_{\hat{r}})\mathbf{e}_{\hat{r}}}{r^3}\right] \tag{21a}$$

$$ds^2 \simeq \left(1 + \frac{2M}{r}\right) dr^2 + r^2(d\theta^2 + \sin^2\theta\, d\varphi^2) \tag{21b}$$

For time-varying bodies (e.g., pulsating or collapsing or exploding bodies) the analogy with electromagnetism remains strong so long as all velocities are small compared with light and gravity is weak enough to be linear. (See Braginsky, Caves, and Thorne 1977 for details.) But large velocities and strong gravity distort the analogy (though they do not fully destroy it; for an extensive discussion see Pollock 1977).

For further insight into the electromagnetic analogy, consider a gyroscope with spin angular momentum **s** in orbit around a rotating body (e.g., the Earth or a black hole) at a large enough distance that the weak-gravity formulas (21) apply. The three aspects of the body's gravity (**g**-field, **H**-field, and space curvature) will each produce a precession of the gyroscope relative to the distant stars.

The interaction of the gyroscope's spin **s** with the earth's gravitomagnetic field **H** is analogous to the interaction of a magnetic dipole $\boldsymbol{\mu}$ with a magnetic field **B**. Just as a torque $\boldsymbol{\mu} \times \mathbf{B}$ acts in the magnetic case, so a torque $\frac{1}{2}\mathbf{s} \times \mathbf{H}$ acts in the gravitational case. [Eqs. (19)–(21) dictate that $\boldsymbol{\mu} \to \frac{1}{2}\mathbf{s}$, $\mathbf{B} \to \mathbf{H}$.] The gyroscope's angular momentum is changed by this torque:

$$\frac{d\mathbf{s}}{dt} = \tfrac{1}{2}\mathbf{s} \times \mathbf{H} \qquad \text{s precesses with the ``gravitomagnetic'' angular velocity } \boldsymbol{\Omega}_{\text{GM}} = -\mathbf{H}/2 \tag{22}$$

Note that Ω_{GM} is independent of the structure of the gyroscope; this is a manifestation of the principle of equivalence, and it permits one to regard the precession as a "dragging of inertial frames" by the rotation of the body. This gravitomagnetic precession is often called the "Lense–Thirring" precession, since Thirring and Lense (1918) were the first to discover it in the equations of general relativity. For a gyroscope in a 500-km-high polar orbit above the earth, as planned in 1990 for NASA's "gravity probe B'' (which will attempt to verify for the first time the existence of gravitomagnetism), $\mathbf{\Omega}_{\text{GM}}$ averages over the orbit to give

$$\langle \mathbf{\Omega}_{\text{GM}} \rangle = \frac{G}{2c^2}\frac{\mathbf{J}}{r^3}$$

$$\simeq 0.05 \text{ arcsec/year} \qquad \text{for GP-B orbiting earth} \tag{23}$$

This is 50 times larger than the design sensitivity of GP-B (Everitt 1974).

The interaction of the orbiting gyroscope with the body's gravitoelectric field $\mathbf{g}$ is analogous to the interaction of a classical spinning electron with the Coulomb electric field $\mathbf{E}$ of an atomic nucleus. Just as motion of the electron through $\mathbf{E}$ induces in the electron's rest frame a magnetic field $\mathbf{B}_{\text{induced}} = -\mathbf{v} \times \mathbf{E}$ and a torque $\boldsymbol{\mu} \times \mathbf{B}_{\text{induced}}$ and a resulting precession of the electron spin (''atomic spin-orbit coupling''), similarly motion of the gyroscope through $\mathbf{g}$ produces in the gyroscope's rest frame a gravitomagnetic field $\mathbf{H}_{\text{induced}} = -\mathbf{v} \times \mathbf{g}$ and a torque $\frac{1}{2}\mathbf{s} \times \mathbf{H}_{\text{induced}}$ and a resulting ''spin-orbit'' precession of the gyroscope with

$$\mathbf{\Omega}_{\text{SO}} = -\tfrac{1}{2}\mathbf{H}_{\text{induced}} = \left(\frac{r_g}{2r}\right)^{5/2} \frac{\mathbf{n}}{r_g/c}$$

$$\simeq 2.3 \text{ (arcsec/year)} \qquad \text{for GP-B orbiting earth} \tag{24}$$

Here $r_g = 2GM/c^2$ is the body's ''gravitational radius'' and $\mathbf{n}$ is the unit normal to the orbital plane.

In the hydrogen atom there is also a ''Thomas precession'' that results from the fact that the product of two ''velocity boosts'' is a combined ''boost and rotation.'' In our gravitational problem the Thomas precession is absent because the gyroscope is presumed to be in a free-fall orbit—that is, it is not accelerated relative to local inertial frames; there are no ''boosts.'' On the other hand, the gravitational field has an aspect, the curvature of space, with no electromagnetic analog; and that space curvature produces an unfamiliar type of precession:

In Figure 3a we see an embedding diagram (cf. Fig. 1) for the curved absolute space of the rotating body. The dashed line is the circular orbit of the gyroscope. The gyroscope can only feel that part of space that is in the immediate neighborhood of its orbit. This allows the pedagogical simplification of replacing the paraboloidal embedding surface of the real curved space by a cone (dotted line) that is tangent to the paraboloid at the gyroscope's orbit. Such a cone can be constructed by drawing a circle on a flat sheet of paper (Fig. 3b), cutting the pie slice (B–A–B') out of

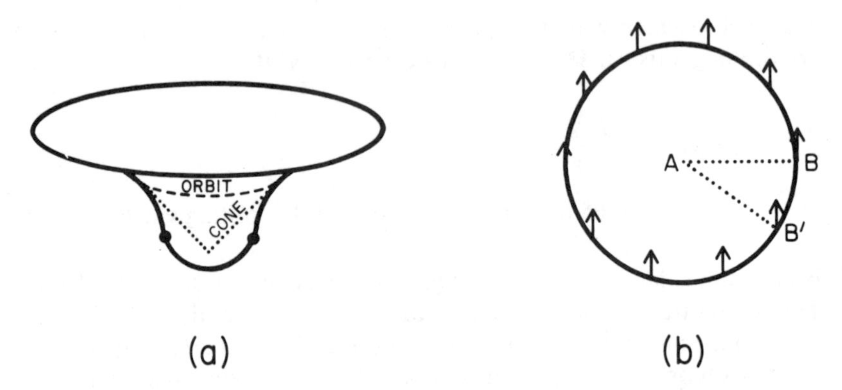

FIGURE 3. Precession of a gyroscope induced by the curvature of absolute space around a body such as the earth or a black hole. (Adapted from Thorne and Blandford 1982.)

it, and joining the edges A–B and A–B' together. As the gyroscope orbits around the cone it always keeps its spin in the same fixed direction on the flat-sheet-of-paper geometry of the cone's surface (arrows in Fig. 3b; no local precession). However, as one easily sees by cutting out the pie slice and pasting the cone together, there will be a net precession of the spin when the gyroscope returns to its starting point $B = B'$. From the shape of the embedding surface [Eq. (8)] and the cone construction one easily finds that the net precession angle after a single Keplerian orbital period $(\pi r_g/c)(2r/r_g)^{3/2}$ corresponds to a precession angular velocity

$$\mathbf{\Omega}_{\text{SC}} = 2\left(\frac{r_g}{2r}\right)^{5/2}\frac{\mathbf{n}}{r_g/c} = 2\mathbf{\Omega}_{\text{SO}} \tag{25}$$

Note that this "space-curvature precession" has twice the angular velocity of the spin-orbit precession. The two together bear the name "geodetic precession":

$$\mathbf{\Omega}_{\text{geo}} = \mathbf{\Omega}_{\text{SO}} + \mathbf{\Omega}_{\text{SC}} = 3\left(\frac{r_g}{2r}\right)^{5/2}\frac{\mathbf{n}}{r_g/c}$$

$$\simeq 6.9 \text{ arcsec/year} \qquad \text{for GP-B orbiting earth} \tag{26}$$

The reader may find it enlightening to compare the preceding derivations of the relativistic precession formulas with the standard derivation in §40.7 of MTW.

(iii) Regions of Strong Gravity Again. Thus far our discussion of the electromagnetic analogy has focused on weak gravity. Turn attention, now, to measurements made by a ZAMO at an arbitrary location outside a rotating black hole. The ZAMO can measure the gravitoelectric field (local acceleration of gravity) $\mathbf{g}$ using a gravimeter or accelerometer in the standard way. Similarly, he can measure the gravitomagnetic field $\mathbf{H}$ using a gyroscope that he carries with himself. So long as he applies at the gyroscope's center of mass the force $-m\mathbf{g}$ required to keep the gyroscope at rest with respect to himself, the gyroscope will precess relative to absolute space with the same gravitomagnetic angular velocity $\mathbf{\Omega}_{\rm GM} = -\frac{1}{2}\mathbf{H}$ as we encountered in regions of weak gravity:

$$\frac{d\mathbf{s}}{d\tau} \equiv \frac{1}{\alpha}\left(\frac{\partial}{\partial t} - \boldsymbol{\beta}\cdot\nabla\right)\mathbf{s} = -\tfrac{1}{2}\mathbf{H} \times \mathbf{s} \tag{27}$$

(law of "Fermi–Walker transport" rewritten in 3+1 language). Here the factor α converts from "per unit universal time" to "per unit ZAMO proper time"; $\partial/\partial t - \boldsymbol{\beta}\cdot\nabla$ is the derivative with respect to universal time moving along the ZAMO's coordinate orbit $d\mathbf{x}/dt = -\boldsymbol{\beta}$ (i.e., at rest in absolute space); and ∇ is the spatial covariant derivative in absolute space.

For further insight into the gravitoelectric and gravitomagnetic fields, including some aspects that are rather different from electromagnetism, see BHMP (Thorne, Price, and Macdonald, eds., 1986).

C. 3+1 Split of Electrodynamics

We turn attention now from the gravitational fields of a black hole (g_{ij}, β_i, α; $\mathbf{g}$, $\mathbf{H}$) to electromagnetic fields in the absolute space around a black hole. By adopting the physical viewpoint of the ZAMOs, we induce a 3+1 split in the laws of electromagnetism (see Macdonald and Thorne 1982 for a derivation, and BHMP for a detailed discussion):

The electromagnetic field tensor $F_{\mu\nu}$ splits up into the electric field $\mathbf{E}$ and magnetic field $\mathbf{B}$ measured by a ZAMO, which like $\mathbf{g}$ and $\mathbf{H}$ are 3-vectors residing in absolute space. These $\mathbf{E}$ and $\mathbf{B}$ are defined physically in terms of the Lorentz force

$$\left(\frac{d\mathbf{p}}{d\tau}\right)_{\rm em} = q(\mathbf{E} + \mathbf{v} \times \mathbf{B}) \tag{28}$$

that a ZAMO sees them exert on a particle that he measures to have charge q, mass m, velocity $\mathbf{v}$, and momentum $\mathbf{p} = m\mathbf{v}/(1 - \mathbf{v}^2)^{1/2}$. Similarly the 4-current J_μ splits up into the charge density ρ_e and current density $\mathbf{j}$, which are a scalar and 3-vector that reside in absolute space and that are defined physically in terms of ZAMO measurements as

$$\rho_e = \left(\begin{matrix}\text{charge per unit} \\ \text{ZAMO-measured volume}\end{matrix}\right),$$

$$\mathbf{j}\cdot\mathbf{n} = \left(\begin{matrix}\text{charge crossing a unit area} \\ \text{orthogonal to } \mathbf{n}\text{, per unit time } \tau\end{matrix}\right). \tag{29}$$

Total charge is conserved in the sense that the charge inside a 3-volume $\mathcal{V}$, whose walls are carried by ZAMOs, changes at a rate given by the integral of the current density over the volume's walls $\partial\mathcal{V}$

$$\frac{d}{dt}\int_{\mathcal{V}} \rho_e \, dV = -\int_{\partial\mathcal{V}} \alpha\mathbf{j}\cdot d\mathbf{A} \qquad \text{if } \partial\mathcal{V} \text{ is carried by ZAMOs} \tag{30}$$

Here dV and dA are volume and area elements as measured by the ZAMOs. Note that the time derivative is with respect to universal time t, not ZAMO time τ. This is because the quantity being differentiated is an integral over some finite region, at different points of which ZAMO time τ ticks at different rates. Such a "global" quantity can only be differentiated with respect to a position-independent time—and the only such time we have is universal time t. Correspondingly, on the right-hand side of (3) the factor $\alpha = d\tau/dt$ must be introduced to convert $\mathbf{j}$ from a "per unit ZAMO time τ" basis to the same "per unit universal time t" basis as is used on the left-hand side. These types of time renormalizations occur extensively in the membrane formalism and may lead to confusion at first. However, because they always have a simple physical or mathematical justification (as above), after a little experience one soon becomes comfortable with them and manipulates them with ease.

Corresponding to the global law of charge conversation (30) there is a local differential law:

$$\alpha\frac{d\rho_e}{d\tau} \equiv \left(\frac{\partial}{\partial t} - \boldsymbol{\beta}\cdot\nabla\right)\rho_e = -\nabla\cdot(\alpha\mathbf{j}_e) \tag{30'}$$

Notice that the time derivative $d/d\tau$ moving with a ZAMO that appears here is the same as that which appeared in the law of gravitomagnetic precession (27).

As in flat space, so also in the curved space around a black hole, the electric and magnetic fields E and B are generated by the charge density ρ_e and current density $\mathbf{j}$ via Maxwell's equations. The $\nabla\cdot\mathbf{E}$ and $\nabla\cdot\mathbf{B}$ Maxwell equations have identically the same form as in flat space

$$\nabla\cdot\mathbf{E} = 4\pi\rho_e \tag{31a}$$

$$\nabla\cdot\mathbf{B} = 0 \tag{31b}$$

except that ∇ is now the covariant divergence in the hole's curved, absolute three-dimensional space rather than the flat-space divergence. As in flat space, so also here, these "Gauss laws" have a simple physical interpretation: One can associate with $\mathbf{E}$ electric field lines whose density (number per unit area in absolute space, as measured by ZAMOs) is proportional to $|\mathbf{E}|$ and whose directions are along $\mathbf{E}$; and one can associate with $\mathbf{B}$ similar magnetic field lines. Then Eq. (31b) says that magnetic field lines never end, and Eq. (31a) says that electric field lines end only on electric charge. We shall find electric and magnetic field lines to be very useful tools for physical insight in our studies below of electromagnetic properties of black holes.

The other two Maxwell equations are most conveniently written in integral rather than differential forms: Consider a closed curve $\mathscr{C}$ which bounds a 2-surface $\mathscr{A}$, that is, $\partial\mathscr{A} \equiv$ (boundary of $\mathscr{A}$) $= \mathscr{C}$; and insist, for simplicity, that $\mathscr{C}$ and $\mathscr{A}$ be carried by ZAMOs (i.e., be at rest in absolute space). (For a rewrite of these Maxwell equations with $\mathscr{C}$ and $\mathscr{A}$ moving in an arbitrary manner, see BHMP.) Then

$$\oint_{\mathscr{C}} \alpha\mathbf{E}\cdot d\mathbf{l} = -\frac{d}{dt}\int_{\mathscr{A}} \mathbf{B}\cdot d\mathbf{A} \tag{31c}$$

$$\oint_{\mathscr{C}} \alpha\mathbf{B}\cdot d\mathbf{l} = +\frac{d}{dt}\int_{\mathscr{A}} \mathbf{E}\cdot d\mathbf{A} + 4\pi\int_{\mathscr{A}} \alpha\mathbf{j}\cdot d\mathbf{A} \tag{31d}$$

Here $d\mathbf{l}$ and $d\mathbf{A}$ are lengths and areas of integration in absolute space as measured by ZAMOs.

Equation (31c) is Faraday's law of induction. It says that a time-changing magnetic flux through the curve $\mathscr{C}$ induces an EMF around $\mathscr{C}$ in virtually the same manner as in flat spacetime. Note that the time derivative in Faraday's law must be d/dt rather than $d/d\tau$ and the time renormalization factor α is present for the same reasons as in the law of charge conservation (30). The fact that the EMF is a line integral of $\alpha\mathbf{E}$ rather than $\mathbf{E}$ goes hand in hand with the fact that, when $\mathbf{B} = 0$ it is $\alpha\mathbf{E}$ rather than $\mathbf{E}$ that is the gradient of a potential [cf. Eq. (37) of Thorne and Macdonald 1982].

Equation (31d) is Ampere's law. It says that the total current through $\mathscr{C}$, measured on a per-unit-universal-time basis ($\alpha\mathbf{j}$ rather than $\mathbf{j}$), plus the displacement current, induces a magnetic field around $\mathscr{C}$ in virtually the same manner as in flat spacetime.

The equivalence principle guarantees that the energy, momentum, and stress in the electromagnetic field as measured by ZAMOs have the familiar forms

$$\begin{pmatrix}\text{energy density}\\ \text{in }\mathbf{E}\text{-field}\end{pmatrix} = \begin{pmatrix}\text{tension along}\\ \mathbf{E}\text{-field lines}\end{pmatrix} = \begin{pmatrix}\text{pressure orthogonal}\\ \text{to }\mathbf{E}\text{-field lines}\end{pmatrix} = \frac{\mathbf{E}^2}{8\pi} \tag{32a}$$

$$\begin{pmatrix}\text{energy density}\\ \text{in }\mathbf{B}\text{-field}\end{pmatrix} = \begin{pmatrix}\text{tension along}\\ \mathbf{B}\text{-field lines}\end{pmatrix} = \begin{pmatrix}\text{pressure orthogonal}\\ \text{to }\mathbf{B}\text{-field lines}\end{pmatrix} = \frac{B^2}{8\pi} \tag{32b}$$

$$(\text{energy flux}) = \frac{\mathbf{E} \times \mathbf{B}}{4\pi} \tag{32c}$$

III. STRETCHING THE HORIZON AND BLACK-HOLE THERMODYNAMICS

Thus far we have focused attention on physics outside a black hole's horizon. In this section we turn attention to the properties of the horizon itself. Our study will begin with a model problem (Section A) that motivates the key step of "stretching the horizon" (Section B), and also motivates the statistical mechanical explanation for the entropy of a hole (Section C). We then turn to an exposition of the laws of thermodynamics and mechanics for the stretched horizon (Section D).

A. Macdonald's Vibrating Magnetic Field Problem

Our motivational model problem, which was developed by Doug Macdonald (Macdonald 1983; Macdonald and Suen 1985), is the vibration of magnetic field lines in the vicinity of a nonrotating (Schwarzschild) black

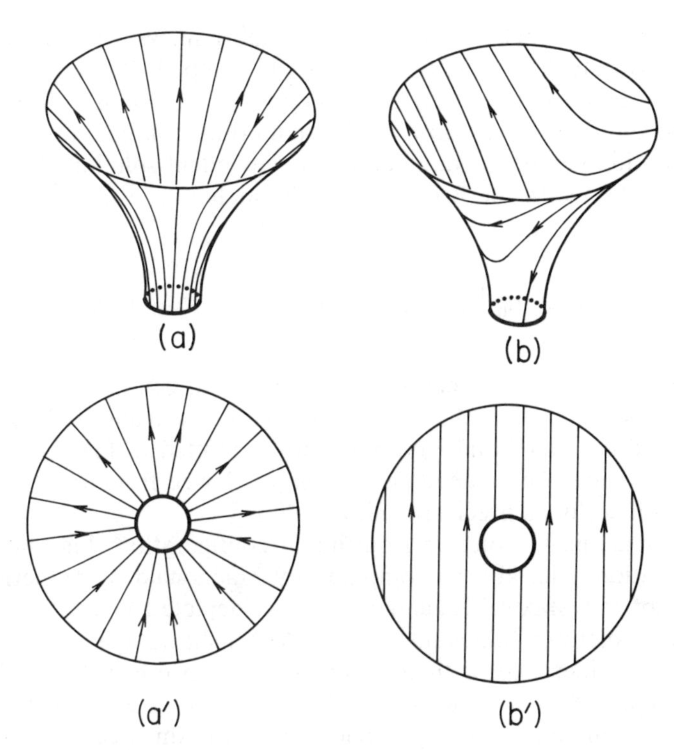

FIGURE 4. The initial and final magnetic field configurations for Macdonald's vibrating magnetic field problem. (a) The initial field lines [Eq. (33)] on an embedding diagram of the "plane" φ = const; (a′) is the same as (a) but viewed from above. (b) and (b′) are the final field lines [Eq. (34)] on an embedding diagram viewed in perspective and from above. [Reproduced from BHMP, i.e., Thorne, Price, and Macdonald, eds. (1986).]

hole; our description of the problem is adapted from BHMP. The magnetic field is anchored in a perfectly conducting spherical shell at radius $r = 10M$, which surrounds the Schwarzchild hole; and the field vibrates in the shell's interior, interacting with the horizon as it moves. The initial magnetic field configuration is shown in Figs. 4a and 4a′. This field is purely radial; but below the equatorial plane the field points into the hole, and above it points out. Mathematically, this initial field is given by

$$\mathbf{B} = B_0 \left(\frac{10M}{r}\right)^2 \cos\theta \mathbf{e}_{\hat{r}} \tag{33}$$

The initial electric field vanishes.

This initial magnetic field is not in equilibrium. When released, it will begin to vibrate. The field's vibrations will dump electromagnetic energy down the hole; and as they lose energy, the vibrations will damp out, bringing the field to rest in a final, equilibrium configuration. The final configuration will be the unique, time-independent solution of Maxwell's equations (31) with the **B**-field lines still anchored in the sphere at $r = 10M$:

$$\mathbf{B} = B_0 \left[\cos\theta \mathbf{e}_{\hat{r}} - \left(1 - \frac{2M}{r}\right)^{1/2} \sin\theta \mathbf{e}_{\hat{\theta}} \right]. \tag{34}$$

Here $\mathbf{e}_{\hat{\theta}}$ and $\mathbf{e}_{\hat{r}}$ are unit vectors in the θ and r directions. This final field [an analytic solution of Maxwell's equations derived and studied by Wald (1974) and by Hanni and Ruffini (1976)] is plotted in Figs. 4b, and 4b′. Note that when the embedding diagram is viewed from above (Fig. 4b′) the final field looks precisely uniform.

The dynamical evolution from initial configuration (Fig. 4a) to final configuration (Fig. 4b), as computed by Macdonald by numerical integration of the Maxwell equations (31), is depicted in Fig. 5. The first column of pictures (times $t/M = 3, 4, 6, 12$) shows a springing outward of the field lines due to their repulsion of each other [Eq. (32b)]. The second column (times $t/M = 20, 24, 28, 32$) shows a springing back in, a reversal, and an outspringing once again; the third column (times $t/M = 40, 44, 48, 52$) shows the next cycle of vibration. Each cycle has smaller amplitude than the preceding because energy is draining down the hole. By time $t/M \sim 200$ (not shown in Fig. 5) the vibrations have become so weak that they are invisible on the scale of these diagrams, and the field lines have settled into the "uniform" equilibrium configuration of Fig. 4b′.

Closer scrutiny reveals that the field lines do not spring on and off the horizon in a simple manner. Rather, near the horizon ($r \lesssim 3M$) the field lines move as though stuck in quicksand: they are sucked inward (by the hole's gravitational pull), but their motion is very sluggish (because of the gravitational redshift embodied in the lapse function α). As a field line (e.g., the dashed line at times $t/M = 3$ and 4) tries to pull away from the horizon, it finds the grip of gravity too strong at $r \lesssim 2.5M$; so it there reconnects itself into two parts, an outer part, which then springs on outward, and an inner part, which is then sucked inward. The reconnected, outspringing field line gets pushed back in by Maxwell pressure at $t/M \sim 12$, gets stuck in the "quicksand" once again at $t/M \sim 24$, and again succeeds in liberating itself at $t/M \sim 28$ only by reconnecting and leaving part of itself behind to fall down the hole.

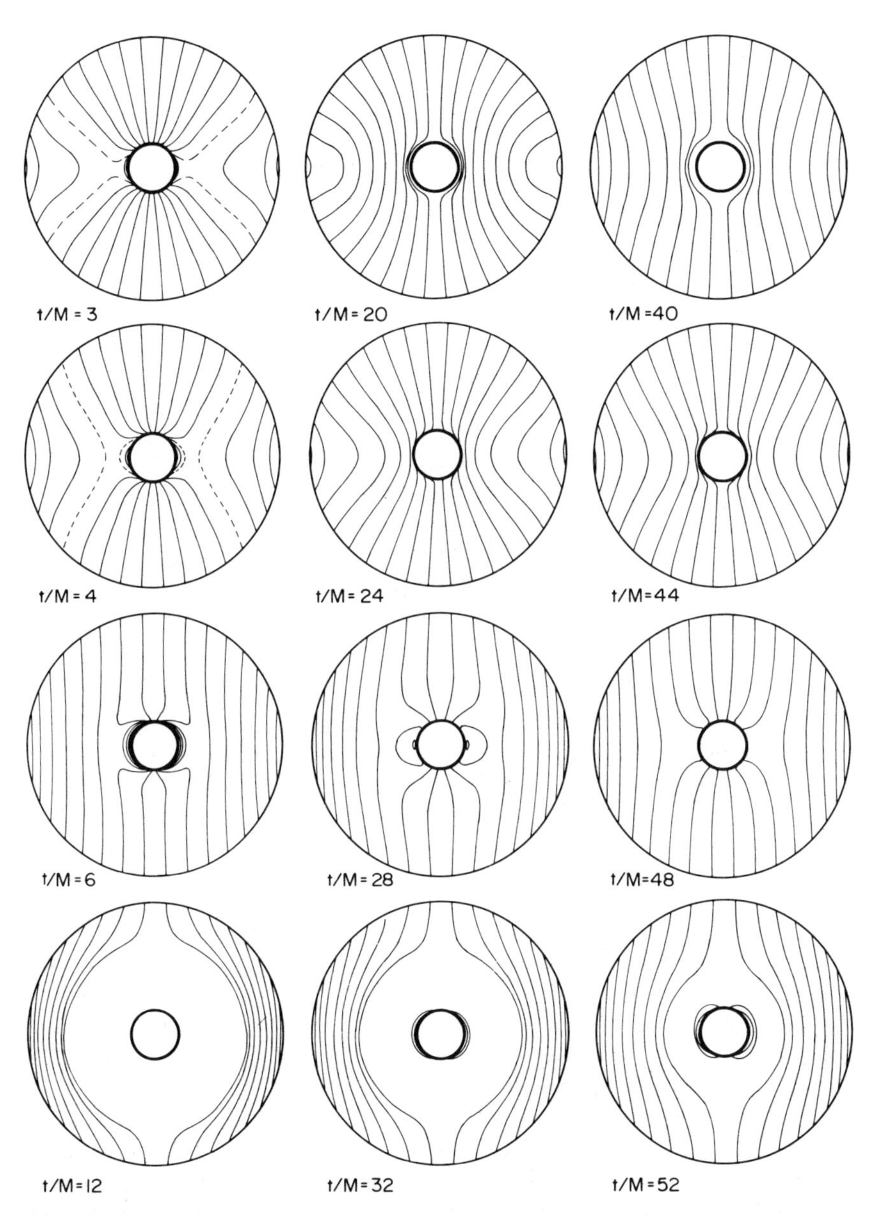

FIGURE 5. The dynamical evolution of the magnetic field lines, beginning with the initial configuration of Fig. 4a′ and leading to the final configuration of Fig. 4b′, as computed by Macdonald. Each drawing is a snapshot of the field lines at a fixed moment of universal time t as viewed on an embedding diagram from above; t is measured in units of $M = GM/c^3 = 4.93 \times 10^{-6}$ sec $(M/M_\odot)$. (Adapted from Macdonald 1983, and Macdonald and Suen 1985.)

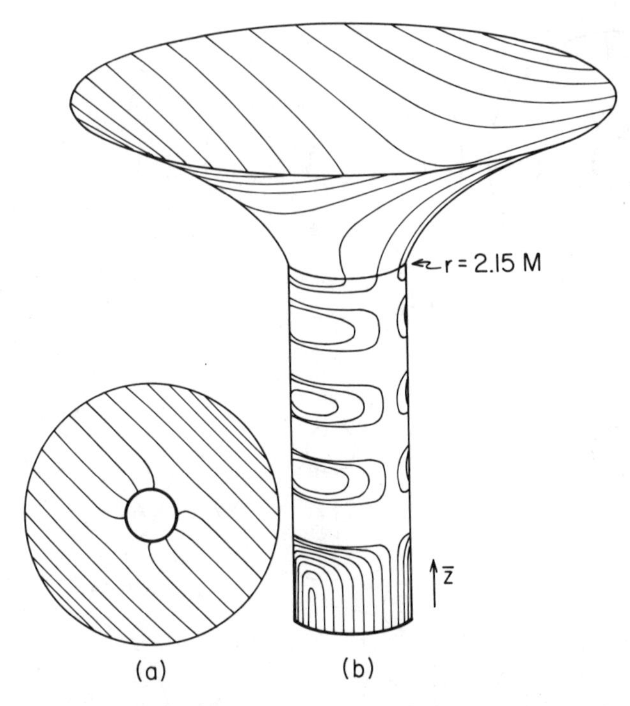

FIGURE 6. Magnetic field lines for the evolution of Fig. 5 at universal time $t = 92M$ as viewed on an embedding diagram from above (a) and in perspective (b). In the perspective drawing the region nearest the horizon $2M < r < 2.15M$ is artifically stretched vertically to exhibit the highly compact boundary-layer structure of the magnetic field. (Adapted from Macdonald 1983, and Macdonald and Suen 1985.)

Because of this episodic reconnection, the field structure near the horizon forms a rather complex "boundary layer." The full details of that boundary layer at time $t/M = 92$ are shown in Fig. 6. Diagram (a) shows the field lines on an embedding diagram viewed from above, with the polar axis rotated to correspond to the chosen orientation of diagram (b). Diagram (b) has two parts: The "horn" part (region above the boundary layer) is an embedding diagram truncated at radius $r = 2.15M$. The "cylinder part" (boundary layer) was obtained by taking the innermost $\Delta r = 0.15M$ of the embedding diagram and stretching it vertically. The amount of stretch is greater the closer one gets to the horizon, thereby permitting a clearer visualization of the very compact magnetic-field structure there. Quantitatively, vertical distance $\bar{z}$ in the cylindrical, boundary-layer, part of the diagram is given by

$$\bar{z} = 2M \ln \left(\frac{r}{2M} - 1 \right) + \text{const} = 4M \ln \alpha + \text{const} \tag{35}$$

so that the horizon is infinitely far down in the diagram ($\bar{z} = -\infty$).

Note that the magnetic field in the boundary layer consists of closed loops. If we had a motion picture of the diagram, we would see the loops being sucked continuously downward toward the horizon as time passes. With each oscillation of the field in the outer (horn) region, a new family of closed loops would form in the inner, boundary-layer region. The bottom of a loop would be created as the outer field springs inward, pushing into the "sluggish" region near the horizon (e.g, at $t/M = 20$ and 40 in Fig. 5, and also at $t/M = 92$ in Fig. 6). The top of the loop would be created as the outer field tries to pull itself off the sluggish region and manages to do so only by reconnection (e.g., at $t/M = 4$ and 28 in Fig. 5). Once created, a loop would sink inward toward the horizon, leaving room above itself for creation of the next loop.

By the time $t/M = 92$ of Fig. 6 the outer field is completing its fourth vibration and is forming its fourth family of loops. Each loop is the relic of one vibration; and the field lines below the first loop, which extend unimpeded radially downward to the horizon ($z = -\infty$), are the relic of the initial configuration. As one might expect, the downward speed of all these relic fields as measured by the ZAMOs is the speed of light:

$$\begin{aligned} v &= \frac{d\,(\text{proper radial distance})}{d\tau} = \frac{(1 - 2M/r)^{-1/2}\, dr}{\alpha dt} \\ &= \frac{dr/dt}{1 - 2M/r} = -1 \end{aligned} \tag{36}$$

By integrating this equation, we find that the radius r and the height $\bar{z}$ of a specific field-line feature as functions of time are

$$r - 2M = \text{const} \times e^{-t/2M}, \qquad \bar{z} = -t + \text{const} \tag{37}$$

Thus, because of the infinite redshift of ZAMO ticking rates ($\alpha \to 0$ at horizon), the field loops never reach the horizon, not even after a time lapse $\Delta t = 10^{100}M$. But they approach the horizon exponentially; and as they do so, their radial thickness becomes exponentially compacted (a compaction that is undone by drawing them on Fig. 6b with unphysical radial coordinate $\bar{z}$).

B. Stretching the Horizon

The preceding description of the vibrating field is made tremendously complex and cumbersome by our insistence on keeping track of the full structure of the sluggishly infalling field in the boundary layer. The description becomes much simpler, and in fact more elegant, if we stretch the horizon—for example, up to the top of the cylinder of Fig. 6b ($r = 2.15M$)—and thereby throw away all the complex details of the boundary layer. (Thorne and Macdonald 1982; Macdonald and Suen 1985). Fortunately, we can discard the boundary layer with its complex field loops without actually losing any information. The field loops are nothing but a fossil record of past history at radius $r \approx 2.2M$, where the field loops were laid down; and we already have the details of that history in the time evolution record of Fig. 5. Discarding the field loops is like ignoring the layers of sediment on the bottom of the ocean when one already knows the history of the ocean and of the processes that laid the sediments down.

The mental deceit of stretching the horizon is made mathematically viable, indeed very attractive, by the fact that it leads to an elegant set of membranelike boundary conditions at the stretched horizon (see Section IV).

The relationship of the stretched horizon to the true horizon is elucidated by adopting, temporarily, the standard four-dimensional "spacetime" view of black holes. Figure 7 is a spacetime diagram of the formation of a Schwarzschild hole by the gravitational collapse of a spherical star. The diagram plots Eddington–Finkelstein time $t_* = t + 2M \ln(r/2M - 1)$ vertically, since that time is well behaved at the horizon while universal time t is not (cf. Box 31.2 of MTW). Two surfaces t = const of constant Schwarzschild time are shown on the diagram. (Although they have a shape similar to those of the embedded surfaces of Figs. 1, 2, 4, and 6, they must not be confused with embedded surfaces. Figure 7 is a spacetime diagram; Figs. 1, 2, 4, and 6 are purely spatial embedding diagrams.) Also shown in Fig. 7 are the time-evolving stretched horizon and true horizon. Note how the true horizon is nested inside the stretched horizon. Both horizons are three-dimensional hypersurfaces in spacetime (but appear two-dimensional because of suppression of one rotational angle from the diagram). The intersection of the stretched horizon with the hypersurface t = const is a two-dimensional surface (one-dimensional

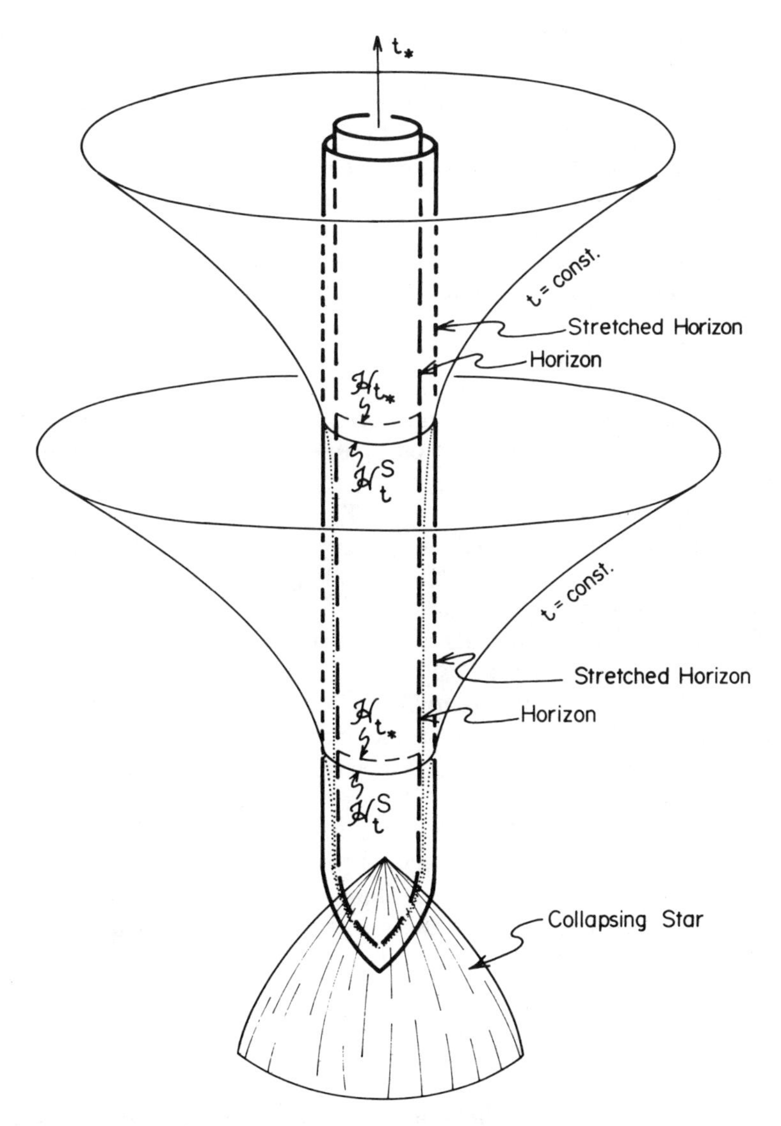

FIGURE 7. Spacetime diagram for the gravitational collapse of a spherical star (bottom of diagram) to form a black hole (upper parts of diagram). This diagram shows the dipping deep into the past of the hypersurfaces of constant universal time t, which is responsible for the formation of the boundary layer in Fig. 6. It also shows the true horizon of the hole (inner cylindrical surface) and the stretched horizon (outer cylindrical surface), which is adopted to get rid of the uninteresting boundary layer. (Reproduced from BHMP.)

in diagram) representing the stretched horizon at a specific moment of time. That "instantaneous stretched horizon" is labeled $\mathcal{H}_t^S$. The instantaneous stretched horizon is very near an instantaneous slice through the true horizon, labeled $\mathcal{H}_{t*}$—"near" as measured by observers who fall freely inward through the horizons. The boundary conditions at the stretched horizon, as developed in Section IV below, attribute to the instantaneous stretched horizon $\mathcal{H}_t^S$ all the physical properties of the instantaneous true horizon $\mathcal{H}_{t*}$.

Figure 7 elucidates the manner in which, inside the stretched horizon, the hypersurfaces t = const contain a fossil record of past history (boundary layer): Inside the stretched horizon those constant-time hypersurfaces (shown dotted) nearly coincide, in a spacetime sense, with the true horizon. As one goes down the true horizon, deeper and deeper into the past, those hypersurfaces asymptote to the true horizon—ultimately wrapping themselves around the true horizon's point of formation inside the star and there almost touching it. By discarding the thin region between true horizon and stretched horizon, one merely discards the past history of the true horizon, which is essentially the same as the past history of the stretched horizon.

This discussion makes clear how much stretching one can perform without much loss of error: One must stretch little enough that physical fields and particles are in essentially the same state on the stretched horizon as on the true horizon—"same state" as seen by physically reasonable, infalling observers, not as seen by the ZAMOs of the membrane viewpoint, who are pathological ($\mathbf{g} \to \infty$) near the horizon. Mathematically, this means (as the relativity aficionado can show by a calculation in Eddington–Finkelstein coordinates) that the lapse function α_H at the stretched horizon and the universal time scale Δt_{SH} on which fields and particles at fixed radius are evolving near the stretched horizon must satisfy

$$\alpha_H{}^2 \ll \frac{\Delta t_{SH}}{M}, \qquad \text{and } \alpha_H{}^2 \ll 1 \tag{38}$$

For the vibrating magnetic field of Fig. 5 the timescale of evolution at all radii is $\Delta t \sim 4M$. Consequently, one can choose the stretched horizon to be outside the true horizon by any amount $\alpha_H^2 \ll 1$. In particular, the amount of stretching suggested by Fig. 6b (stretched horizon at top of cylindrical, boundary-layer region; $\alpha_H^2 = 1 - 2M/2.15M = 0.070$) is marginally acceptable.

Of course, one must also choose the stretched horizon small enough that nothing which enters it ever reemerges. Usually criterion (38) will be adequate for this; but in pathological situations (e.g., when an object is lowered on an infinitely strong rope to very near the true horizon and is then pulled back up), a more stringent criterion than (38) may be needed. In any given situation a little judicious thought should produce a wise choice for the amount of stretching. The judicious thought needed is very similar to that used in choosing a grid size for the numerical solution of a differential equation.

C. The Entropy of a Black Hole

The stretching of the horizon is intimately tied to the statistical mechanical origin of a black hole's entropy (Zurek and Thorne 1985): Considered a quiescent hole that exhibits, outside and at its stretched horizon, a mass M and angular momentum J. These can be thought of as macroscopic variables analogous to the energy $\mathscr{E}$, volume V, and number of molecules N in a chamber of monomolecular gas; and the precise configuration of matter and fields in the boundary layer below the stretched horizon can be thought of as analogous to the precise configuration of molecules inside the chamber. Just as the entropy S in the gas chamber can be regarded as Boltzman's constant k times the logarithm of the number of quantum mechanically independent molecular configurations corresponding to fixed $\mathscr{E}$, V, N, so the entropy S_H of the hole can be regarded as k times the logarithm of the number of quantum mechanically independent configurations beneath the stretched horizon that could produce the given M and J. Moreover, since each configuration in the hole's boundary layer corresponds to a unique past history by which the hole was made, the entropy S_H can also be regarded as k times the logarithm of the number of possible past histories for the black hole.

A detailed evaluation of the number W of quantum mechanically independent configurations beneath the stretched horizon gives

$$S_H = k \ln W = \frac{k}{4\hbar} A_H, \qquad A_H = 4\pi({r_H}^2 + a^2) = 8\pi M r_H \qquad (39)$$

where A_H is the hole's surface area. [The second equation can be derived by integrating $\rho\tilde{\omega}\, d\theta\, d\varphi$ over the horizon, $r = r_H$, Eq. (6).] This value for black-hole entropy was originally guessed, to within a factor of order unity, by Bekenstein (1972a,b, 1973, 1980) and was first derived exactly by Hawking (1975, 1976) using thermodynamic arguments (see below). Its statistical mechanical origin, as described above, was discovered by Zurek and Thorne (1985).

D. The Thermodynamics and Mechanics of a Black Hole

Quantum field theory reveals that a ZAMO very close to the horizon sees himself bathed in thermal radiation with a temperature proportional to his gravitational acceleration $g \equiv |\mathbf{g}|$:

$$T = \frac{\hbar}{2\pi k} g \tag{40}$$

where $\hbar$ and k are $1/2\pi$ times Planck's constant and Boltzmann's constant (Unruh 1976). In the membrane formalism one regards this radiation as emitted by the stretched horizon. Those quanta which have almost zero angular momentum travel outward from the stretched horizon, through the ZAMO's feet, and on to infinity; while those with larger angular momenta fly out some distance, then get pulled back by the hole's gravity, shower down on the ZAMO's head, and sink into the stretched horizon. As it travels, each quantum gets gravitationally redshifted (or blueshifted) in energy by an amount inversely proportional to the gravitational shift of ZAMO proper time τ, and the temperature of the thermal spectrum gets similarly shifted, $T \propto 1/\alpha$. Equations (16b) and (40) reveal the proportionality constant in this temperature shift:

$$T = \alpha^{-1} T_H, \qquad T_H \equiv \frac{\hbar}{2\pi k} g_H \tag{41}$$

$$g_H \equiv \lim_{\alpha \to 0} \alpha g = \frac{r_H - M}{2Mr_H} \tag{42}$$

The quantity g_H is called the hole's "surface gravity," since it is the gravitational acceleration measured by a ZAMO, renormalized with one factor of α to make it finite. The quantity T_H is the temperature of the fully redshifted thermal radiation when it has finally escaped from the hole's neighborhood, that is, at $\alpha \to 1$. Equivalently, T_H is the locally measured temperature T at the stretched horizon renormalized with a factor of α to make it finite. By analogy with the name "surface gravity" for g_H, T_H is called the hole's "surface temperature."

Prior to Hawking's (1974, 1975) discovery of this "Hawking radiation," Bekenstein (1972a,b, 1973) and Bardeen, Carter, and Hawking (1973) had derived, classically, the first law of black hole mechanics

$$dM = (1/8\pi) g_H \, dA_H + \Omega_H \, dJ \tag{43}$$

for the change in a quiescent hole's mass M which accompanies changes in its surface area A_H and angular momentum J; and Hawking (1971a) had derived, classically, the second law of black hole mechanics, that the surface area A_H of a black hole can never decrease. Hawking's discovery that a hole radiates as though it had a surface temperature $T_H = (\hbar/2\pi k)g_H$ made it immediately obvious (Hawking 1975, 1976) that these first and second laws of black hole mechanics are really the first and second laws of thermodynamics in disguise: A black hole must be endowed with an entropy proportional to its surface area with proportionality constant $k/4\hbar$ [Eq. (39)]; and the first and second laws of thermodynamics must read

$$dM = T_H\, dS_H + \Omega_H\, dJ \tag{44a}$$

$$dS_H \geq 0 \qquad \text{for classical processes} \tag{44b}$$

The emission of Hawking radiation (a quantum process) causes the hole's surface area and entropy to decrease, but only by an amount less than the entropy deposited in the external universe by the outflowing radiation

$$dS_H + dS_{\text{universe}} \geq 0 \qquad \text{for all processes}$$

(second law of thermodynamics for universe plus black hole; Hawking 1976; Zurek and Thorne 1985).

To within factors of order unity a hole's surface temperature and entropy are

$$T_H \simeq (6 \times 10^{-8}\text{K}) \left(\frac{M_\odot}{M}\right) \tag{45a}$$

$$S_H \simeq (1 \times 10^{77}k) \left(\frac{M}{M_\odot}\right)^2 \tag{45b}$$

where $M_\odot$ is the mass of the sun. For comparison, the surface temperature and entropy of a typical star of mass $M \sim M_\odot$ are $T \sim 10^4$K, $S \sim (10^{57}k)(M/M_\odot)$. The enormous entropy of a hole [$S_H \sim 10^{20}(M/M_\odot)$ times larger than that of a star of the same mass] reflects the enormous number of possible ways the hole might have been made and the corresponding enormous number of possible microconfigurations that could exist in the boundary layer beneath its stretched horizon.

Throughout this paper we shall restrict attention to the only kinds of holes that can form in the present universe: holes with masses larger than about $2M_\odot$. (Smaller holes may have formed in the big bang; Hawking 1971b.) For such holes the surface temperature is so low that Hawking radiation is of no practical importance; and so henceforth we shall largely ignore the Hawking radiation.

All the thermodynamic properties of a system in equilibrium can be computed from a knowledge of just one function: the system's total mass-energy M expressed in terms of its other "extensive" variables. For a Kerr black hole we can choose as the other extensive variables the hole's entropy S_H and angular momentum J; and from Eqs. (11) and (39) with $a = J/M$ we can derive the fundamental function

$$M(S_H, J) = \left(\frac{\hbar}{4\pi k} S_H + \frac{\pi k}{\hbar} \frac{J^2}{S_H}\right)^{1/2} \tag{46}$$

From this fundamental function and the first law of thermodynamics (44a) one can compute the hole's surface temperature $T_H = (\partial M/\partial S_H)_J$ and angular velocity $\Omega_H = (\partial M/\partial J)_{S_H}$. After some algebra they agree of course, with equations (41) and (12).

The thermodynamic formalism for black holes differs in one important respect from that for most systems which one encounters in textbooks: The black hole mass M is *not* homogeneous of first degree in the extensive variables S_H and J; rather, it is homogeneous of degree $\frac{1}{2}$ [Eq. (46)]. One consequence of this is the fact that $T_H S_H + \Omega_H J$ is equal to $M_H/2$ rather than M_H [Euler's theorem; "Smarr's (1973) formula"]. For other consequences see Tranah and Landsberg (1980).

If we take a black hole of mass M and angular momentum J and spin it down in the most efficient manner possible (entropy S_H and surface area A_H held fixed), we can thereby extract all its rotational energy and leave behind a nonrotating hole with mass

$$M_{\text{irr}} = \left(\frac{A_H}{16\pi}\right)^{1/2} = \tfrac{1}{2}(r_H{}^2 + a^2)^{1/2} \tag{47}$$

This mass is called the "irreducible mass" of the original hole (Christodolou 1970) because no classical process can reduce the hole's mass below this amount. The rotational energy extracted during the spindown is

$$M_{\text{rot}} \equiv M - M_{\text{irr}} = M[1 - \sqrt{\tfrac{1}{2} + \tfrac{1}{2}(1 - a^2/M^2)^{1/2}}] \tag{48}$$

Since no black hole can ever have an angular momentum larger than $J_{max} = M^2$, [i.e. $a_{max} = M$ (Bardeen, Carter, and Hawking 1973)], the rotational energy of a hole can never exceed $(M_{rot})_{max} = M(1 - 1/\sqrt{2}) = 0.293M$. In the real universe the fastest spinning holes probably have $a \simeq 0.998M$, corresponding to $M_{rot} \simeq 0.271M$ (Thorne 1974).

The moment of inertia I_H of a rotating hole can be defined in the usual manner as the ratio of its angular momentum $J = Ma$ to its angular velocity $\Omega_H = a/(2Mr_H)$:

$$I_H = \frac{J}{\Omega_H} = 2M^2[M + (M^2 + a^2)^{1/2}]$$
$$\simeq 4M^3 \qquad \text{for a slowly rotating hole} \tag{49}$$

Notice that

$$M_{rot} \simeq \tfrac{1}{2} I_H \Omega_H{}^2 \qquad \text{for a slowly rotating hole} \tag{50}$$

just as one might expect.

These thermodynamic and mechanical relations will play important roles in the model problems of Section V.

IV. ELECTRODYNAMICS OF THE STRETCHED HORIZON

In studies of physics around a black hole using the membrane viewpoint, one must impose boundary conditions at the stretched horizon, which we shall take to be a surface of constant lapse function, $\alpha = \alpha_H \ll 1$. We have already met one example of such boundary conditions: the stretched horizon must be regarded as a perfect thermal emitter (as well as a perfect absorber), with locally measured temperature $T = T_H/\alpha_H$. In this section we shall study the membranelike boundary conditions that one must impose on external electromagnetic fields, including Gauss's law, Ampere's law, Ohm's law, and the law of charge conservation (Section A); and the Lorentz force law and law of ohmic dissipation (Section B).

A. The Laws of Gauss, Ampere, Ohm, and Charge Conservation

The boundary conditions on classical electric and magnetic fields **E** and **B** at the stretched horizon can be derived from a 3 + 1 split of the standard

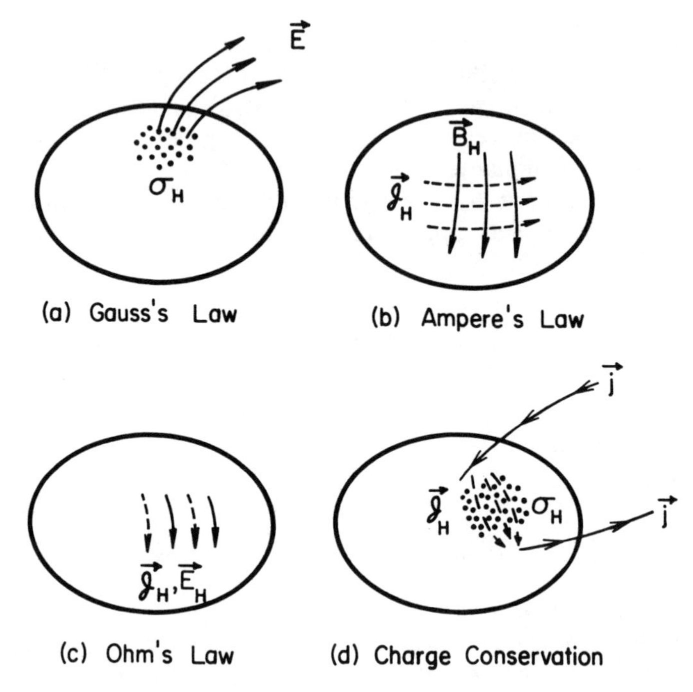

FIGURE 8. Electromagnetic boundry conditions at the stretched horizon. (Adapted from Thorne 1986.)

general relativistic laws of electrodynamics. As first shown by Znajek (1978) and by Damour (1978) (with augmentation of the 3+1 formalism by Thorne and Macdonald 1982), it is useful to regard these boundary conditions as arising from physical properties of a fictitious membrane that resides at the location of the stretched horizon.

More specifically, it is useful to pretend that the stretched horizon is endowed with a surface density of electric charge σ_H so defined that it terminates the normal component of any electric field that pierces the stretched horizon; see diagram (a) of Fig. 8:

$$\sigma_H \equiv \begin{pmatrix} \text{charge per unit area} \\ \text{on stretched horizon} \end{pmatrix} \equiv (E_n/4\pi)_{\text{stretched horizon}} \tag{51}$$

$$E_n \equiv \mathbf{E}\cdot\mathbf{n},\ \mathbf{n} \equiv \mathbf{e}_{\hat{r}} = \text{unit radial vector} \tag{52}$$

Similarly, it is convenient to pretend that there exists a surface current on the stretched horizon, so defined that it terminates any tangential magnetic field in accordance with Ampere's law; see Fig. 8b. But in making this idea precise, we must be careful about the type of time used in defining the surface current. The proper time of a ZAMO must not be used because it ticks at a pathologically slow rate at the stretched horizon, and its ticking rate even depends on just where we choose the stretched horizon to be. The only reasonable choice is a time that is independent of the stretched horizon's location: universal time t. Thus we define

$$\mathcal{J}_H \equiv \begin{pmatrix} \text{charge crossing a unit length per unit} \\ \text{universal time } t \text{ in the stretched horizon} \end{pmatrix} \tag{53}$$

Since this surface current is smaller by a factor $\alpha_H = d\tau/dt$ than that which would be measured by a ZAMO at the stretched horizon, Ampere's law must equate it to 4π times a tangential magnetic field which is smaller by α_H than that, $\mathbf{B}_{\|}$, measured by the ZAMO, that is, to 4π times the "horizon field"

$$\mathbf{B}_H \equiv \alpha_H \mathbf{B}_{\|}. \tag{54}$$

Thus Ampere's law—which is actually the definition of $\mathcal{J}_H$—says

$$\mathbf{B}_H \equiv 4\pi \mathcal{J}_H \times \mathbf{n}; \tag{55}$$

see Fig. 8b. It turns out (see below) that, in the limit as $\alpha_H \to 0$, $\mathbf{B}_H$ and $\mathcal{J}_H$ remain finite while $\mathbf{B}_{\|}$ diverges as $1/\alpha_H$.

The horizon's Gauss's law (51) and Ampere's law (55) are merely definitions. Not so for the remaining two boundary conditions, which are consequences of these definitions, of the general relativistic Maxwell equations (31) and of the demand that the electromagnetic field as seen by freely falling observers be well behaved at the horizon. The first of these boundary conditions, Ohm's law, states that the tangential electric field, renormalized in the same manner as for $\mathbf{B}$

$$\mathbf{E}_H \equiv \alpha_H \mathbf{E}_{\|}, \tag{56}$$

is proportional to the stretched horizon's surface current

$$\mathbf{E}_H = R_H \mathcal{J}_H \tag{57}$$

with a proportionality constant

$$\begin{aligned} R_H &\equiv (\text{surface resistivity of stretched horizon}) \\ &= 4\pi \simeq 377 \text{ ohms} \end{aligned} \tag{58}$$

(Fig. 8c). It should not be surprising that the horizon displays a surface resistivity of $4\pi = 4\pi/c \simeq 377$ ohms. After all, that is the impedance at the end of an open wave guide; it is the "impedance of the vacuum" into which waves propagate freely.

Notice, in fact, that Ohm's law (57) and Ampere's law (55) together imply that the tangential electric and magnetic fields look to ZAMOs at the stretched horizon like a divergently strong ingoing electromagnetic wave

$$\mathbf{E}_{\|} \propto \frac{1}{\alpha_H}, \qquad \mathbf{B}_{\|} \propto \frac{1}{\alpha_H}, \qquad \mathbf{B}_{\|} = \mathbf{E}_{\|} \times \mathbf{n} \tag{59a}$$

Physically this is the "fault" of the ZAMOs, not of the fields: In order to prevent themselves from falling into the hole, the ZAMOs at the stretched horizon must move outward relative to physically reasonable infalling observers with nearly the speed of light; and this motion converts the nonwavelike tangential fields seen by infalling observers into the divergently large wavelike field (59a), while leaving the normal fields E_n and B_n finite and unchanged. In the special case where infalling observers see no tangential field, Maxwell's equations (31) demand that $\mathbf{E}_{\|}$ and $\mathbf{B}_{\|}$ go to zero at the stretched horizon

$$\mathbf{E}_{\|} \propto \alpha_H, \qquad \mathbf{B}_{\|} \propto \alpha_H \qquad \text{when their } O(1/\alpha_H) \text{ parts vanish.} \tag{59b}$$

The remaining boundary condition is the law of charge conservation at the stretched horizon. Definitions (51) and (55) of σ_H and $\mathcal{J}_H$, together with the general relativistic Maxwell equations (31), imply that

$$\frac{\partial \sigma_H}{\partial t} + {}^{(2)}\nabla \cdot \mathcal{J}_H + (\alpha j_n)_{\text{at stretched horizon}} = 0 \tag{60}$$

Here ${}^{(2)}\nabla \cdot \mathcal{J}_H$ is the two-dimensional divergence of the two-dimensional surface current in the stretched horizon; $j_n \equiv \mathbf{j} \cdot \mathbf{n}$ is the normal component of the volume current (charge per unit area per unit time) leaving the stretched horizon as measured by a FIDO; and αj_n is that departing current converted to the same per-unit-universal-time basis as is used in the first two terms of equation (60). Of course, universal time t appears in this law of charge conservation rather than ZAMO time τ because ZAMO time ticks at a pathologically slow and α_H-dependent rate at the stretched horizon, whereas universal time ticks at a rate that is physically reasonable.

The law of charge conservation (60) says that, whenever electric charge falls into the stretched horizon, it does not pass on through and into the hole's interior. Rather, it stays on the stretched horizon, moving around and always being conserved, until such a time as charge of the opposite sign falls into the stretched horizon to annihilate it. Put differently, but equivalently, electric current can flow both into the stretched horizon (infalling positive charges) and out of the stretched horizon (infalling negative charges); and any current that flows in then stays on the stretched horizon, with its charge being conserved, until it flows back out again (Fig. 8d).

The four electromagnetic laws of the stretched horizon—Gauss, Ampere, Ohm, and charge conservation—are approximations to these same laws for the true horizon, as originally formulated by Damour (1978) and (in slightly different form) by Znajek (1978). They are approximations because of the errors inherent in stretching the horizon—errors of order α_H, which become arbitrarily small if the stretched horizon is squeezed arbitrarily close to the true horizon. Thorne and Macdonald (1982) and Price and Thorne (1986) give formal proofs that these laws are equivalent to the Damour–Znajek laws and also give derivations of these laws.

The four electromagnetic laws, written in the precise forms of Eqs. (51) (Gauss), (55) (Ampere), (57) (Ohm), and (60) (charge conservation), together with the renormalization definitions of $\mathbf{B}_H$ and $\mathbf{E}_H$ [Eqs. (54) and (56)] and the radiative boundary condition [Eq. (59a)], are valid not only for quiescent, Kerr holes, but also highly dynamical holes—indeed for any kind of black hole whatsoever. The proof at the true horizon is given by Damour (1978); the proof at the stretched horizon is given by Price and Thorne (1985). We shall explore these laws for quiescent holes in the model problems of Section V.

The four electromagnetic laws of the stretched horizon, together with Maxwell's equations for the external universe in 3 + 1 form [Eqs. (31)], constitute a basis for computing the interaction of a black hole with arbitrary electromagnetic fields, charges, and currents. It is not necessary, in studying such interactions, to think of the horizon charge and current as real; but in my own research I have found it very helpful to do so.

B. Lorentz Force and Ohmic Dissipation in the Stretched Horizon

When a black hole endowed with surface charge σ_H and surface current $\mathcal{J}_H$ is immersed in external electric and magnetic fields, those fields exert a Lorentz force on the horizon. This Lorentz force can be thought of as producing a rate of change of horizon momentum density

$$\begin{aligned}\mathbf{\Pi}_H &\equiv \text{(momentum per unit area in stretched horizon)} \\ &= \text{(a vector tangential to the horizon)}\end{aligned} \qquad (61)$$

(For detailed discussions of the horizon momentum density see Damour 1979, 1982; Price and Thorne 1986; and BHMP.)

Not surprisingly, the Lorentz force (rate of change of momentum density, measured on a per-unit-universal-time basis) is

$$\frac{d\mathbf{\Pi}_H}{dt} = \sigma_H \mathbf{E}_H + \boldsymbol{\mathcal{J}}_H \times \mathbf{B}_n \qquad (62)$$

(Damour 1978, 1979, 1982). It is the normal component of the magnetic field $\mathbf{B}_n \equiv B_n \mathbf{n}$ that enters because the Lorentz force must be tangential to the horizon and the tangential magnetic field produces a normal force. The component of the Lorentz force along the generator of rotations $\partial/\partial\varphi = \tilde{\omega}\mathbf{e}_{\hat{\varphi}}$ (i.e., the $\mathbf{e}_{\hat{\varphi}}$ component of the force multiplied by the "lever arm" $\tilde{\omega}$) is the rate of change of the hole's angular momentum:

$$\begin{aligned}\frac{dJ}{dt} &= \int_{\text{stretched horizon}} \left(\frac{d\mathbf{\Pi}_H}{dt}\right) \cdot \left(\frac{\partial}{\partial\varphi}\right) dA = \int \left(\frac{d\Pi_{H\hat{\varphi}}}{dt}\right) \tilde{\omega}\, dA \\ &= \int (\sigma_H E_{\hat{\varphi}} - \mathcal{J}_{H\hat{\theta}} B_n)\tilde{\omega}\, dA\end{aligned} \qquad (63)$$

As the current $\boldsymbol{\mathcal{J}}_H$ flows in the stretched horizon it produces ohmic dissipation at a rate given by the standard relation

$$T_H \frac{dS_H}{dt} = \int R_H \mathcal{J}_H^2\, dA \qquad (64)$$

(Znajek 1978; Damour 1978, 1979, 1982). By virtue of Ohm's law (57) and Ampere's law (55) for the stretched horizon we can rewrite this ohmic dissipation as

$$T_H \frac{dS_H}{dt} = \alpha_H^2 \int \left(\frac{1}{4\pi}\mathbf{E} \times \mathbf{B}\right) \cdot (-\mathbf{n}\, dA) \qquad (65)$$

Since $(1/4\pi)\mathbf{E} \times \mathbf{B}$ is the energy flux measured by ZAMOs [Eq. (32c)], the right-hand side of (65) is the redshifted energy (one factor α_H for redshift) carried into the stretched horizon per unit universal time (other factor α_H to convert from ZAMO time to universal time), as measured by ZAMOs at the stretched horizon. Thus the energy required for ohmic dissipation comes from the external electromagnetic field. The reader may wish to go back and think about this for the model problem of Macdonald's vibrating magnetic field (Section III.A).

The ohmic dissipation (65) always increases the hole's mass, $dM = T_H\, dS_H > 0$, at the expense of external, electromagnetic energy. However, the Lorentz force (62) can decrease the hole's mass, $dM = \Omega_H\, dJ < 0$ if the force acts in the opposite direction to the hole's rotation; as a result, the hole's rotational energy can be fed into the external electromagnetic field. We shall explore this possibility in the model problem of Section V.E.

V. SOME ELECTROMAGNETIC MODEL PROBLEMS

We shall now illustrate the membrane formalism by a series of pedagogical model problems, including the charge separation induced in the stretched horizon of a nonrotating hole by the electric field of an external point charge (Section A); the EMF induced in the stretched horizon of a rotating hole by a threading magnetic field ("magnetic $\times$ gravitomagnetic battery"; Section D); and the role of a black hole as a lumped circuit element in a DC electric circuit—specifically, as a resistor (Section B), the rotor of a motor (Section C), and a battery (Section E). These model problems illustrate the power of the membrane formalism to predict, quickly and easily, the behaviors of matter and fields near a black hole.

These model problems are restricted to nondynamical electric and magnetic fields. The only dynamical model problem considered in this paper is Macdonald's vibrating magnetic field (Section III.A). After studying the nondynamical problems of this section, the reader may find it instructive to go back and apply to Macdonald's problem the stretched-horizon concepts of Section IV. For other dynamical model problems and the insights they give, see Macdonald and Suen (1985) and BHMP.

This paper's presentation of the model problems is adapted from BHMP.

A. Charge Separation in the Stretched Horizon

As a first model problem (due to Hanni and Ruffini 1973) consider a point particle with positive charge Q that is lowered very quickly down the

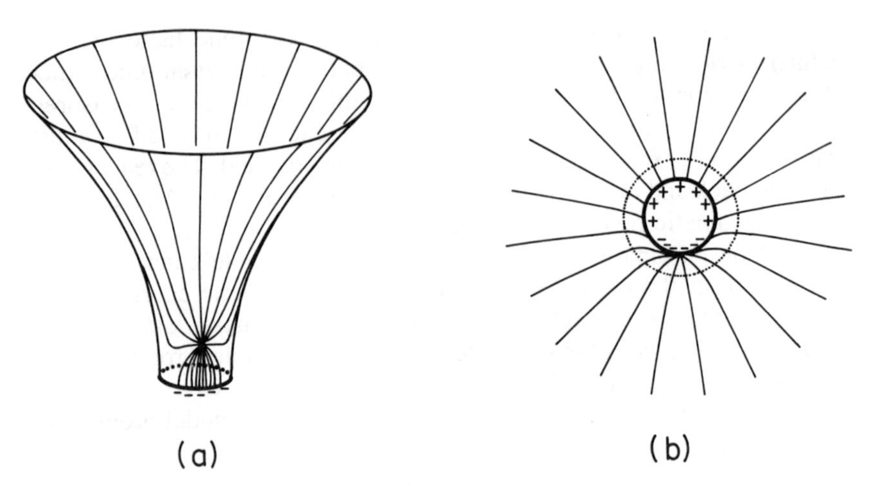

FIGURE 9. Electric field of a point charge at rest outside a nonrotating black hole, and the charge separation it produces on the stretched horizon. (a) A perspective view of the embedding diagram; (b) a view from above. In (b) the dotted circle is a possible location for the stretched horizon if one wishes to regard the charge as lost into the hole. (Reproduced from BHMP.)

polar axis ($\theta = 0$) of a nonrotating black hole and then comes to rest near the stretched horizon. As the particle descends, it begins to produce a tangential electric field $\mathbf{E}_H$ at the stretched horizon, and that $\mathbf{E}_H$ begins to drive horizon currents $\mathcal{J}_H$ and ohmic dissipation. Soon after the particle comes to rest, the horizon currents will have produced a substantial charge separation in the horizon, which influences the form of the external field. The charge separation, in fact, must drive $\mathbf{E}_H = R_H \mathcal{J}_H$ to zero, since the particle at rest cannot provide energy for continuing ohmic dissipation. The final, equilibrium configuration, with $\mathbf{E}_H$ vanishing and thus $\mathbf{E}$ normal to the stretched horizon [Eq. (59b)], must look something like Fig. 9. In fact, Fig. 9 is the final, stationary solution to Maxwell's equations as derived by Cohen and Wald (1971), Hanni and Ruffini (1973), and Linet (1976).

This dynamical evolution and final equilibrium configuration are qualitatively the same as occur in the laboratory when one places a charged particle near a metal sphere of finite conductivity. In the black hole case, as for a laboratory sphere, the closer the particle is to the horizon, the more localized becomes the region of negative horizon charge below the particle, and the more uniformly the positive charge gets spread over the rest of the horizon (cf. Fig. 9b). In the limit as the particle's radius r_0

approaches $r_H = 2M$, with the stretched horizon always kept well below the particle, the charge density approaches

$$\sigma_H = -Q\delta_2(\mathbf{x} - \mathbf{x_0}) + \frac{Q}{4\pi r_H^2} \qquad (66)$$

where δ_2 is the two-dimensional Dirac delta function on the stretched horizon and $\mathbf{x_0}$ is the point on the stretched horizon directly below the particle. The uniformly spread positive charge, in the limit $r_0 \rightarrow 2M$, gives rise to a radial electric field that, if the stretched horizon were chosen outside the particle (dashed circle of Fig. 9b) instead of inside, would be attributed to the particle having gone down the hole and its charge having been spread over the horizon (Cohen and Wald (1971).

B. Black Hole as a Resistor in an Electric Circuit

Having gained insight into the horizon's surface charge, we now turn to a model problem designed for insight into the surface current and resistivity. In this problem (due to Damour 1978) a nonrotating black hole is hooked up to a simple DC circuit containing a battery, the hole, and connecting wires (Fig. 10). The battery produces a voltage V, it has negligible internal resistance, and it is located far from the hole. The wires are perfectly conducting, and they descend to the hole along the polar axes. The hole's horizon provides the only nonzero resistance in the circuit. The total resistance is the surface resistivity divided by the cross-sectional length $2\pi r_H \sin\theta$ across which the current flows, and integrated over the distance $r_H\, d\theta$ through which it flows:

$$R_T = \int_{\text{north pole}}^{\text{south pole}} R_H \left(\frac{r_H\, d\theta}{2\pi r_H \sin\theta}\right) \qquad (67)$$

If the wires have vanishing radius, the integration must be taken from the precise north pole ($\theta = 0$) to the precise south pole ($\theta = \pi$), and the integral is then infinite. With an infinite resistivity, the circuit can carry no current. More interesting is the case of wires with finite radius. To make the analysis analytically simple, we take them to have the form of perfectly conducting conical surfaces with opening half-angles θ_0 (Fig.

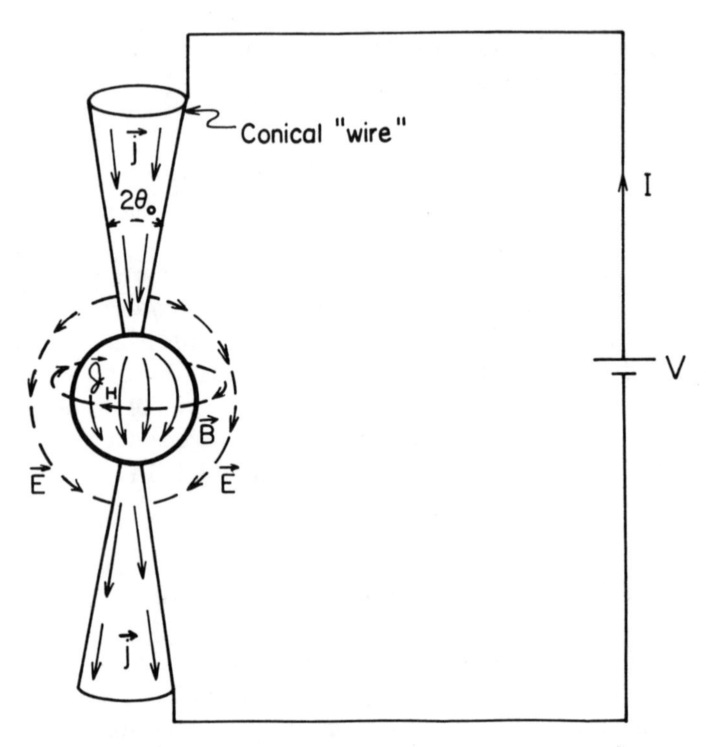

FIGURE 10. Nonrotating black hole as the resistor in an electric circuit. (Reproduced from BHMP.)

10). Then the integral in Eq. (67′) is from θ_0 to $\pi - \theta_0$, and the resulting total resistance presented to the circuit by the hole's horizon is

$$R_T = 4 \ln \left[\cot \left(\frac{\theta_0}{2} \right) \right] = (120 \text{ ohms}) \ln \left[\cot \left(\frac{\theta_0}{2} \right) \right] \tag{67′}$$

since in geometrized units $1 = 1/c = 30$ ohms.

The total current I that flows in the circuit, measured on a per-unit-universal-time basis (since it is a global quantity)

$$I = \int_{\text{any cross section}} \alpha \mathbf{j} \cdot d\mathbf{A} \tag{68}$$

is given by the standard circuit equation

$$I = \frac{V}{R_T} \tag{69}$$

By symmetry this current must flow uniformly on the horizon, where the cross-sectional length is $2\pi r_H \sin\theta$. Thus the horizon current must point in the θ direction and have magnitude $I/2\pi r_H \sin\theta$:

$$\mathcal{J}_H = \frac{I}{2\pi r_H \sin\theta}\,\mathbf{e}_{\hat{\theta}} \tag{70}$$

Ohm's law and Ampere's law for the stretched horizon tell us that this horizon current must be driven by a horizon electric field and must produce a horizon magnetic field given by

$$\mathbf{E}_H = \mathcal{J}_H R_H = \frac{2I}{r_H \sin\theta}\,\mathbf{e}_{\hat{\theta}} \tag{71a}$$

$$\mathbf{B}_H = -\frac{2I}{r_H \sin\theta}\,\mathbf{e}_{\hat{\varphi}} \tag{71b}$$

The tangential electric and magnetic fields $\mathbf{E}_{\|}$ and $\mathbf{B}_{\|}$ measured by the ZAMOs at the stretched horizon are the same as these $\mathbf{E}_H$ and $\mathbf{B}_H$, renormalized by $1/\alpha_H$.

The ZAMO-measured electric and magnetic fields at larger radii, as computed from the full Maxwell equations (31), turn out to be (Damour 1978)

$$\mathbf{E} = \frac{2I}{\alpha r \sin\theta}\,\mathbf{e}_{\hat{\theta}} \tag{72a}$$

$$\mathbf{B} = -\frac{2I}{\alpha r \sin\theta}\,\mathbf{e}_{\hat{\varphi}} \tag{72b}$$

Note that the electric field $\mathbf{E}$ has the properties, as it must, that it is equal to $\mathbf{E}_H/\alpha_H$ at the stretched horizon, and it produces a potential drop equal to the battery voltage between the upper conducting cone and the lower conducting cone

$$\int_{\text{upper conducting cone}}^{\text{lower conducting cone}} \alpha \mathbf{E}\cdot d\mathbf{l} = IR_T = V \tag{73}$$

Note, similarly, that the magnetic field $\mathbf{B}$ of Eq. (72b) is equal to $\mathbf{B}_H/\alpha_H$ at the stretched horizon, and when integrated in a closed loop around the hole and wires, it gives 4π times the current in the circuit [Ampere's law, Eq. (31d)]

$$\int_{\text{closed curve around hole}} \alpha \mathbf{B} \cdot d\mathbf{l} = 4\pi I \tag{74}$$

The current (71a) flowing in the stretched horizon produces ohmic dissipation

$$T_H \frac{dS_H}{dt} = \int R_H \mathcal{J}_H{}^2 \, dA = \int \mathcal{J}_H \cdot \mathbf{E}_H \, dA$$

$$= \int (\mathcal{J}_{H\hat{\theta}} 2\pi r_H \sin\theta) E_{H\hat{\theta}} r_H \, d\theta = \int I \mathbf{E}_H \cdot d\mathbf{l} = IV = I^2 R_T \tag{75}$$

Thus the dissipation rate is given by the familiar lumped-circuit expression "total current squared times total resistance." A black hole is identical to an ordinary resistor, so far as its DC circuit properties are concerned!

This model problem also illustrates the ingoing-wave nature of the tangential fields at the horizon, and the manner that Poynting flux provides the energy for the ohmic dissipation: The electric and magnetic fields outside the hole, as measured by the ZAMOs [Eqs. (72)], are precisely radiative in form (**E** and **B** are orthogonal and are equal in magnitude). The DC Poynting flux $\mathbf{E} \times \mathbf{B}/4\pi$, as measured by the ZAMOs, is directed radially inward and has magnitude $\alpha^{-2} I^2/(\pi r_H{}^2 \sin^2\theta)$. To convert from a flux of locally measured energy to a flux of redshifted energy ("energy-at-infinity") we must multiply by one factor of α. To convert from a flux measured on a per-unit-ZAMO-time basis to a flux measured on a per-unit-universal-time basis we must multiply by a second factor of α. These two renormalizations remove the α-dependence from the Poynting flux and give a result that, when integrated over the stretched horizon, is precisely equal to the ohmic dissipation rate of Eq. (75).

C. Black Hole as the Rotor in an Electric Motor

As an example of slowly rotating holes and electromagnetic torques on them, we shall study a model problem in which a slowly rotating hole acts as the rotor in a simple electric motor (Fig. 11).

It is instructive to recall, first, the construction of an analogous motor in the laboratory: Take a conducting sphere of radius r_H that is free to turn about its vertical axis and immerse it in a homogeneous, vertical

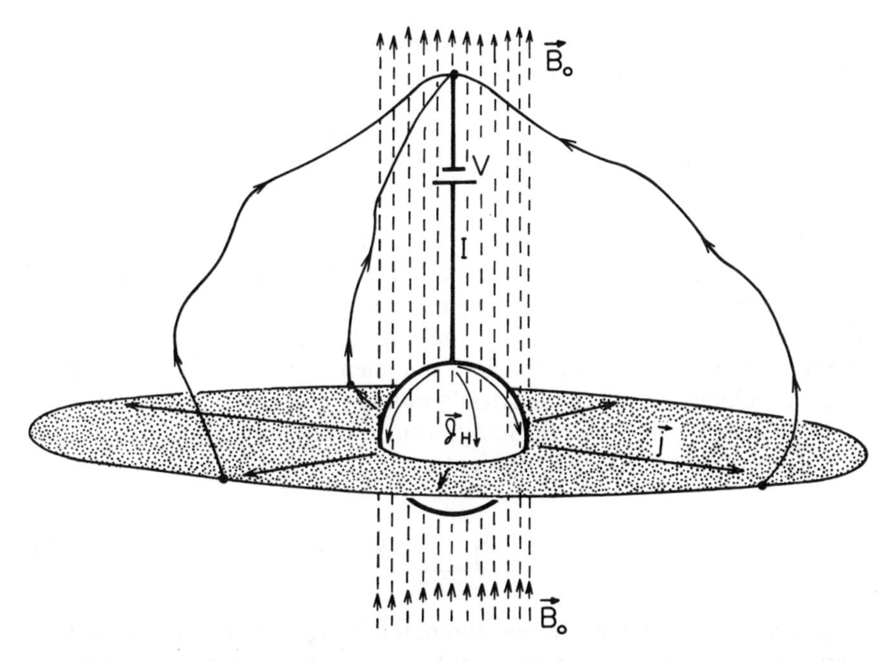

FIGURE 11. Slowly rotating black hole as the rotor in an electric motor. (Reproduced from BHMP.)

magnetic field of strength B_0. Then attach a wire to the sphere's top pole and attach a flat conducting disk to its equator. (The attachments are usually made via wire brushes that slide freely over the sphere as it turns.) Run wires from the outer edge of the disk up to the polar wire, and insert a battery in the circuit, thereby causing a total current I to flow. The upper hemisphere of the sphere will then carry a surface current

$$\mathcal{J}_H = \frac{I}{2\pi r_H \sin\theta}\,\mathbf{e}_{\hat{\theta}} \tag{76}$$

which will interact with the magnetic field to produce a torque on the sphere and spin it up:

$$I_H \frac{d\Omega_H}{dt} = \frac{dJ}{dt} = \int_{\text{sphere}} (\mathcal{J}_H \times \mathbf{B}_0 \cdot \mathbf{e}_{\hat{\varphi}}) r_H \sin\theta \, dA = -\tfrac{1}{2} I B_0 r_H{}^2 \tag{77}$$

It is possible, in principle, to replace the conducting sphere by a Schwarzschild black hole. The immersing, vertical magnetic field B_0 is then identical to the final state of Macdonald's vibrating-magnetic-field problem [Eq. (34); Fig. 4b]:

$$\mathbf{B}_0 = B_0 \left[\cos\theta\, \mathbf{e}_{\hat{r}} - \left(1 - \frac{2M}{r}\right)^{1/2} \sin\theta \mathbf{e}_{\hat{\theta}} \right] \tag{78}$$

and the horizon current has precisely the same form (76) as for our laboratory motor. Remarkably, the equation for the torque also has precisely the same form as in the laboratory case: The Lorentz force law (63) for the stretched horizon guarantees that Eq. (77) is unchanged except for the reinterpretation of I_H as the hole's moment of inertia, Ω_H as the hole's angular velocity, r_H as the horizon's radius ($r_H = 2M$), and so on.

In this black hole motor the horizon current not only produces a torque, but also produces an ohmic dissipation

$$\frac{dM_{\text{irr}}}{dt} = \frac{T_H\, dS_H}{dt} = \int_{\text{horizon}} (\mathcal{J}_H{}^2 R_H)\, dA = I^2 R_T \tag{79}$$

which is identical in form to that computed in Eq. (75) for the same sphere without an immersing magnetic field (except for a factor one half in Eq. (67′) for the horizon's total resistance due to the current then covering both hemispheres of the hole but now covering only the upper hemisphere). The mass of the hole changes as a result both of ohmic dissipation (growth of irreducible mass) and of changing angular velocity (change of rotational mass-energy); cf. the first law of thermodynamics, Eq. (44a):

$$\frac{dM}{dt} = \frac{dM_{\text{irr}}}{dt} + \frac{dM_{\text{rot}}}{dt} = T_H \frac{dS_H}{dt} + \Omega_H \frac{dJ}{dt} = I^2 R_T - \tfrac{1}{2}\Omega_H I B_0 r_H{}^2 \tag{80}$$

D. Rotating Hole Immersed in a Time-Independent, Vacuum Magnetic Field

Consider next a rapidly rotating Kerr black hole immersed in a static magnetic field [(i.e., a field with $(\partial\mathbf{B}/\partial t)_{x^j} = 0$]); see Fig. 12. We initially do not require that the field be symmetric about the hole's rotation axis, but we do require that some of the field lines thread through the stretched horizon.

We can gain insight into the interaction of the **B**-field with the hole's rotation by applying Faraday's law of induction to a carefully chosen closed curve $\mathscr{C}$ (Fig. 12). The curve begins at some arbitrary point $\mathscr{Q}$ on

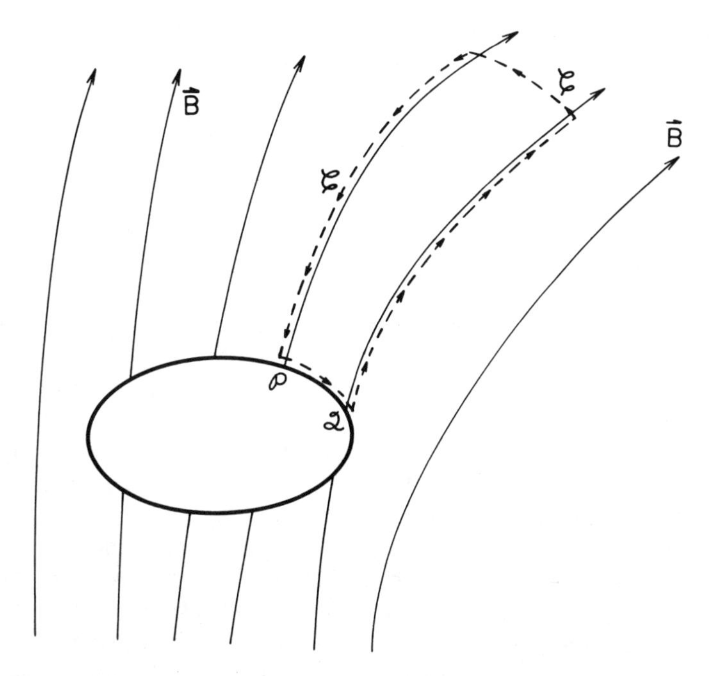

FIGURE 12. Rapidly rotating black hole immersed in a time-independent magnetic field. Interaction of the hole's gravitomagnetic potential **β** with the immersing magnetic field **B** induces an EMF around the closed circuit $\mathscr{C}$. This EMF can be thought of as due to a "battery" that resides in the stretched horizon. (Reproduced from BHMP.)

the stretched horizon, extends out of the horizon and up a magnetic field line to a point far from the hole, then crosses over to some other field line and descends down it to the stretched horizon at point $\mathscr{P}$, then extends along the stretched horizon back to its starting point $\mathscr{Q}$. The instantaneous EMF around this closed curve

$$\text{EMF} = \int_{\mathscr{C}} \alpha \mathbf{E} \cdot d\mathbf{l} \tag{81}$$

can be computed from Faraday's law (31c). Since the magnetic field is time-independent ($\partial \mathbf{B}/\partial t = 0$), one might have thought that the magnetic flux linked by the curve is constant in time. However, the curve $\mathscr{C}$ in Faraday's law is required to be at rest in absolute space, which means it is dragged into orbital motion relative to the spatial coordinates x^j by the rotation of the hole: $dx^j/dt = -\beta^j$ [Eq. (10)]. The resulting shearing motion of $\mathscr{C}$ ($|\,d\mathbf{x}/dt\,|$ big near the stretched horizon and small far away) causes $\mathscr{C}$ to link a time-changing magnetic flux:

$$\text{EMF} = -\frac{d}{dt}\int_{\mathscr{A}} \mathbf{B}\cdot d\mathbf{A} = -\int_{\mathscr{C}} \mathbf{B}\cdot(-\boldsymbol{\beta}\times d\mathbf{l})$$

$$= -\int_{\mathscr{C}} \boldsymbol{\beta}\times\mathbf{B}\cdot d\mathbf{l} = -\int_{\mathscr{P}}^{\mathscr{Q}} \boldsymbol{\beta}\times\mathbf{B}\cdot d\mathbf{l} \qquad (82)$$

Note, as indicated by the last integral, that because $\mathbf{B}$ is parallel to the outgoing and ingoing portions of $\mathscr{C}$, and $\boldsymbol{\beta}$ vanishes ($|\,\boldsymbol{\beta}\,| \sim J/r^2$) on the distant portion of $\mathscr{C}$, the EMF is produced entirely between points $\mathscr{P}$ and $\mathscr{Q}$ on the hole's horizon.

One way to view this EMF is in terms of an analogy with a unipolar inductor (Blandford and Znajek 1977). An alternative way, which we shall adopt, is to think of the hole's gravitomagnetic potential $\boldsymbol{\beta}$ and the immersing magnetic field $\mathbf{B}$ as jointly producing a "surface battery" in the horizon of the hole—a battery whose EMF between the two points $\mathscr{P}$ and $\mathscr{Q}$ is given by Eq. (82). However, we must be a bit careful with this battery viewpoint: Only when the closed curve $\mathscr{C}$ rises out of the horizon parallel to the magnetic field and extends to large distances does the EMF get produced solely in the horizon.

Let us now specialize to an axisymmetric, time-independent magnetic configuration such as that in Fig. 13. We wish to determine the physical consequences of the "batterylike" EMF in the hole's horizon. We shall do so with the help of Ampere's law (31d) and Faraday's law (31c) applied to the circle $\mathscr{C}'$ in the stretched horizon of Fig. 13a: By axisymmetry Ampere's law says

$$2\pi\tilde{\omega}_H B_{H\hat{\phi}} = \int_{\mathscr{C}'} \alpha\mathbf{B}\cdot d\mathbf{l} = \frac{d}{dt}\int_{\mathscr{A}} \mathbf{E}\cdot d\mathbf{A} + 4\pi\int_{\mathscr{A}} \alpha\mathbf{j}\cdot d\mathbf{A} \qquad (83)$$

The first term on the right-hand side vanishes because any electric field $\mathbf{E}$ generated by interaction of the time-independent $\mathbf{B}$ field and the hole's time-independent rotation ($\boldsymbol{\beta}$-field) must itself be time independent; and the second term vanishes because there are no currents $\mathbf{j}$ present in the space outside the hole. Thus the integral of $\alpha\mathbf{B} = \mathbf{B}_H$ around the curve $\mathscr{C}'$ vanishes, which means that the $\hat{\phi}$-component ("toroidal component") of $\mathbf{B}_H$ must vanish at the stretched horizon. But the vanishing of the toroidal $B_{H\hat{\phi}}$ implies (by Ampere's law for the stretched horizon) that the $\hat{\theta}$-component ("poloidal component") of the horizon current $\mathscr{J}_H$ must

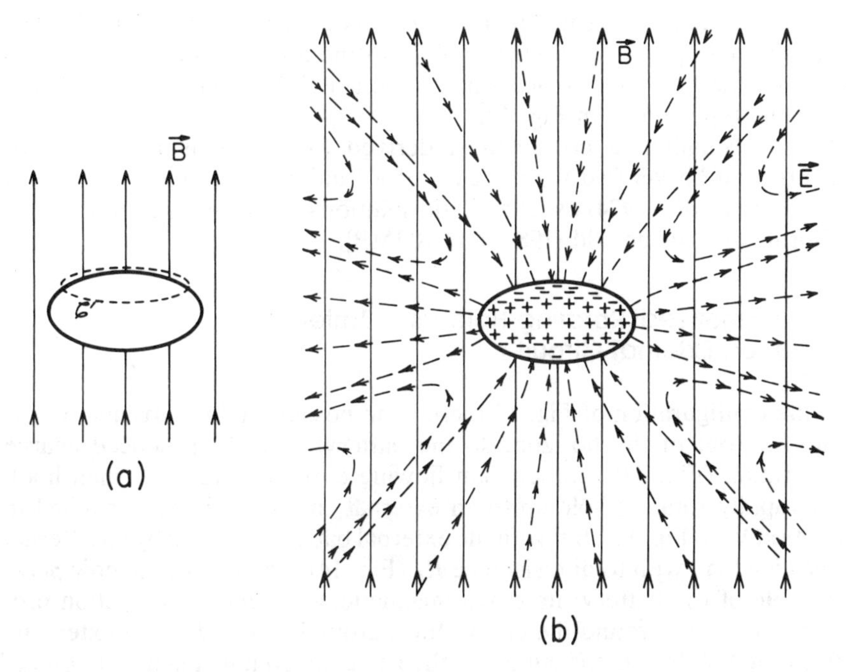

FIGURE 13. Rapidly rotating black hole immersed in a time-independent, axisymmetric magnetic field. Diagram (a) is used to infer that no currents are flowing in the stretched horizon. Diagram (b) shows the charge separation produced by the horizon's "gravitomagnetic × magnetic battery," and the electric field generated by that charge separation. (Reproduced from BHMP.)

vanish, which in turn implies (by Ohm's law) that the poloidal component of the horizon electric field $E_{H\hat{\theta}}$ must vanish. An analogous argument based on Faraday's law (31c) applied to the curve $\mathscr{C}'$ shows that the toroidal component of $\mathbf{E}_H$ and $\mathscr{J}_H$ and the poloidal component of $\mathbf{B}_H$ all vanish. Thus there are no tangential fields $\mathbf{E}_H$ or $\mathbf{B}_H$ whatsoever and no surface current $\mathscr{J}_H$.

The vanishing of the horizon's poloidal $\mathbf{E}_H$ might seem peculiar in light of the poloidal, "battery-like" EMF (82) in the stretched horizon. Something must be counteracting that EMF to make the tangential poloidal electric field vanish. The only thing that can do so is "charge separation" in the stretched horizon; Fig. 13b: When the hole was first immersed in the magnetic field, the "batterylike" EMF must have pushed positive charges in the horizon toward the equator and negative charges toward the poles, until the charge separation created sufficient electric field of its own to counteract the EMF and leave the horizon free of tangential electric field and current. The resulting charge distribution (Fig. 13b) obviously must produce an electric field in the absolute space around the hole—a field required to account for the net EMF around closed curves $\mathscr{C}$ of the type shown in Fig. 12.

These qualitative conclusions, derived by simple membrane arguments, can be verified for the case of a "uniform" magnetic field by a study of Wald's (1974) exact analytic solution of Maxwell's equations [cf. Thorne and Macdonald (1982) and BHMP].

E. Magnetized, Rotating Hole as a Battery for an External Circuit

In the configuration of Fig. 13 there is no possibility for currents to flow into and out of the horizon, so the batterylike EMF produced charge separation (Fig. 13b) rather than flowing currents. Now we shall hook the rapidly rotating hole up to an external circuit of the type studied in Section V.C (Fig. 11), but with the external battery replaced by an external resistive load with total resistance R_L (Fig. 14). The black hole now plays the role of the battery; its gravitomagnetic $\times$ magnetic interaction produces an EMF around the circuit [i.e., around a closed curve extending from the hole's equator out along the perfectly conducting disk to one of the distant wires, up the wire and through the resistive load to the top of the conical conductor, down the conical conductor to its tip at $\theta = \theta_0 \ll 1$ on the horizon, and poloidally (in θ-direction) down the horizon to the equatorial starting point]:

$$V \equiv \text{EMF} \equiv \oint \alpha \mathbf{E}\cdot d\mathbf{l} = \int_{\text{tip of cone}}^{\text{equator}} (-\boldsymbol{\beta} \times \mathbf{B})\cdot d\mathbf{l}$$

$$= \Omega_H \int_{\theta_0}^{\pi/2} B_n \tilde{\omega}_H \rho_H \, d\theta \quad (84)$$

The second integral is the "batterylike" EMF of Eq. (82); and the last expression is this EMF rewritten using $\boldsymbol{\beta} = -\Omega_H \partial/\partial\varphi = -\Omega_H \tilde{\omega}_H \mathbf{e}_{\hat{\varphi}}$ [Eq. (4a) evaluated at the horizon using (12)], and $d\mathbf{l} = (\rho_H \, d\varphi)\mathbf{e}_{\hat{\theta}}$ [cf. Eq. (4)]. Here ρ_H and $\tilde{\omega}_H$ are the spatial metric coefficients ρ and $\tilde{\omega}$ [Eq. (4d)] evaluated on the horizon.

The hole's voltage (84) causes a total current $I = d(\text{charge})/dt$ to flow through the circuit. On the horizon this current is carried by a poloidal (θ-directed) surface current

$$\mathscr{J}_H = \frac{I}{2\pi\tilde{\omega}_H} \mathbf{e}_{\hat{\theta}} \quad (85)$$

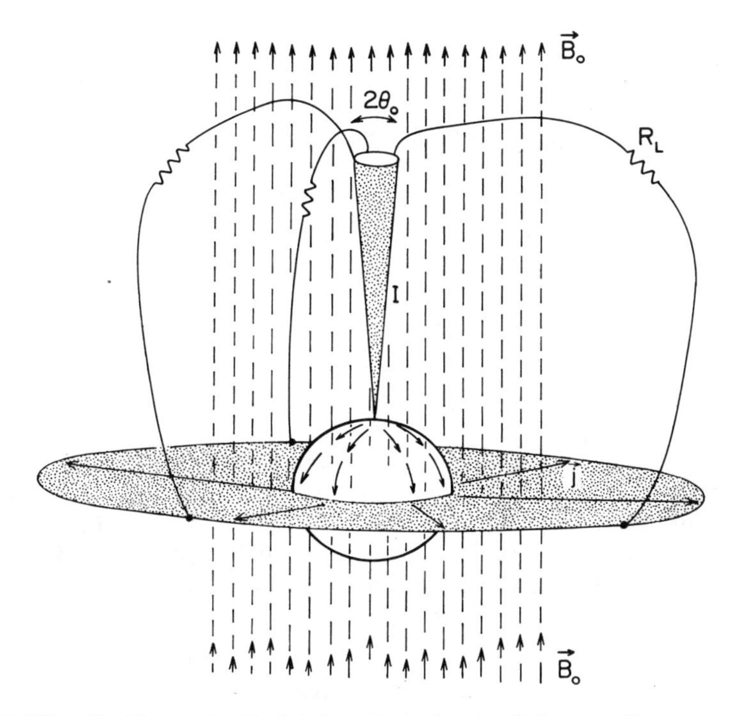

FIGURE 14. Rapidly rotating black hole as the battery in a DC circuit. (Reproduced from BHMP.)

[One can verify, by applying Faraday's and then Ohm's law to the loop $\mathscr{C}'$ of Fig. 13a, that the toroidal component (φ-component) of the surface current vanishes.]

One can compute the total current I in the circuit by evaluating the integral $\oint \alpha \mathbf{E} \cdot d\mathbf{l} = V$ around the circuit using ohm's law. The portion of the integral that lies in the horizon gives

$$V_H = \int_{\theta_0}^{\pi/2} \alpha_H E_{\hat{\theta}} \rho_H \, d\theta = \int_{\theta_0}^{\pi/2} R_H \mathscr{J}_{H\hat{\theta}} \rho_H \, d\theta$$

$$= I \int_{\theta_0}^{\pi/2} R_H \frac{\rho_H \, d\theta}{2\pi\tilde{\omega}_H} \equiv IR_T \quad (86)$$

The first equality follows from $d\mathbf{l} = (\rho_H \, d\theta)\mathbf{e}_{\hat{\theta}}$; the second follows from Ohm's law; the third follows from Eq. (85) for the surface current; and the fourth is the usual definition of the total resistance R_T as the integral, along the conductor (horizon), of the surface resistivity R_H divided by the cross-sectional length through which the current is carried; cf. Eq. (67). For simplicity we shall assume that the equatorial disk and the narrow polar cone are perfect conductors and are so constructed that each bit of disk or cone is at rest in absolute space (i.e., in the reference frame of a local ZAMO). Then the tangential electric field must vanish on the disk and cone, and they thus contribute nothing to $\oint \alpha \mathbf{E} \cdot d\mathbf{l} = V$. The wires leading from the outer edge of the disk to the upper edge of the cone are at rest in the asymptotically flat region ($\alpha \simeq 1$, $\beta \simeq 0$) of absolute space; so they can be analyzed using the usual flat-space version of circuit theory. They contribute

$$V_L = IR_L \tag{87}$$

to the integral $\oint \alpha \mathbf{E} \cdot d\mathbf{l}$, where R_L is the total resistance of the load computed in the usual way ($R_L^{-1} = R_1^{-1} + R_2^{-1} + R_3^{-1} + \cdots$) for a parallel circuit. By adding the battery and load contributions (86) and (87) we obtain the familiar result

$$I = \frac{V}{R_T + R_L} \tag{88}$$

for the circuit's current I in terms of the EMF V of the horizon's "battery" (Eq. 84) and the total horizon and load resistances R_T and R_L.

Just as the voltage-current relation (88) is identical to that of flat-space DC circuit theory, so also the energy balance of the circuit is the same: Ohmic heating deposits energy in the resistive load at a rate

$$\frac{dE_L}{dt} = I^2 R_L \tag{89}$$

Similarly, Ohmic heating increases the hole's area A_H, entropy S_H, and irreducible mass $M_{\text{irr}} = (A_H/16\pi)^{1/2}$ at a rate

$$\frac{g_H}{8\pi}\frac{dA_H}{dt} = T_H \frac{dS_H}{dt} = \int_{\text{Horizon}} R_H \mathcal{J}_H^2 \, dA$$

$$= \int_{\theta_0}^{\pi/2} E_{H\hat{\theta}} \mathcal{J}_{H\hat{\theta}} 2\pi\tilde{\omega}_H \rho_H \, d\theta = I \int_{\theta_0}^{\pi/2} E_{H\hat{\theta}} \rho_H \, d\theta = I^2 R_T \tag{90}$$

The first equality follows from $T_H \propto g_H$ and $S_H \propto A_H$ [Eqs. (41) and (39)];

the second equality is the law of ohmic heating (64); the third follows from Ohm's law $R_H \mathcal{J}_H = \mathbf{E}_H$ and from $dA = 2\pi\tilde{\omega}_H \rho_H \, d\theta$ [cf. the spatial metric (4b)]; the fourth follows from equation (85) for $\mathcal{J}_H$; and the fifth follows from Eq. (86).] The energy for these ohmic losses comes, ultimately, from the rotation of the hole; so as the power is dissipated the hole must spin down:

$$\begin{aligned} -\Omega_H \frac{dJ}{dt} &= -\Omega_H \int_{\text{Horizon}} (\mathcal{J}_H \times \mathbf{B}_n) \cdot \frac{\partial}{\partial \varphi} \, dA \\ &= \Omega_H \int_{\theta_0}^{\pi/2} \mathcal{J}_{H\hat{\theta}} B_n \tilde{\omega}_H 2\pi \tilde{\omega}_H \rho_H \, d\theta = I \Omega_H \int_{\theta_0}^{\pi/2} B_n \tilde{\omega}_H \rho_H \, d\theta \\ &= IV = I^2 (R_T + R_L) \end{aligned} \tag{91}$$

[Here the first equality is the horizon's Lorentz force law (63); the second equality follows from $\partial/\partial\varphi = \tilde{\omega}_H \mathbf{e}_{\hat{\varphi}}$ and $dA = 2\pi\tilde{\omega}_H \rho_H \, d\theta$; the third equality follows from Eq. (85) for $\mathcal{J}_H$; the fourth follows from Eq. (84) for the horizon's batterylike voltage; and the fifth follows from the lumped-circuit Eq. (88).] Note that Eqs. (89)–(91) say that the mass of the hole decreases as a result of spindown plus ohmic heating at precisely the same rate as energy gets deposited in the resistive load:

$$\frac{dM}{dt} = T_H \frac{dS_H}{dt} + \Omega_H \frac{dJ}{dt} = -\frac{dE_L}{dt} \tag{92}$$

It would be nice if we could exhibit an analytic solution for the electric and magnetic fields in the absolute space around the hole and circuit. Unfortunately, no analytic solution is known. On the other hand, none is really needed if all we wish to understand is the behavior of the circuit. Just as in laboratory electronics (where one rarely asks about the **E** and **B** fields in the space between the circuit elements), so also here, we have succeeded in analyzing the circuit in full detail without detailed knowledge of **E** and **B**.

VI. ASTROPHYSICAL APPLICATIONS OF THE MEMBRANE FORMALISM

The membrane formalism is a useful tool in building models for the power sources of quasars and active galactic nuclei. This section will sketch its

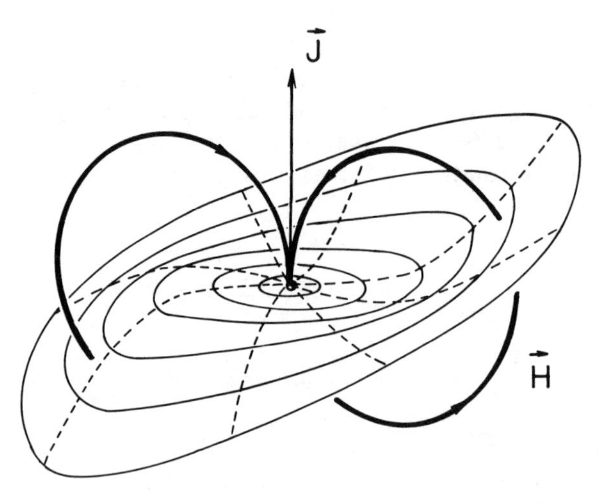

FIGURE 15. An accretion disk around a black hole is driven into the hole's equatorial plane at small radii by the coupling of viscosity to gravitomagnetic (**H**-induced) precession ("Bardeen–Petterson effect"). (Reproduced from Thorne and Blandford 1982.)

applications very briefly. For details the reader should consult the papers cited.

It is now clear that quasars and other strong extragalactic radio sources are fed power ($dE/dt \gtrsim 10^{44}$ erg sec^{-1}) by jets of gas and magnetic field and that each jet is generated by a compact supermassive object ($M \gtrsim 10^7$ solar massses) in the nucleus of a galaxy [see the chapter by Rees in this volume; also see Begelman, Blandford, and Rees (1984)]. A supermassive black hole is the prime candidate for the compact object.

In some radio sources the compact object must hold the jet direction constant for times as long as 10 million years. The only way a black hole can do this is by the gyroscopic action of its spin; and the only way it can communicate the direction of its spin to the jet is via its gravitomagnetic field **H**. In fact, **H** should produce a gravitomagnetic precession of any accretion disk encircling the hole [analog of Eq. (22)]; and it turns out (Bardeen and Petterson 1975) that this gravitomagnetic precession should couple to the disk's viscosity to produce forces that drive the inner region of the disk into the hole's equatorial plane (Fig. 15). The resulting configuration has only two preferred directions along which to send jets: the north and south poles of the hole.

Some observed jets precess with precession periods $\gtrsim 10^4$ years. If the jets are attached to the spin axis of a black hole, this must be caused by a precession of the central hole's spin. The most promising way that the spin can be made to precess is by orbital motion of the hole (now acting as a "test gyroscope") around a companion hole or other massive object (which acts as a source of **g** and **H** fields and of space curvature) (see Begelman, Blandford, and Rees 1980). The resulting geodetic precession [Eq. (26)] will be faster than the gravitomagnetic precession (22) due to the companion's spin and will have a period

$$\frac{2\pi}{\Omega_{\rm geo}} \sim 10^4 \text{ years} \left(\frac{\text{distance between holes}}{0.01 \text{ parsecs}}\right)^{5/2} \left(\frac{\text{mass of companion}}{10^8 \text{ solar masses}}\right)^{-3/2} \tag{93}$$

There are several plausible models for generation of the jets. One of the most attractive (Blandford and Znajek 1977) relies on the hole's rotational energy as the power source, and relies on the horizon's "gravitomagnetic × magnetic battery" as that power source's agent (astrophysical variant of the model problem in Section V.E and Fig. 14).

More specifically, consider a rotating black hole surrounded by a magnetized accretion disk (Fig. 16). As the disk's plasma accretes, it drags magnetic field lines with itself, depositing them on the horizon. Although the B fields in the disk may be very chaotic, when deposited on the horizon they quickly slide around ("imperfect MHD"; "finite conductivity of the horizon"; time scale of sliding $\sim 4M$) until they become very orderly; that is, the horizon "cleans" the magnetic field (Macdonald and Thorne 1982; Macdonald and Suen 1985). The ultimate, nearly uniform magnetic-field configuration on the horizon, as depicted in Fig. 16, is that which minimizes the horizon's ohmic dissipation (Macdonald and Thorne 1982). These quasiuniform **B**-field lines are held on the hole by Maxwell pressure from the surrounding chaotic field lines, which in turn are anchored in the disk by currents. If the disk were suddenly removed, the field would slide off the horizon, convert itself into radiation, and fly away in a time $\sim 4M$ (cf. Macdonald's vibrating magnetic field, Section III.A and Fig. 5).

Near the horizon the magnetic field **B** will be so strong that currents cannot flow across it; but far from the horizon, where **B** is weaker, they can. Consequently, the **B**-field acts like a DC transmission line: the horizon's "gravitomagnetic × magnetic battery" [Eq. (82)] drives currents around closed loops, such as curve $\mathscr{C}$ of Fig. 16: up the **B**-field from the horizon to a weak-**B** region, across the **B**-field there, and then back down

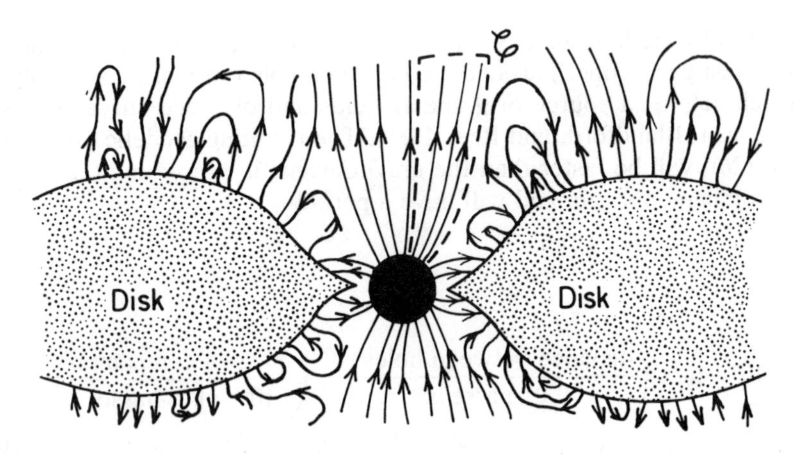

FIGURE 16. Black hole surrounded by a magnetized accretion disk. The closed loop $\mathscr{C}$ plays a role analogous to the DC circuit of Fig. 14; currents flowing around it transport power from the hole's rotation into the magnetosphere, where the power may be used to drive jets ("Blandford–Znajek process"). (Adapted from Thorne and Blandford 1982.)

the **B**-field to the horizon and along the horizon through the "battery" to the starting point (distributed-current analog of model problem of Section V.E and Fig. 14). These currents transmit power in the form of Poynting flux from the horizon to the weak-B region, where it is deposited into charged particles and accelerates them to ultrarelativistic energies. Both the horizon with its battery, and the weak-B "acceleration region" with its particles soaking up power, possess total electric resistances of order 30 ohms (Znajek 1978; Damour 1978; Lovelace, MacAuslan, and Burns 1979; Macdonald and Thorne 1982; Phinney 1983a,b). Thus the battery and its load are impedance matched, and this "Blandford–Znajek" (1977) process has the optimum possible efficiency for depositing the hole's rotational energy into ultrarelativistic charged particles—particles that might well generate the jets observed in quasars and galactic nuclei. Numbers for typical jet models are

$$\begin{pmatrix}\text{Battery}\\ \text{voltage}\end{pmatrix} \sim \left(\frac{J}{J_{\max}}\right) B_n r_g \sim (10^{20}\,\text{V}) \left(\frac{J}{J_{\max}}\right)\left(\frac{B_n}{10^4\,\text{G}}\right)\left(\frac{M}{10^8 M_\odot}\right) \tag{94}$$

$$\begin{pmatrix}\text{Power}\\ \text{output}\end{pmatrix} \sim \left(\frac{(\text{Voltage})^2}{4R_T}\right)$$
$$\sim (10^{45}\,\text{erg sec}^{-1}) \left(\frac{J}{J_{\max}}\right)^2 \left(\frac{B_n}{10^4\,\text{G}}\right)^2 \left(\frac{M}{10^8 M_\odot}\right)^2 \tag{95}$$

where $J/J_{\max} = a/M$ is the black hole's angular momentum ("gravitomagnetic dipole moment") in units of the maximum possible angular momentum GM^2/c of a hole, $M/10^8 M_{\odot}$ is the hole's mass in units of 10^8 solar masses, B_n is the strength of the ordered magnetic field threading the hole, and R_T = (length/circumference) $\times R_H \sim 30$ ohms is the total electrical resistance between the horizon's polar regions and its equatorial regions.

For further discussion of the details of this Blandford–Znajek process see Blandford and Znajek (1977), Macdonald and Thorne (1982), Macdonald (1984), and Phinney (1983a,b); and for a discussion of related astrophysical issues see the chapter by Rees in this volume, and Begelman, Blandford, and Rees (1984).

VII. CONCLUSION

In this paper we have restricted attention to electromagnetic aspects of the membrane formalism and to quiescent black holes. In doing so we have largely ignored such features of a black hole's horizon as its viscosity and viscous stress, its surface pressure, its surface energy density, and its surface momentum density. Also ignored have been deformations of the horizon, both stationary and dynamical, by the gravitational pulls of external and infalling matter and by black hole vibrations; the general formulations of the laws of dissipation, energy conservation, and momentum conservation (Navier–Stokes equation) for the hole's horizon; and analyses of the evolution of the hole's mass, entropy, and angular momentum due to interactions with external objects and infalling matter. A pedagogical review of these is given in BHMP (Thorne, Price, and Macdonald, eds. 1986), along with a more detailed review of electromagnetic features of the formalism than that given here.

ACKNOWLEDGMENTS

For extensive discussions that underlie the work reported here, I thank R. Price, T. Damour, R. Znajek, D. Macdonald, W.-M. Suen, and R. Blandford. Portions of this chapter were adapted from BHMP (Thorne, Price, and Macdonald, eds. 1986). I thank my coauthors of the relevant chapters of that book (R. Price, D. Macdonald, W.-M. Suen, and X.-H. Zhang) for permission to do this adaption and thank D. Macdonald for careful preparation of many of the figures. This work was supported in part by the National Science Foundation, Grant AST 82-14126.

REFERENCES

Arnowitt, R., Deser, S., and Misner, C. W., 1962, in *Gravitation: An Introduction to Current Research*, L. Witten (ed.), Wiley, New York, p. 227.

Bardeen, J. M., 1973, in *Black Holes*, C. DeWitt and B. S. DeWitt (eds.), Gordon and Breach, New York, p. 215.

Bardeen, J. M., Carter, B., and Hawking, S. W., 1973, *Commun. Math. Phys.*, **31**, 161.

Bardeen, J. M., and Petterson, J. A., 1975, *Ap. J.* (*Lett.*), **195**, 65.

Begelman, M. C., Blandford, R. D., and Rees, M. J., 1980, *Nature*, **287**, 307.

Begelman, M. C., Blandford, R. D., and Rees, M. J., 1984, *Rev. Mod. Phys.*, **56**, 255.

Bekenstein, J. D., 1972a, unpublished Ph.D. thesis, Princeton University.

Bekenstein, J. D., 1972b, *Nuovo Cimento Lett.*, **4**, 737.

Bekenstein, J. D., 1973, *Phys. Rev. D*, **7**, 2333.

Bekenstein, J. D., 1974, *Phys. Rev. D*, **9**, 3292.

Bekenstein, J. D., 1980, *Phys. Today*, **33**, January issue, 24.

Blandford, R. D., and Znajek, R. L., 1977, *M.N.R.A.S.*, **179**, 433.

Braginsky, V. B., Caves, C. M., and Thorne, K. S., 1977, *Phys. Rev. D*, **15**, 2047 (and esp. 2054).

Chandrasekhar, S., 1983, *The Mathematical Theory of Black Holes*, Clarendon Press, Oxford.

Christodolou, D., 1970, *Phys. Rev. Lett.*, **25**, 1596.

Cohen, J., and Wald, R., 1971, *J. Math. Phys.*, **12**, 1845.

Damour, T., 1978, *Phys. Rev. D*, **18**, 3598.

Damour, T., 1979, unpublished Ph.D. thesis, University of Paris VI.

Damour, T., 1982, in *Proceedings of the Second Marcel Grossmann Meeting on General Relativity*, R. Ruffini (ed.), North Holland, Amsterdam.

Everitt, C. W. F., 1974, in *Experimental Gravitation:* Proceedings of Course 56 of the International School of Physics "Enrico Fermi," B. Bertotti (ed.), Academic Press, New York, p. 331.

Forward, R. L., 1961, *Proc. I.R.E.*, **49**, 892.

Hanni, R. S., and Ruffini, R., 1973, *Phys. Rev. D*, **8**, 3259.

Hanni, R. S., and Ruffini, R., 1976, *Lettre al Nuovo Cimento*, **15**, 189.

Hartle, J. B., 1973, *Phys. Rev. D*, **8**, 1010.

Hartle, J. B., 1974, *Phys. Rev. D*, **9**, 2749.

Hawking, S. W., 1971a, *Phys. Rev. Lett.*, **26**, 1344.

Hawking, S. W., 1971b, *M.N.R.A.S.*, **152**, 75.

Hawking, S. W., 1974, *Nature*, **248**, 30.

Hawking, S. W., 1975, *Commun. Math. Phys.*, **43**, 199.

Hawking, S. W., 1976, *Phys. Rev. D*, **13**, 191.

Hawking, S. W., and Hartle, J. B., 1972, *Commun. Math. Phys.*, **27**, 283.

Linet, B., 1976, *J. Phys. A*, **9**, 1081.

Lovelace, R. V. E., MacAuslan, J., and Burns, M., 1979, in *Proceedings of La Jolla Institute*

Workshop on Particle Acceleration Mechanisms in Astrophysics, American Institute of Physics, New York.

Macdonald, D., 1983, unpublished Ph.D. thesis, California Institute of Technology.

Macdonald, D., 1984, *M.N.R.A.S.*, **211,** 313.

Macdonald, D., and Suen, W.-M., 1985, *Phys. Rev. D,* (in press).

Macdonald, D., and Thorne, K. S., 1982, *M.N.R.A.S.*, **198,** 345.

Misner, C. W., Thorne, K. S., and Wheeler, J. A., 1973, *Gravitation*, University of Chicago Press, Chicago, cited in text as MTW.

Phinney, E. S., 1983a, in *Proceedings of the Torino Workshop on Astrophysical Jets*, A. Ferrari and A. G. Pacholczyk (eds.), Reidel, Dordrecht.

Phinney, E. S., 1983b, unpublished Ph.D. thesis, University of Cambridge.

Pollock, M. D., 1977, unpublished Ph.D. thesis, University of Cambridge.

Price, R. H., and Thorne, K. S., 1986, *Phys. Rev. D*, **33,** 915.

Smarr, L., 1973, *Phys. Rev. Lett.*, **30,** 71.

Smarr, L., and York, J. W., Jr., 1978, *Phys. Rev. D*, **17,** 2529.

Suen, W.-M., Price, R. H., and Redmount, I. H., 1986, *Phys. Rev. D*, submitted.

Thirring, H., and Lense, J., 1918, *Phys. Z.*, **19,** 156.

Thorne, K. S., 1974, *Ap. J.*, **191,** 507.

Thorne, K. S., 1986, in *Near Zero*, Festschrift for William M. Fairbank, C. W. F. Everitt (ed.), Freeman, San Francisco.

Thorne, K. S., and Blandford, R. D., 1982, in *Extragalactic Radio Sources*, Proceedings of IAU Symposium No. 97, D. S. Heeschen and C. M. Wade (eds.), Reidel, Dordrecht, p. 255.

Thorne, K. S., and Macdonald, D., 1982, *M.N.R.A.S.*, **198,** 339.

Thorne, K. S., Price, R. H., and Macdonald, D., eds., 1986, *Black Holes: The Membrane Paradigm*, Yale University Press, New Haven, Conn.; cited in text as BHMP.

Tranah, D., and Landsberg, P. T., 1980, *Collective Phenomena,* **3,** 81.

Unruh, W. G., 1976, *Phys. Rev. D,* **14,** 870.

Wald, R. M., 1974, *Phys. Rev. D,* **10,** 1680.

York, J. W., Jr., 1979, in *Sources of Gravitational Radiation*, L. Smarr (ed.), Cambridge University Press, Cambridge, p. 83.

Znajek, R., 1976, unpublished Ph.D. thesis, University of Cambridge.

Znajek, R., 1978, *M.S.R.A.S.*, **185,** 833,

Zurek, W. H., and Thorne, K. S., 1985, *Phys. Rev. Lett.*, **54,** 2171.

6.

Jets and Galactic Nuclei

MARTIN J. REES

Martin J. Rees has been the Plumian Professor of Astronomy and Experimental Philosophy at the University of Cambridge since the ripe old age of 31. He has also served as the Director of the prestigious Institute for Astronomy there. He is a leading theoretical astrophysicist who has done fundamental research in the fields of cosmology, X-ray astronomy, galaxies, and quasars.

I. EVIDENCE FOR DIRECTED OUTFLOW FROM ACTIVE GALAXIES

Most galaxies are basically just self-gravitating assemblages of stars and gas, but in 1954 radioastronomers provided the first clues that some galaxies might be more than this. Baade and Minkowski showed that the radio source Cygnus A, the second most intense object in the radio sky, was associated with a remote galaxy with a redshift of 0.05. This immediately indicated that some peculiar galaxies might be detectable by radio techniques even if they were so far away that the integrated light from 10^{11} stars failed to register optically. Radio maps (Fig. 1) tell us that the emission from a source like Cygnus A comes from two blobs symmetrically disposed on either side of the central galaxy. This double structure, in which the overall separation of the components may be a million light-years (or even more), seems characteristic of the strongest radio sources.

The (polarized) radio emission was quickly recognized as synchrotron radiation, implying that the radio lobes must contain relativistic electrons and magnetic fields. The field strength is unknown, but the magnetic energy content of the lobes must be $(B^2/8\pi) \times$ (volume). The energy in relativistic electrons depends on the field strength, being proportional to (radio power output) $\times$ $B^{-3/2}$ (the B-dependence arises because more particles, of higher individual energy, would be needed to produce the observed synchrotron power in a weak field). The total energy is mini-

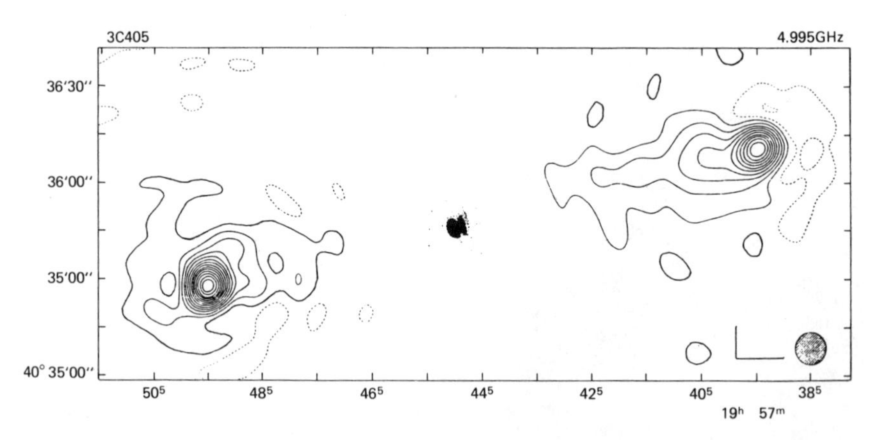

FIGURE 1. A historic radio map (at 5 GHz) of Cygnus A, made with the Cambridge "one-mile" telescope (Mitton and Ryle 1969). This illustrates the best data available on any strong radio source in the 1960s, which showed the characteristic double structure but no evidence of any energetic link between the central galaxy and the lobes.

mized when there is rough equipartition between these two forms. Burbidge (1958) was the first to do such a calculation: he showed that the energy stored in the giant radio sources—in magnetic fields and relativistic plasma—must exceed the rest mass of a million suns. This implied that galaxies could somehow generate power in this particular form at a level vastly exceeding a single supernova.

The major contribution of optical astronomy came in 1963, when searches for the optical counterparts of some radio sources led to the discovery of an unsuspected new class of objects, which looked like stars on photographic plates, but whose spectra displayed emission lines with large redshifts (Hazard et al. 1963; Schmidt 1963). These objects, the quasars, have optical luminosities far exceeding normal galaxies, even though they are more compact. Some are strongly variable: the luminosity may change by $>10^3$ galactic luminosities in only a few days. The optical emission in these highly variable objects is strongly polarized and is probably synchrotron radiation. The majority of quasars, however, are less variable and only weakly polarized; in these, the central synchrotron source may be surrounded and shrouded by gas, so that we only see "secondary" radiation that has been reprocessed and thermalized further out.

During the last 21 years, a vast and bewildering body of accumulated data has borne out the general concept (adumbrated in the pioneering paper by Burbidge, Burbidge, and Sandage (1963)) of "violent activity" in galactic nuclei. Radio sources and quasars are the prime examples of this phenomenon, but the Seyfert galaxies and BL Lac objects are generally interpreted as instances of the same thing.

Optical observations offer a wealth of information on the spectrum, polarization, and variability of galactic nuclei. From such data, physical conditions in the emission regions can be inferred. However, there is essentially no direct information about the spatial structure of the central regions where the visible light originates. Radio astronomers, on the other hand, *can* provide structural information. This is because the radio-emitting regions are often very extended, and also because the angular resolution of radio interferometers surpasses anything optical imaging can yet achieve.

We consequently now have important clues to why radio sources have their characteristic double morphology, and to the nature of the energetic link between the nucleus and the radio lobes. These features were entirely mysterious in the 1960s, when the best maps available resembled Fig. 1, but have been clarified in the last decade, thanks mainly to the improved resolution and sensitivity of the Very Large Array (VLA) in New Mexico.

Figure 2 shows a well-known double source, Hercules A (3C 348),

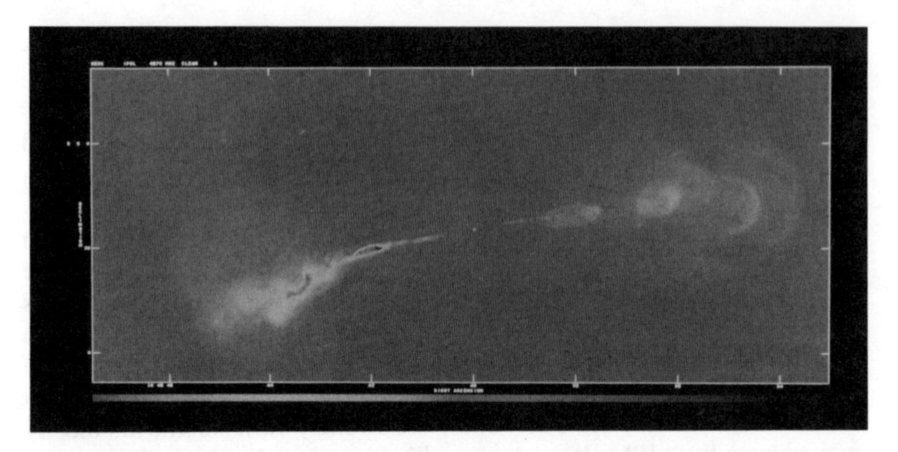

FIGURE 2. VLA map of the giant double source Hercules A, showing narrow straight "jets" linking the galaxy to the lobes (Dreher and Feigelson 1984). This is a high-power source that is atypical in showing such conspicuous jets: normally the jets are only as strong (relative to the lobes) in sources of lower power.

displaying conspicuous bridges of radio emission stretching almost all the way from the central galaxy to the lobes. Jetlike features are now detected in more than 100 double sources (Bridle and Perley, 1984). Some of the jets are rather inconspicuous: the recently discovered jet in Cygnus A (Perley, Dreher, and Cowan, 1984) is barely detected even with the VLA's dynamic range.

The smallest angular scales resolvable by the VLA (0.1 arcsec) correspond to linear dimensions as large as several hundred parsecs in a remote extragalactic source; for finer resolution we must resort to Very Long Baseline Interferometry (VLBI). The montage of NGC 6251 (Fig. 3) is specially interesting in showing direct continuity between a plasma "blowtorch" one parsec long and the large-scale jets and giant radio struc-

FIGURE 3. Montage, adapted from Bridle and Perley (1984), showing the radio source associated with the galaxy NGC 6251 over a wide range of angular scales. The top panel shows the large-scale structure: a double source ~2 Mpc in extent. The second panel shows the jet and the (much weaker) counterjet; lower panels show the high-surface-brightness inner parts of the jet at increasing resolution. The large brightness asymmetry between jet and counterjet and the straightness of the jet are characteristic of moderately high-power radio sources. The bottom panel, obtained with milliarcsecond resolution via the VLBI technique, shows that the jet emanates from a "nozzle" less than 1 pc in scale at the galactic nucleus. The primary power supply probably comes from a region about 5 powers of 10 smaller still. (From Begelman et al. 1984.)

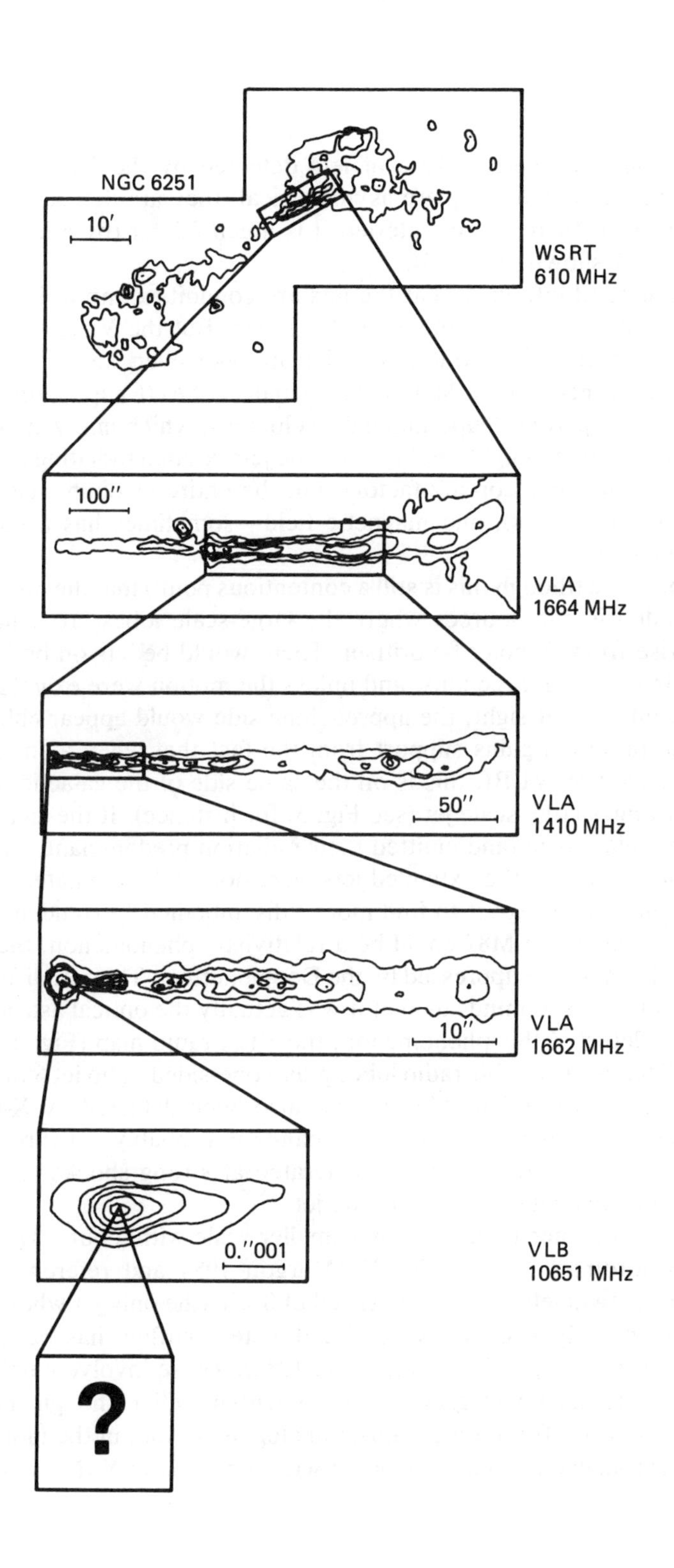
NGC 6251
10′
WSRT
610 MHz
100″
VLA
1664 MHz
50″
VLA
1410 MHz
10″
VLA
1662 MHz
0.″001
VLB
10651 MHz
?

ture. A common characteristic of jets detected by the VLA in strong double sources is that they are asymmetrical: they are detected on only one side; or, if there is a counterjet, it is much fainter (by a factor 60:1 in NGC 6251 shown in Fig. 3).

It is a natural inference that the jets are conduits along which energy and momentum flow into the extended lobes. But the VLA maps offer no direct evidence for motion. VLBI maps, however, sometimes show dramatic evidence of this: Marshall Cohen describes (Chapter 10) several instances of apparent "superluminal" velocities, which imply an outflow speed close to that of light. Not only do the jets contain radiating electrons with high individual Lorentz factors, but the entire medium—electrons, protons (or positrons), and magnetic field—sometimes has a bulk Lorentz factor $\gtrsim 5$.

It is possible (though this is still a contentious point) that the prevalence of one-sided jets in sources where the large-scale lobes are symmetric could arise from Doppler favoritism: There would be jets on both sides, ejected in opposite directions, and unless the motion were exactly transverse to our line of sight, the approaching side would appear enhanced. This interpretation gains support from the fact that when a small-scale jet is detected by VLBI, this is on the same side of the galactic nucleus as the stronger large-scale jet (see Fig. 3, for instance). If the small-scale jets were relativistic (and emitted their radiation predominantly in a forward cone) whereas the extended jets were not (and so radiated isotropically), one would expect to find more "disembodied" extended jets.

The famous jet in M87 could be a relativistic phenomenon, the counterjet perhaps being suppressed by the Doppler effect. This jet, discovered by Curtis in 1918, reminds us that it was actually the optical astronomers who first detected this phenomenon; the VLA radio map (Fig. 4) shows that M87 has weak double radio lobes plus a one-sided radio jet (coinciding with the optical jet). The M87 jet has also been detected by X-ray astronomers; the emission in all wave bands is probably synchrotron radiation, produced by electrons accelerated at strong shocks associated with the conspicuous "knots" in the jet.

Jetlike phenomena are seen on a smaller scale within our own galaxy. The extraordinary object SS 433 (Margon 1984 and references cited therein) has twin jets with a flow speed of $0.27c$ (the only jet whose speed is unambiguously known). Recently, directed outflow has been found from some protostars (Bally and Lada 1982): These involve much lower energies (and shallower gravitational potential wells), though the collimation may arise from a mechanism analogous to that in the more spectacular extragalactic jets. Associated with the galactic X-ray source Sco

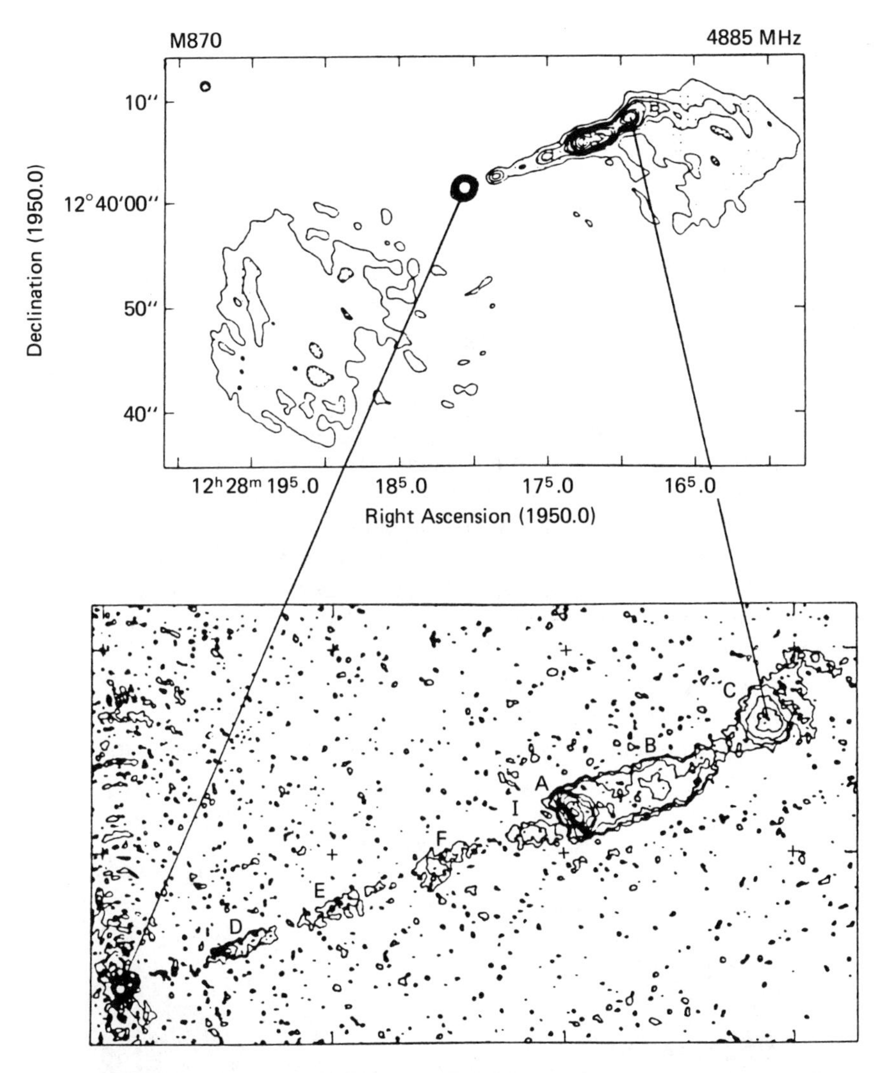

FIGURE 4. Two radio maps of M87. The bottom picture (from Biretta et al. 1983) shows a 15 GHz VLA map of the jet, with 0.12″ resolution. The high brightness features correspond with the optical knots (the emission being synchrotron radiation in both bands). The top picture (from Owen et al. (1980)) shows a more extended radio view at lower resolution, which reveals that M87 is a miniature double source, with roughly symmetrical lobes ~2 kpc in size and a one-sided jet.

X1 are double radio components, resembling a miniature version of an extragalactic double source (Fomalont et al. 1983).

II. INTERPRETATION OF JETS

Any theory of radio sources has, in effect, three ingredients: (i) a source of relativistic plasma in the center; (ii) some bifurcation and collimation mechanism, that is, a way in which relativistic plasma can be squirted out preferentially in two opposite directions; and (iii) a place far away where the relativistic plasma is stopped by interaction with the intergalactic medium, in a shock front (see Fig. 5).

The speed of advance V of the "working surface," where the jet is stopped by the external medium, is governed by ram pressure balance—the balance between the momentum density in the beam and the $\rho_{ext}V^2$ pressure force exerted by the surrounding medium. (Here ρ_{ext} is the density of the surrounding medium.) The beam energy is randomized by shocks when it impinges on the external medium; particles here are accelerated and these regions are identified with the "hot spots" in the radio source components. Even if the beams are relativistic, V itself is not; we therefore do not expect the same Doppler asymmetry in the lobes and hot spots as in the emission from the jets themselves. The relativistic plasma then accumulates in a cocoon of lower energy density and lower

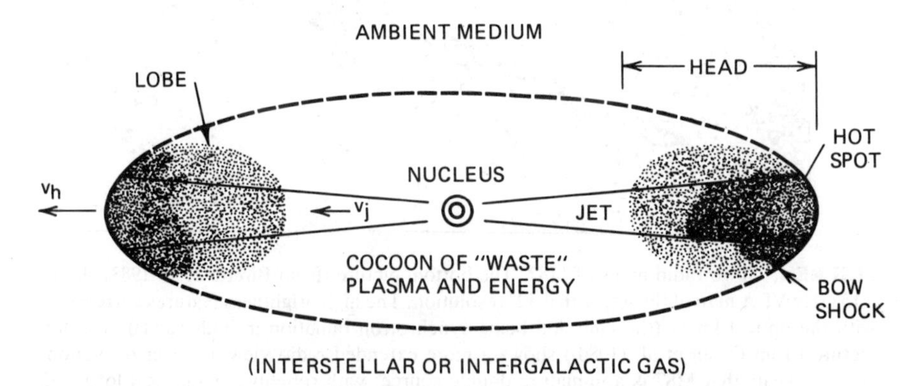

FIGURE 5. Schematic diagram of a strong double radio source, illustrating the nomenclature according to the "twin exhaust" theory (Blandford and Rees 1974). The entire head advances into the intergalactic medium at a speed V given by ram pressure balance (see text), which is much smaller than the jet speed. (From Begelman et al. 1984.)

radio emissivity. Many jets and cocoons, on scales 1—100 kpc, are now mapped beautifully with the VLA.

We are justified in thinking of jets as basically fluid phenomena and using fluid-dynamical analogies, the reason being that the gyroradius $m_e\gamma_e v_e/eB$ for the particles in the jet, and the Debye length $(kT_e/n_e e^2)^{1/2}$, are both *much* less than the jet dimensions, so that charge neutrality is closely satisfied, unless the particles have energies of 10^{19} eV (cf. Lovelace 1976); also the relative mean velocities of the electrons and ions is small. In effect, the flow is fluidlike and the MHD approximation is valid (as in the solar wind).

The data on jets pose a whole range of questions (all discussed more fully by Begelman et al. 1984).

A. What Are the Speeds of the Flow?

The small-scale superluminal sources certainly indicate outflow at a speed of about c, the speed of light, but it is unclear whether the nuclei of all radio sources generate relativistic jets; nor is it clear whether the high speed would persist over the jets' whole length or whether frictional or dissipative effects would gradually slow them down. A tenable viewpoint is that in the strong double sources (Cygnus A, for instance) the jets have high Mach number and low internal dissipation, and maintain speeds of approximately c out to distances of several hundred kiloparsecs, thereby transporting energy to the extended components in an almost loss-free way. The fact that often only one jet is seen could then arise merely from Doppler favoritism. In lower-luminosity sources (where the jets often appear two-sided and are more conspicuous relative to the extended lobes), the flow is presumably slower and more dissipative.

B. Do the Jets Primarily Contain Electron–Positron Plasma or Do They Consist of Ordinary (Entrained) Material?

A slow-moving jet would presumably be made of ordinary swept-up material. However, there are reasons (discussed in Section IV) for conjecturing that the "central engine" generates an e^+–e^- plasma. The kinetic energy requirements of relativistic jets are then somewhat reduced: each electron need be neutralized by just one positron (0.5 MeV of rest mass) rather than a proton (936 MeV).

C. How Is the Collimation of Narrow Jets Maintained?

There are basically three possibilities:

1. The internal pressure may be balanced by external gas pressure. This requires

$$P_{\text{ext}} = \frac{1}{\mathcal{M}^2} \quad \text{(momentum flux density along jet)} \tag{1}$$

where $\mathcal{M}$ is the jet's internal Mach number. As the jet travels outward through regions of decreasing P_{ext} it would widen in accordance with a Bernoulli-type equation (modified, however, by the effects of the internal dissipation, whose occurrence is implied by the presence of freshly accelerated relativistic electrons in the jet).

2. If the jets were "heavy," so that their internal sound speed was small compared with the flow velocity, they could stay narrow even if they flowed outward ballistically without confinement. This requires an internal Mach number of $\mathcal{M} > \theta^{-1}$, where θ is the collimation angle.

3. If the jet were *magnetically confined* by a field with a toroidal component B_T, whose strength varied roughly inversely with distance d from the jet axis for d in the range $d_{\text{min}} < d < d_{\text{max}}$, then the pressure in the core of the jet (i.e., for $d < d_{\text{min}}$) could exceed the external pressure by a factor

$$P_{\text{jet core}} \cong \left(\frac{d_{\text{max}}}{d_{\text{min}}}\right)^2 P_{\text{ext}} \tag{2}$$

For stability there would need to be a component of magnetic field aligned with the jet, maintained by a velocity gradient (shear).

X-ray observations can in some cases set an upper limit to the density of hot gas in a galaxy or cluster potential well, and hence place a limit on P_{ext}. If this exceeds the minimum internal pressure of a jet-emitting synchrotron radiation, then alternative 1 can be ruled out. Alternative 2 can also be excluded in some cases, because the required amount of material in the jet would produce too much Faraday rotation to be compatible with the observed polarization. A jet encounters a wide range of external conditions as it propagates out from the galactic nucleus, through the interstellar medium of the host galaxy, and eventually into intergalactic space; different mechanisms could well be responsible for collimating jets on different scales.

D. How Do Jets Propagate Stably Out to Such Vast Distances?

Clues to this question can come from simulations. Aerodynamical experiments may provide valuable insights into purely hydrodynamic aspects of jet physics. They have limitations, however: they cannot demonstrate the dynamical effects of magnetic fields and relativistic bulk velocity, and only restricted ranges of Mach number, density ratio, and adiabatic index are practicable in the laboratory. Nevertheless, through the judicious choice of different jet gases and the use of wind tunnels and vacuum chambers, it should be possible to spot some of the trends accompanying changes in the various dimensionless quantities that characterize astrophysical jets. The experiments might include firing a supersonic jet into undisturbed gas (or head on into a wind, so that the "working surface" is stationary in the laboratory frame).

A different type of experiment that could achieve higher Mach numbers than wind tunnels would involve the propagation of intense particle beams—or, alternatively, laser beams—into an ambient gas (cf. Bekefi et al. 1980); although the internal dynamics of such beams differ crucially from those in the cosmic-scale beams, they could provide a much higher momentum density than an ordinary gas jet. The interaction with the external medium as such a beam advances may simulate the structure of "hot spots" and cocoons in very strong sources.

The greatest progress will surely come, however, from use of increasingly sophisticated and powerful hydrodynamical codes. These have already uncovered some gas-dynamical properties of supersonic flows that were unanticipated by analytical models and may have counterparts in radio maps (Norman et al. 1983; Williams and Gull 1985). Within a few years, high-resolution three-dimensional computations, incorporating electromagnetic effects (MHD) should be feasible. We can then test whether it is plausible that jets are confined magnetically, and whether the polarization patterns observed in jets can be explained in terms of the kinematics of expanding shear flows.

In the extended radio components the energy deposited by the jets is dissipated via complex interactions with interstellar and intergalactic media: to model their intricate and environment-dependent morphology will be difficult, just as weather prediction is difficult. But the jets are plainly initiated on a scale of $\lesssim 1$ pc. So a key characteristic of active galactic nuclei (AGNs) is that they can not only generate a vast "in situ" luminosity (as in the quasars), but sometimes squirt out these relativistic jets. The jets, however, are just one aspect of the general range of phenomena that encompasses quasars, blazars, BL Lacs, Seyferts, and the rest of the zoo.

III. THE PRIMARY POWER SUPPLY

Galactic nuclei display activity in many forms, in all wave bands, and on scales spanning many powers of 10 (see Fig. 6). The central engines may nevertheless be similar in all the most highly active nuclei—the primary power being reprocessed in a variety of ways, depending on details of the galactic environment.

What then is the central engine? My favored view—the "bottom line" of the diagram reproduced as Fig. 7—is that a variety of runaway processes might be expected in galactic nuclei, but that the almost inevitable endpoint is the collapse of a large fraction of the matter involved to a massive black hole. A black hole offers a more efficient power source than any conceivable progenitor (see Chapter 5 by Kip Thorne). This suggests that we should associate the most powerful objects—quasars and strong radio sources—with black holes, and then consider, as a secondary issue, whether some other categories of AGN (e.g. starburst nuclei, Seyferts, etc.) could be precursor stages in any of the evolutionary pathways in Fig. 7.

The various precursors—supermassive stars, dense star clusters, etc.—received perfunctory attention in the 1960s, but all merit further study. Spherical supermassive stars are fragile: they would be supported almost entirely by radiation pressure, which varies as $\rho^{4/3}$, rendering them only neutrally stable in Newtonian theory; general relativistic corrections, even when these are small, are then destabilizing. However, not much is known about supermassive stars with relativistic (differential) rotation. The stars in an idealized cluster can be treated as "point masses," interacting with each other only gravitationally; but in real systems stellar collisions and interactions with gas would be important, and we do not know whether a dense cluster of ordinary stars would turn itself into a supermassive star or into a cluster of compact stars (neutron stars) that would then evolve further via dynamical processes.

Before "homing in" on black holes, let me venture a historical digression, by quoting from a paper in *Phil. Trans. Roy. Soc.* for 1784 by John Michell, an underappreciated polymath of eighteenth-century science.

> If the semidiameter of a sphaere of the same density with the sun were to exceed that of the sun in the proportion of 500 to 1, a body falling from

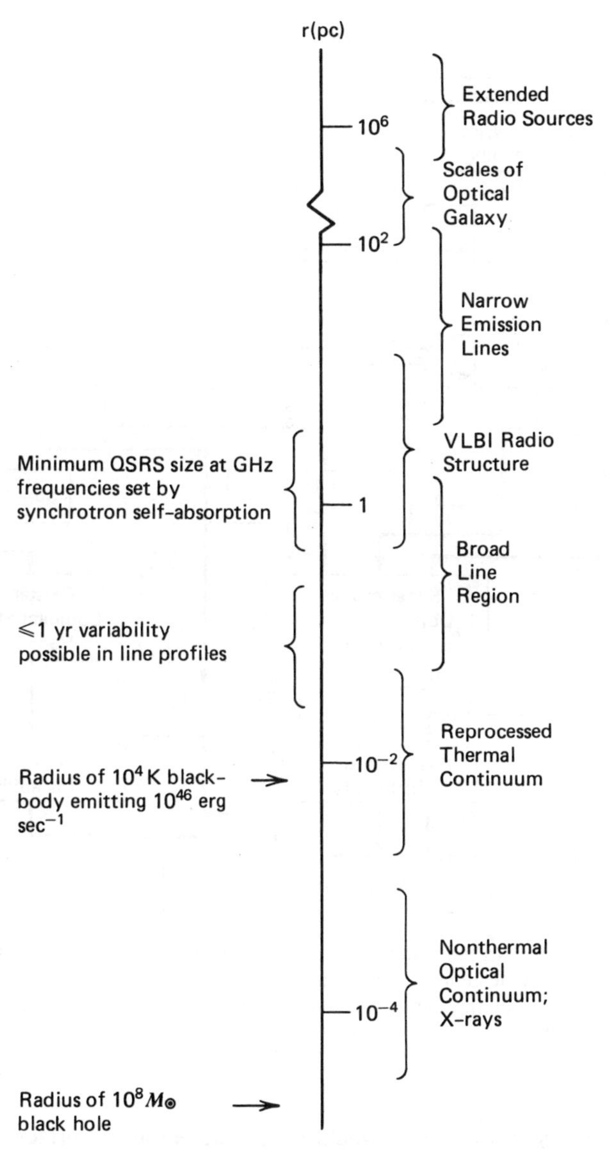

FIGURE 6. A rough indication of the scales involved in the various forms of activity energized by AGNs.

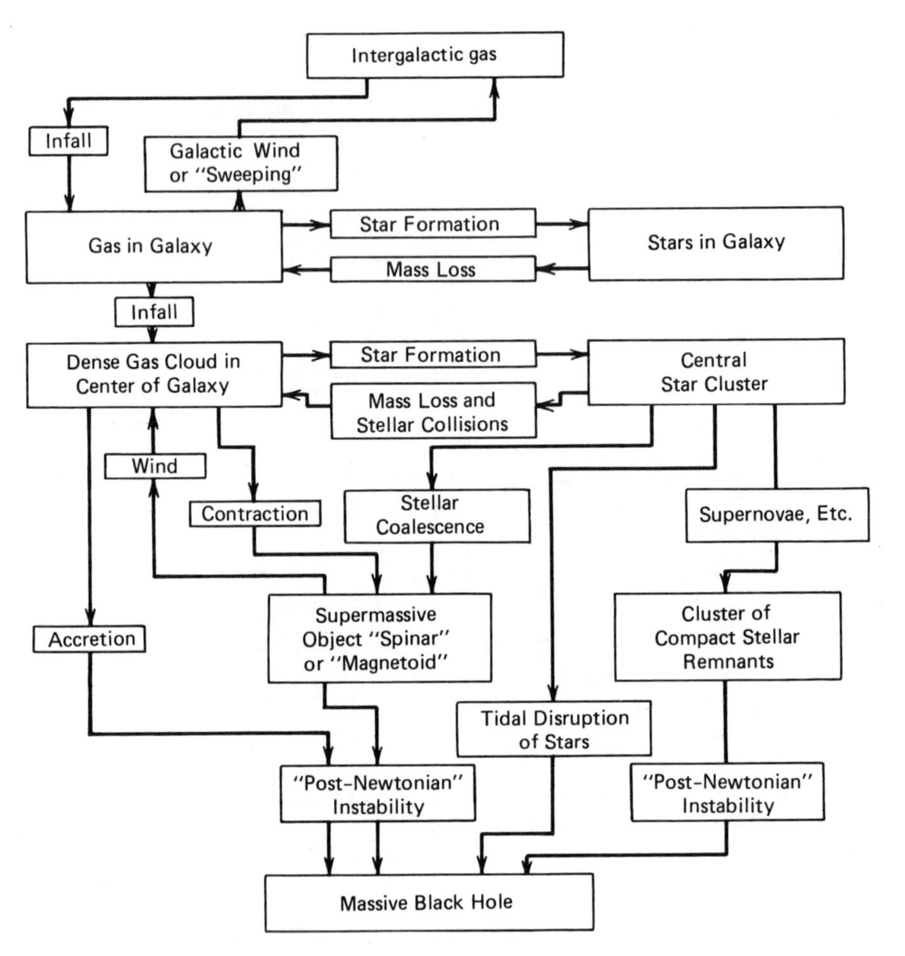

FIGURE 7. A flow chart indicating the ways whereby gas and stars may accumulate near the center of a galaxy, leading eventually to a runaway catastrophe. This may occur via a variety of pathways but the almost inevitable endpoint is a massive black hole. (From Begelman et al. 1984.)

> an infinite height towards it, would have acquired at its surface a greater velocity than that of light, and consequently supposing light to be attracted by the same force in proportion to its "vis inertia" with other bodies, all light emitted from such a body would be made to return towards it, by its own proper gravity.

He thus notes, as did Laplace a decade later, that the most massive objects in the universe might therefore be unobservable except by their gravitational influence on material orbiting near them. If we accord Michell his due priority, 1984 can be celebrated as the black hole bicentennial.

Michell based his argument on Newtonian gravity and the ballistic theory of light. I shall later discuss some astrophysical effects that depend on black holes having the specific properties inferred from Einstein's general relativity; but there are some important order-of-magnitude statements that are insensitive to black hole physics, but apply in any gravitational potential well of "$1/r$" form around a compact mass M. A characteristic luminosity is the so-called Eddington limit, at which radiation pressure on free electrons balances the force of gravity on the associated protons:

$$L_{\mathrm{Ed}} = \frac{4\pi G M m_p c}{\sigma_T} \cong 1.3 \times 10^{38} \left(\frac{M}{M_\odot}\right) \text{erg sec}^{-1} \tag{3}$$

In this formula m_p is the proton mass and σ_T the Thompson scattering cross section. The relevance of L_{Ed} in an actual AGN model depends on the details. However, insofar as L_{Ed} is a fiducial limiting luminosity in accretion-powered sources, it suggests that the relevant mass range for quasars (with $L \cong 10^{46}$ erg sec^{-1}) is $10^8 M_\odot$. Strong radio sources offer a second argument pointing to a similar mass: whatever produced the overall energy content of the extended lobes must have involved a mass of at least this order. The Schwarzschild radius of a black hole of mass $10^8 M_\odot$ solar masses is

$$r_s = \frac{2GM}{c^2} = 3 \times 10^{13}\, M_8\ cms \tag{4}$$

Note that $10^8 M_\odot$ is roughly the mass envisaged by Michell: it is about the mass of a hole whose "density," defined as $\sim M/r_s^3$, equals that of the Sun (which is similar to the density of terrestrial solids). Whereas stellar-mass black holes cannot form until after the matter has been squeezed to supernuclear density, the formation of these supermassive holes does not entail these physical complications. An experimenter could pass through the Schwarzschild horizon of one of these holes before being discomforted by tidal gravitational forces and would still have an hour or so for leisured study of the region $r < r_s$ (which is invisible to external observers) before being destroyed by the central singularity.

A massive black hole sitting isolated in empty space would be passive and quiescent. To release any power, it must be surrounded by plasma (and/or magnetic fields). This would indeed be expected in a galactic nucleus. Gas expelled from ordinary stars via stellar winds and supernovae could fall in from the body of the galaxy; gas could even come from intergalactic clouds captured by the galaxy. [Relevant here is the evidence that galaxies are more likely to be active if they are interacting with a neighbor (Balick and Heckman 1982), and that quasars may be in interacting galaxies (Hutchings and Campbell 1983).] Alternatively, the gas supply may originate in the central parts of the galaxy, coming from, for example, (1) debris from stars tidally disrupted by the hole, (2) debris from stellar collisions in a compact star cluster around the hole, or (3) a positive feedback process whereby stars are induced to lose mass (and thereby provide further fuel) by irradiation from a luminous central source.

There are several powers of 10 (cf. Fig. 6) between the relativistic domain (where $r \cong r_s$) and the scales ($\gtrsim$1 pc) probed directly by VLBI, and whence the emission lines come. Within this range of scales, flow patterns would be controlled by a potential with the familiar Newtonian "$1/r$" form. However, because this potential well is much deeper than that ordinarily encountered in stellar or galactic physics, it can compress or confine gas at extraordinarily high temperatures. Let us examine the inner regions in more detail.

A. The Transrelativistic Region

The virial temperature is given by

$$kT_{\text{virial}} = \frac{GMm_p}{r} = m_p c^2 \left(\frac{2r}{r_s}\right)^{-1} \tag{5}$$

Within a radius $\sim 10^3 r_s$, kT_{virial} exceeds the rest mass energy of an electron $m_e c^2$ (~0.5 Mev, corresponding to $T = 6 \times 10^9$ K). However, cooling processes (especially pair production and Compton cooling) are so efficient that electrons cannot in general be maintained at such elevated temperatures that they are moving highly relavitistically. So pressure gradients cannot balance gravity in this "transrelativistic" regime unless either (1) *radiation* pressure balances gravity, which requires $L \cong L_{\text{Ed}}$, or (2) *ion* pressure balances gravity. This requires $T_{\text{ion}} \cong T_{\text{virial}}$ ($>T_e$) and is possible only if the ion $\rightarrow$ electron transfer is slower than the inflow time scale. This kind of "two-temperature plasma" (cf. Eardley, Lightman, and Shapiro 1975) can exist only at low densities where Coulomb collisions between ions and electrons are rare and in the absence of collective plasma effects coupling the two species [see Rees et al (1982) for fuller discussion of ion-supported equilibria in contexts where $kT_e \gtrsim m_ec^2$].

If neither alternative 1 nor 2 is fulfilled, then material at $r \lesssim 10^3 r_s$ must be either in almost free fall or in a rotationally supported thin disk.

Another surprising property of compact luminous sources is that they naturally give rise to electron–positron plasma. I shall discuss below how a rather exotic model for the central engine in radio galaxies can bring this about, but in fact one can give a more general argument that applies to any compact source of high-energy photons. Suppose that a source of radius r_* emits a luminosity L_γ in the form of MeV photons. These photons can interact with each other to produce electron–positron pairs, the interaction cross section being of the order of the Thompson cross section. The MeV photons (each carrying energy $\sim m_ec^2$) will collide with each other before escaping if $n_\gamma \sigma_T r_*$ exceeds unity, where $n_\gamma \cong (L_\gamma / 4\pi m_e c^3 r_*^2)$ is the photon density. This implies that no γ-rays can escape from any source whose "compactness parameter" L_γ / r_* exceeds a certain threshold value (see Fig. 8). This condition is equivalent to

$$L_\gamma > 10(m_e/m_p)\,(r_*/r_s)\,L_{\rm Ed} \qquad (6)$$

and when it is written this way we see that it is readily fulfilled by nonthermal sources associated with black holes. If (6) were to hold, the primary source would shroud itself in an optically thick "false photosphere"

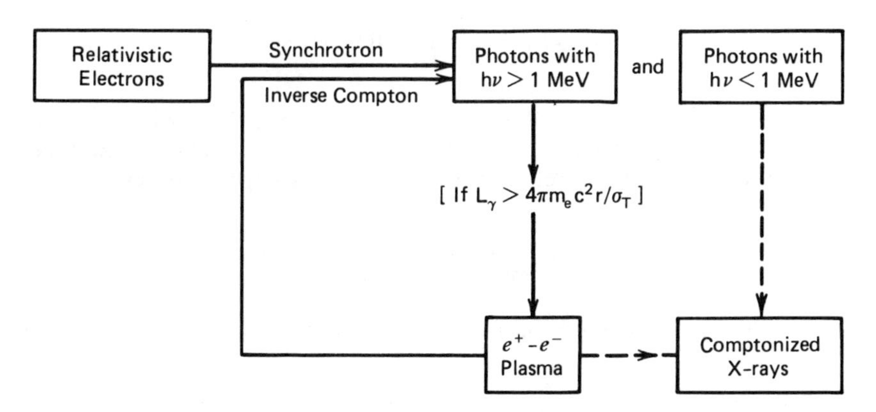

FIGURE 8. Processes that lead inevitably to the production of an optically thick e^+–e^- plasma in sources with a hard spectrum and a high "compactness parameter" L/r.

of e^+–e^- pairs, which would scatter *all* radiation, not just the part with $h\nu > m_e c^2$ (Guilbert et al. 1983).

B. The General Relativistic Domain

Material falling inward will eventually reach $r \simeq r_s$ where a Newtonian "$1/r$" potential no longer applies. Indeed it is here—in the deepest part of the potential well—that the main power output is generated. We must therefore take cognizance of the inherently relativistic features of the gravitational field.

The physics of dense star clusters and of supermassive objects is complex and poorly understood. In contrast, the final state of such a system—if indeed gravitational collapse occurs—is comparatively simple, at least if we accept general relativity. According to the so-called no-hair theorems, the endpoint of a gravitational collapse, however messy and asymmetric it may have been, is a standardized (uncharged) black hole characterized by just two parameters—mass and spin—and described exactly by the Kerr metric. If the collapse occurred in a violent or sudden way, it would take only a few dynamical time scales for the black hole to settle down: during that period gravitational waves would be emitted. But the final stationary state would be described by the Kerr solution, provided only that the material left behind outside the hole did not provide a strong perturbation.

Simple order-of-magnitude considerations tell us that the material participating in an accretion flow does not significantly perturb a Kerr metric. The dynamical (or free-fall) time scale near the Schwarzschild radius is

$$t_{\text{free fall}} \cong \left(\frac{r_s}{c}\right) \cong 1000 M_8 \text{ sec} \tag{7}$$

The mass inflow rate required to yield a luminosity L_{Ed}, with efficiency ϵ, is $\epsilon^{-1}\dot{M}_{\text{Ed}}$, where $\dot{M}_{\text{Ed}} \equiv L_{\text{Ed}}/c^2$. The time it would take for the hole's mass to double, if it accreted at a rate $\dot{M}_{\text{Ed}}$, is (Salpeter 1964)

$$t_{\text{Ed}} \cong \frac{c\sigma_T}{4\pi G m_p} \cong 4 \times 10^8 \text{ years} \tag{8}$$

So the ratio of the mass of inflowing material to the mass of the hole itself is

$$\frac{\dot{M}_{\text{Ed}} t_{\text{free fall}}}{M} \sim (t_{\text{free fall}}/t_{\text{Ed}})\, \epsilon^{-1}\, (L/L_{\text{Ed}})\, (v_{\text{free fall}}/v_{\text{inflow}})$$

Since $t_{\text{free fall}} \cong 10^{-13} M_8 t_{\text{Ed}}$, this ratio is negligibly small for all plausible values of the other terms.

The expected spin of the hole depends on the route by which it formed (see Fig. 7). A precursor spinning fast enough to be significantly flattened by rotational effects when its radius was $\gg r_s$ would probably have *more* angular momentum per unit mass than the critical value GM/c. A massive black hole that forms "in one go" is thus likely to have been fed with as much spin as it can accept and to end up near the top of the range of angular momentum permitted by the Kerr metric; the same is true for black holes that grow by accretion of infalling galactic gas (though the expectation is less clear if they grow by tidal disruption of stars). We should therefore take full account of the distinctive properties of *spinning* black holes, which Kip Thorne has reviewed for us in Chapter 5.

Fortunately, there is one inherently relativistic effect that simplifies things. The Lense–Thirring "dragging of inertial frames" can create a flow pattern near the hole that is *axisymmetric* (with respect to the hole's spin axis) irrespective of the symmetry and orientation of the flow at larger distances.

The accretion flow pattern depends on two things:

1. The value of $\dot{m} = \dot{M}/\dot{M}_{\text{Ed}}$, which determines (for a given radiative efficiency) the dynamical importance of radiation pressure
2. The inflow time scale, which depends on the effective viscosity

The cooling time scale depends inversely on the density and is therefore proportional to v_{inflow} for a given $\dot{m}$. The ratio of inflow and cooling time scales, an important parameter that controls the temperature of the gas, depends on $\dot{m}(v_{\text{inflow}}/v_{\text{free fall}})^{-2}$. When $\dot{m} \gtrsim 1$, this quantity is large, cooling is efficient, and the accretion-powered luminosity would be of order L_{Ed}. But for sufficiently low $\dot{m}$, cooling via thermal processes is ineffective; the infalling material will attain "transrelativistic" temperatures, and the only significant radiation emitted during the inflow would be synchrotron or Compton emission.

There is an extensive literature on AGN models powered by accretion onto massive black holes [see Rees (1984) Wiita (1985) for reviews]. Such models can, in broad terms, provide acceptable models for quasars. However, one cannot reliably predict the spectrum, nor whether the radiation is thermal or nonthermal: the hardest thing to estimate is what fraction of the power dissipated by viscous friction would go into relativistic particles (via shocks, magnetic reconnection, etc.) rather than being shared among all the particles. Nor do we know how steady or stable the inflow pattern might be. This is a topic where detailed numerical simulations would be worthwhile, particularly if these allowed us to treat unsteady accretion, nonaxisymmetric instabilities, and realistic radiative emission and transfer processes.

IV. THE "PRIME MOVER" IN STRONG RADIO SOURCES

Despite the lack of quantitative understanding of AGNs in general, the strong radio galaxies (e.g., Cygnus A) have a distinctive property that offers a clue to their central mechanism. The remarkable feature of these particular AGNs is that the "kinetic" power required to energize the extended radio lobes (transmitted by the jets in the form of relativistic particles or Poynting flux) exceeds the radiative luminosity of the nucleus itself. Is there a mechanism that could generate an intense plasma outflow, even if the accretion rate and nuclear luminosity were low?

There is indeed another possible source of power over and above the gravitational energy released by infalling matter. The part of the rest mass of a spinning black hole associated with its spin (up to 29% of the total) can in principle be extracted, as was first emphasized by Penrose (1969). By exploiting the analogy between a black hole's horizon and an electrical conductor (see Chapter 5) Blandford and Znajek (1977) suggested a realistic astrophysical context whereby *electromagnetic torques* can extract this energy, rather as the unipolar inductor mechanism brakes a spinning conductor. Three conditions are necessary, all of which can be fulfilled if the hole is surrounded by a small amount of plasma (such as could result from accretion with low $\dot{m}$):

1. *Magnetic fields threading the hole must be maintained by an external current system.* The requisite flux could have been advected in by slow accretion; even if the field within the inflowing matter were tangled, in the magnetosphere it would nevertheless be well ordered. The surrounding plasma would be a good enough conductor to maintain surface currents that could confine such a field within the hole's magnetosphere. The only obvious upper limit to the field is set by the requirement that its total energy should not exceed the gravitational binding energy of the infalling gas.

2. *There must be a current flowing into the hole.* Although the trans-relativistic plasma expected around the hole when $\dot{m}$ is low radiates very little, it emits some bremsstrahlung gamma rays. Some of these will interact in the funnel to produce a cascade of electron–positron pairs, yielding more than enough charge density to "complete the circuit" and carry the necessary current—enough, indeed, to make the magnetosphere essentially charge-neutral, in the sense that $(n^+ + n^-) \gg |(n^+ - n^-)|$, so that relativistic MHD can be applied.

3. *The proper "impedance match" must be achieved between the hole and the external resistance.* Phinney (1983) has explored the physics of the relativistic wind whose source is the pair plasma created by $\gamma + \gamma \rightarrow e^+ + e^-$ in the hole's magnetosphere, and that flows both outward along the funnel and into the hole. He finds consistent wind solutions in which about one-half of the hole's spin energy is transformed into Poynting flux and a relativistic electron–positron outflow.

The general scheme is depicted in Fig. 9. Even a low-level and inefficient accretion flow can "anchor" a magnetic field that threads the hole, and thereby tap the hole's spin energy; in these conditions the extracted power naturally goes predominantly into a relativistic bifurcated outflow. The power extracted is of order $B^2 r_s^2 c$: for a field $\sim 10^4$ G, which can be

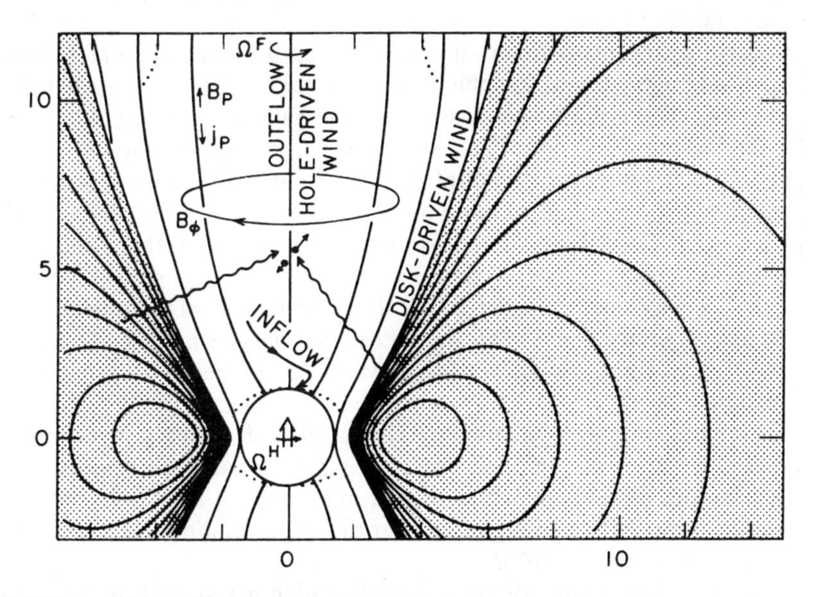

FIGURE 9. Schematic illustration of the hypothetical "central engine" in radio sources (Rees et al. 1982; Phinney 1983). Interaction between a Kerr hole and the magnetic field generates a hydromagnetic wind. Stippled area: external matter confining a poloidal magnetic field B_p (of strength 10^3–10^4 G) to the hole. The precise geometry is unimportant; that shown is appropriate for a pressure-supported torus with constant specific angular momentum. Wavy lines: γ-rays generated in the external matter create pairs in the otherwise empty "zone of nonstationarity" from which accreting material is excluded. On field lines which cross the event horizon, these pairs carry a current that extracts rotational energy from the hole in the form of a direct-current Poynting flux. (From Begelman et al. 1983.)

confined by plasma of density only 10^{-11} g cm^{-3}, this can be $\sim 10^{45}$ erg sec^{-1}. This mechanism seems specially appropriate for strong radio galaxies such as Cygnus A (Rees et al. 1982)—objects where the energy flowing along the jets dominates the radiative output of the AGN itself. Electron–positron pairs moving with Lorentz factors of approximately 100 would transport some kinetic energy, but most of the power outflow would initially be in the form of Poynting flux associated with the magnetic field coiled round the jet axis, and "frozen in" to the pair plasma. This Poynting flux may be converted into fast particles where the jet encounters ambient material (perhaps on the scale of the VLBI radio components). The expected magnetic field in the jet has the kind of configuration that could cause magnetic confinement and collimation (see Section II). The plasma around the hole that supplies the currents and anchors the field is just a catalyst—in principle, the power output of a radio galaxy could be sustained with zero accretion rate if some of the hole's spin energy were channeled into the surrounding plasma to compensate for its (small) radiative losses.

According to this idea, radio galaxies harbor massive black holes formed long ago via catastrophic collapse (maybe during a quasar phase of activity). The holes lurked quiescent, the galaxy being swept clean of gas, for billions of years. Then some event, maybe interaction with a companion, triggered renewed infall—maybe at a low rate but sufficient to reactivate the nucleus by applying a magnetic field. This "engages the clutch," tapping the hole's latent spin energy, and converting it into nonthermal directed outflow—Poynting flux and e^+–e^- plasma—which ploughs its way out to scales $\gtrsim 10^{10}$ times larger. If this is indeed what happens in Cygnus A and M87, then these very large-scale manifestations of AGN activity could offer the most direct evidence for inherently relativistic effects.

How much energy is available? Up to 29% of a hole's mass can be extracted in principle; models for relativistic winds with realistic efficiency factors allow conversion of rest mass with at least a few percent efficiency—more than enough to power the largest radio galaxy, provided that the hole mass is $\gtrsim 10^8 M_\odot$.

So there are two quite distinct ways in which massive black holes can generate a high luminosity: straightforwardly by accretion or via the electromagnetic process just described, where the energy comes from the hole itself. The latter process tends to give purely nonthermal phenomena, whereas accretion yields an uncertain mixture of thermal and nonthermal power. The properties of an AGN must depend, among other things, on the relative contributions of these two mechanisms, which are functions of $\dot{m}$ and the spin of the hole. The properties of AGNs must depend on other parameters—the nuclear mass M, the orientation and properties of the host galaxy, and so on. Ideally, one would like a unified model that explains the multifarious types of AGN in the same way that our theories for the Hertzsprung–Russell diagram do this for stars.

Still more ambitiously one would like to place quasars in the general context of galactic evolution. Do quasars die and get resurrected as radio galaxies? How and by what route does a condensed mass first accumulate in the centers of galaxies? What are the masses involved? How do the pyrotechnics in the nucleus react back on the rest of the galaxy? How common is it for massive holes, remnants of past activity, to lurk quiescent in the centers of normal galaxies? Already we have some clues from quasar demography, and from the study of nearby galaxies.

V. STATISTICS OF AGNs AND THEIR DEFUNCT REMNANTS

Even if AGNs are precursors on the route toward black hole formation rather than involving black holes that have already formed, massive black holes should exist in profusion as remnants of past activity; they would be inconspicuous unless infall onto them recommenced and generated a renewed phase of accretion-powered output or catalyzed the extraction of latent spin energy.

Estimates of the masses of AGNs and numbers of "dead" AGNs are bedeviled by uncertainty about how long individual active objects live, and the evolutionary properties (i.e., the dependence on redshift z) of the AGN population. It has long been known that the evolution is *strong*, amounting to a factor up to 1000 in comoving density for the strongest sources; and that it is *differential*, being less steep for lower-luminosity objects of all kinds. We are still a long way from understanding *why* the luminosity function evolves in this way. Anyway, at the epoch $z = 2$ the population of powerful AGNs was declining on a time scale $t_{\rm Evo} \cong 2 \times 10^9\ h_{100}^{-1}$ years (h_{100} being the Hubble constant in units of 100 km sec^{-1} Mpc^{-1}); this is of course an *upper limit* to the active lifetime t_Q of a typical individual AGN, since there may be many generations of objects within the period $t_{\rm Evo}$.

The individual remnant masses would be

$$8 \times 10^7\ h_{100}^{-3} \epsilon^{-1} \left(\frac{t_Q}{t_{\rm Evo}}\right) M_\odot \tag{9}$$

ϵ being the overall efficiency with which rest mass is converted into electromagnetic radiation over a typical quasar's active lifetime t_Q.

Clearly t_Q cannot exceed t_{Evo}. On the other hand, it could be that $t_Q \ll t_{\mathrm{Evo}}$, so that many successive short-lived generations of quasars are born and die in the period during which the overall population declines. Because of this uncertainty, quasar data alone would be compatible with either a small number of very massive remnants [10^9–$10^{10} M_\odot$, obtained by writing $t_Q \cong t_{\mathrm{Evo}}$ in (9)], or a larger number of less massive remnants.

Another issue is the nature of the lower luminosity AGNs (e.g., Seyfert galaxies). Are these quasar precursors (see Fig. 7)? Are they dying quasars with high M and low $\dot{m}$ or are they "miniquasars" with $L \cong L_{\mathrm{Ed}}$ but lower M? Answers to these questions depend on the complex interactions between evolving galaxies and their nuclei and must await theoretical progress along a broad front.

What can be said about the possible presence of massive black holes—"starved" remnants of some early active phase—in nearby galaxies? Young et al. (1978) advanced evidence that there was a dark central mass in the center of M87. They plotted the surface brightness as a function of distance and found a high brightness "cusp," and also anomalously high stellar velocities, near the center. They claimed on this basis that there was a central dark mass of 3–5 $\times$ 10^9 $M_\odot$. This conclusion has been subject to various doubts [see Richstone and Tremaine (1984) and references cited therein]. If the stellar velocities are not isotropic, one can explain the observations without invoking a large central dark mass at all; and even if there is one, it does not have to be a black hole, but could be a star cluster with a large mass-to-light ratio. We have to await the Space Telescope to get data with better resolution. Perhaps more convincing is Tonry's (1984) evidence for a dark central mass in M32. The inferred mass is "only" 5 $\times$ $10^6 M_\odot$; but M32 is a small nearby elliptical galaxy, so this result suggests that such objects may not be uncommon.

When the jets on a strong radio source switch off, the extended lobes expand adiabatically and their radio luminosity fades, on a time scale of a few times 10^8 years. The nearby galaxy Centaurus A (Fig. 10) is a rather weak radio source but has giant diffuse lobes: it may be a dying source which, in its prime, rivaled Cygnus A in luminosity. The nucleus is shrouded by dust, so we do not have the kind of optical data that Young et al. obtained for M87. But Centaurus A is the nearest galaxy we have reason to think spewed out a great deal of energy in the past. It has a compact radio and X-ray nucleus, with a jet; so there is some residual activity. The only reason for believing that there is some large ($>10^8 M_\odot$) mass at the center is that Centaurus A's extended radio lobes (if one applies the classic Burbidge argument) pose extremely high energy requirements.

There is a compact (and variable) radio source in our own Galactic

FIGURE 10. Optical photograph of Centaurus A with radio contours superposed. The contours shown indicate only the regions of most intense radio emission from Centaurus A. Radio emission can be traced out to scales more than ten times larger than those shown here.

center that is still almost unresolved. Its dimensions are $\lesssim 10^{14}$ cm. Another peculiar phenomenon toward the Galactic center is a strong electron–positron annihilation line, which there is strong reason to believe is variable. The variability is claimed on the basis of obtaining different results on balloon flights 6 months apart. This can be modeled in terms of low-level accretion onto a $\sim 10^6 M_\odot$ hole (Rees 1982).

There cannot, however, be a *monster* black hole in our Galactic center. Infrared data on the motions of gas in the central parsec imply that if there is a black hole there at all, its mass cannot be more than $\sim 10^6 M_\odot$. That is the mass inferred if one interprets the velocities as virial velocities; but it is not mandatory to interpret them that way—they could be ejection velocities—and therefore this is just an upper limit (Lacy et al. 1982). There is really *no* firm evidence for a black hole in the Galactic center. There are however two phenomena that are *unique* to the Galactic center: the radio source and the large fraction of the γ-ray energy coming out in the e^+–e^- annihilation line. If unique phenomena occur in a unique location in the Galaxy, it is not necessarily "ad hoc" to invoke a special kind of object; and black hole models can account for both these things. But this is the only reason to suspect a collapsed object of more than a stellar mass in the Galactic center.

VI. CONCLUDING REMARKS

It is now more than 20 years since quasars were discovered. When we look back over that period, progress seems depressingly slow. Sometimes one has the illusion that the subject is advancing rapidly, but we have really had a rather slow advance, with "sawtooth" variations superposed on it as fashions come and go. Many of the bizarre ideas current in the early days can now be abandoned; but if we look back at the insightful early work of, for instance, Salpeter (1964) and Zeldovich and Novikov (1964), who were already arguing that supermassive collapsed objects were implicated in the quasar phenomenon, progress indeed seems slow. This is partly because quasars were, in a sense, discovered too early. If they had been discovered in (say) the early 1970s, when Seyferts and other lower-level manifestations of AGN were already familiar, and when pulsars and compact X-ray sources had familiarized us with the likely efficiency of gravitational power sources, one suspects that a consensus would have developed more quickly, and there would have been less disposition to regard quasars as demanding "new physics" and so on.

The central engine, less than 1 light day across, involves plasmas under much more extreme conditions than those normally studied (e.g., electron–positron pairs). Inherently relativistic effects may be crucial. If so, the study of active galactic nuclei acquires another motivation: we would like some diagnostic to *test* strong-field general relativity—to learn whether the spacetime around a collapsed object does indeed resemble the Kerr metric.

But although the physics of the central engine is exotic, the key problem is at least well posed: axisymmetric plasma dynamics in a specified gravitational field, the aim being to calculate how much power is derived from accretion and extracted from the hole's spin, and to find the form in which these respective contributions emerge. Such calculations play the same part in our modeling of AGNs that nuclear physics does in our understanding of stellar structure and evolution. The evidence that black holes have anything to do with AGNs is purely circumstantial. Some theorists might brazenly assert that this is no less true for other cherished beliefs in astrophysics: the almost unquestioned idea that stars are powered by nuclear energy is supported by evidence that is also "merely" circumstantial. However, the confrontation of models with observations—indirect even for stars—is more ambiguous still for AGNs. This is because in stars the energy percolates to the observable surface in a relatively steady and well-understood way; in AGNs, on the other hand, it is reprocessed into all parts of the electromagnetic spectrum on scales spanning many powers of 10 (cf. Fig. 6), in a fashion that depends on ill-understood environmental and geometrical effects within the host galaxy. Indeed, the phenomena of quasars and radio galaxies cannot be understood until they are placed in the general context of galactic evolution. The massive black hole hypothesis is, however, vulnerable in various ways. It would, for instance, have to be abandoned (or at least severely modified) if very regular periodicities were found in AGNs, or if there were observational upper limits $\ll 10^8 M_\odot$ to the central masses in strong radio galaxies.

To conclude, I should like to recall something from the report of the Field Committee on the prospects for U.S. astronomy and astrophysics. Its panel on extragalactic astronomy was foolhardy enough to list the problems for the 1980s in order of importance. And at the top of the list was "the nature of the energy source in galactic nuclei." We have not yet got very far in interpreting these phenomena—quasars, cosmic jets, and the rest—nor in meshing the theoretical "toy" models with the real data. Theorists still lag far behind the observers. But, if this is indeed a "number one problem," maybe even the meager progress already achieved (and that we can realistically hope for in the next decade) is worth something.

ACKNOWLEDGMENTS

I am grateful for discussions with many colleagues and collaboration with M. C. Begelman, R. D. Blandford, A. C. Fabian, and E. S. Phinney on topics touched on in this chapter.

REFERENCES

Baade, W., and Minkowski, R., 1954, *Astrophys. J.*, **119**, 206.

Balick, B., and Heckman, T. M., 1982, *Ann. Rev. Astrophys.*, **20**, 431.

Bally, J., and Lada, C. J., 1982, *Astrophys. J.*, **265**, 824.

Begelman, M. C., Blandford, R. D., and Rees, M. J., 1984, *Rev. Mod. Phys.*, **56**, 255.

Bekefi, G., Field, B. T., Parmentola, J., and Tsipis, K., 1980, *Nature*, **248**, 219.

Biretta, J. A., Owen, F. N., and Hardee, P. E., 1983, *Astrophys. J. (Lett.),* **274,** L27.

Blandford, R. D., and Rees, M. J., 1974, *MNRAS,* **169,** 395.

Blandford, R. D., and Znajek, R. L., 1977, *MNRAS,* **179,** 433.

Bridle, A. H., and Purley, R. A., 1984, *Ann. Rev. Astr. Astrophys.,* **22,** 319.

Burbidge, G. R., 1958, *Astrophys. J.,* **129,** 841.

Burbidge, G. R., Burbidge, E. M., and Sandage, A. R., 1963, *Rev. Mod. Phys.,* **35,** 947.

Curtis, H. D., 1918, *Lick Obs. Publications,* **13,** 11.

Dreher, J. W., and Feigelson, E. D., 1984, *Nature,* **308,** 43.

Eardley, D. M., Lightman, A. P., and Shapiro, S. L., 1975, *Astrophys. J. (Lett.),* **199,** L153.

Fomalont, E. B., Geldzahler, B. J., Hjellming, R. M., and Wade, C. M., 1983, *Astrophys. J. (Lett.),* **275,** 802.

Guilbert, P. W., Fabian, A. C., and Rees, M. J., 1983, *MNRAS,* **205,** 593.

Hazard, C., Mackay, M. B., and Shimmins, A. J., 1963, *Nature,* **197,** 1037.

Hutchings, J. B., and Campbell, B., 1983, *Nature,* **303,** 584.

Lacy, J. H., Townes, C. H., and Hollenbach, D. J., 1982, *Astrophys. J.,* **262,** 120.

Lovelace, R. V. E., 1976, *Nature,* **262,** 649.

Margon, B., 1984, *Ann. Rev. Astr. Astrophys.,* **22,** 507.

Michell, J., 1784, *Phil. Trans. R. Soc.,* **74,** 35.

Mitton, S., and Ryle, M., 1969, *MNRAS,* **146,** 221.

Norman, M. L., Smarr, L. L., Winkler, K-H. A., and Smith, M. D., 1982, *Astron. Astrophys.,* **113,** 285.

Owen, F. N., Hardee, P. E., and Bignell, R. L., 1980, *Astrophys. J. (Lett.),* **239,** L11.

Penrose, R., 1969, *Nuovo Cim.,* **1,** 252.

Perley, R. A., Dreher, J. W., and Cowan, J. J., 1984, *Astrophys. J. (Lett.),* **285,** L35.

Phinney, E. S., 1983, Ph.D. thesis, Cambridge University.

Rees, M. J., 1982, in *The Galactic Centre,* R. Blandford and G. Riegler (eds.), American Institute of Physics, New York, p. 166.

Rees, M. J., 1984, *Ann. Rev. Astr. Astrophys.,* **22,** 471.

Rees, M. J., Begelman, M. C., Blandford, R. D., and Phinney, E. S., 1982, *Nature,* **295,** 17.

Richstone, D. O., and Tremaine, S., 1984, *Astrophys. J.,* **286,** 27.

Salpeter, E. E., 1964, *Astrophys. J.,* **140,** 796.

Schmidt, M., 1983, *Nature,* **197,** 1040.

Tonry, J. L., 1984, *Astrophys. J. (Lett.),* **283,** L27.

Wiita, P., 1985, *Phys. Rep.,* **123,** 117.

Williams, D., and Gull, S., 1985, *Nature,* **303,** 39.

Young, P. J., Westphal, J. A., Kristian, J., Wilson, C. P., and Landauer, F. P., 1978, *Astrophys. J.,* **221,** 721.

Zeldovich, Y. B., and Novikov, I. D., 1964, *Dokl. Acad. Nauk. SSSR,* **158,** 811.

7.

Probes of the Universe

ROBERT V. WAGONER

Robert V. Wagoner is a Professor of Physics at Stanford University. He has made outstanding contributions in relativistic astrophysics, gravitation theory, cosmology, and nucleosynthesis.

I. INTRODUCTION

Much is being learned about our cosmic environment by the clever application of well-understood physics to the increasing variety of astronomical data. It is especially important in cosmology, where we are dealing with the most remote objects and the most extreme physical conditions, that we have confidence in the physical models that form the basis of our understanding of the universe. In this chapter I try to emphasize this requirement in discussing the foundations of cosmology and various methods of increasing our understanding of the structure and evolution of the universe.

It is probably safe to say that most lists of the most important problems in cosmology would include the following:

1. Validity of the standard Big-Bang model.
2. Determination of the parameters of the model via the properties of the recent universe.
3. Nature of the very early universe.
4. Origin and evolution of structure.
5. Nature of the dark matter that appears to dominate the visible matter.

In this chapter we are mostly concerned with the first two problems. In other words, we focus on the properties of the universe as a whole during those epochs in which we have experimentally induced confidence in the relevant laws of physics. Readers desiring more information on the other topics are directed to the following sources. The very early universe and the growth of structure have been discussed in many recent workshops. Those proceedings edited by Gibbons, Hawking, and Siklos (1982) and Audouze and Tran Thanh Van (1983) are representative. A comprehensive review of the inflationary model of the early universe is presented by Brandenberger (1985), while a standard reference on large-scale structure remains the book by Peebles (1980a). The observational basis for the now widely held belief in a significant component of matter so far detected only gravitationally has been reviewed by Faber and Gallagher (1979), Peebles (1980b), and Rubin in Chapter 9. The most up-to-date reference on all three of these topics is probably the Proceedings of the Fermilab Inner Space/Outer Space Workshop, held in 1984 (Kolb and Turner 1985).

A particular problem we discuss in most detail is the determination of cosmological distances. Significant progress toward obtaining the expansion rate, deceleration, and other related properties of the universe seems possible through the use of better distance indicators and the availability of more powerful telescopes. In keeping with our general philosophy, and supported by the lessons of history, we will advocate the use of only well-understood physical systems as probes of the universe.

II. THE STANDARD MODEL: EVIDENCE AND STRUCTURE

A. Evidence

Six types of observational evidence presently form the basis for the adoption of the standard Big-Bang class of models as the simplest and most natural description of the general features of our universe. They are the observations of isotropy, expansion, ages, the microwave background spectrum, nuclear abundances, and the evidence for the validity of classical general relativity. We briefly discuss each in turn, reserving more detailed comments on some of them to subsequent portions of this chapter.

The uniformity of the large-scale angular distribution of galaxies over the sky, first noted by Hubble, has been subsequently made more precise and extended to related objects such as X-ray and radio sources (Peebles 1980a). But the most stringent limits come from the microwave background radiation, with the latest data (Uson and Wilkinson 1984) depicted in Fig. 1. Note that with the dipole anisotropy due to our motion removed, the upper limits on the fractional anisotropy in intensity are $\sim 10^{-4}$ at large angular scales and are approaching 10^{-5} on scales of a few minutes of arc. It is important to remember that these limits are also the most powerful because they refer to the greatest distances (independent of the particular source of the background). Based on the lack of any contrary evidence, we shall adopt the Copernican Principle that we occupy a *typical* position in the universe. The observations of isotropy then imply the Cosmological Principle: large-scale homogeneity as well.

The expansion velocity of the universe, as deduced from the Doppler shift, has been verified to be consistent with the Hubble relation $v = H_0 D$ within its realm of validity $v_* \ll v \ll c$, where v_* is the random velocity (induced mainly by concentrations of mass relatively nearby the galaxy observed). However, the value of the Hubble constant remains uncertain within a range $40 \lesssim H_0 \lesssim 110$ km sec^{-1} Mpc^{-1}. We will address the major problem, the determination of the distance D, in later sections.

The "age" of the universe implied by the Hubble constant in the absence of significant deceleration (or acceleration) is then $H_0^{-1} = 9 - 24 \times 10^9$ years. An independent lower limit to the age of the universe is

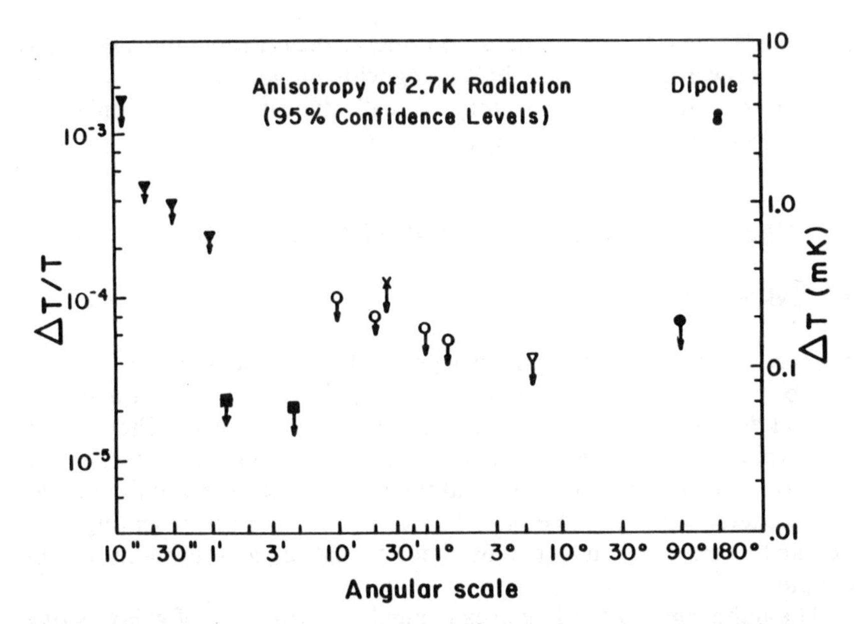

FIGURE 1. Upper limits on the anisotropy of the intensity of the 2.7 K background radiation (updated from Uson and Wilkinson 1984). Anisotropy on large angular scales (≳6°) is measured from balloons and satellites because of atmospheric affects, while that at smaller angular scales comes from ground-based radiotelescopes. (The limits for angles ≤1′ are from the VLA.)

obtained from the calculated age of the oldest globular clusters. The latest value is $(16 \pm 3) \times 10^9$ years (Iben and Renzini 1985). Improved observations with the Hubble Space Telescope (HST) should increase our confidence in any such number (see Chapter 11), as would a resolution of the solar neutrino problem (see Chapter 4). Nevertheless, the consistency of these two ages is an important check on any model of the universe. The ages of radioactive nuclei are typically determined from their observed and (theoretical) initial abundances relative to their decay product. Although the results are sensitive to the chemical evolution of our galaxy, a variety of chronometers (such as U^{235}/U^{238} and Th^{232}/U^{238}) give ages in the same range as H_0^{-1} (Audouze 1980).

The spectrum of the microwave background radiation remains the strongest evidence that the universe was much denser and hotter in the past [even if it was subsequently scattered or was produced by stars and thermalized by dust as in nonstandard models (Carr, Bond, and Arnett 1984)]. The latest observations (Smoot et al. 1985) are consistent with a blackbody of temperature $T = 2.7$–2.8 K, at least at wavelengths greater than 0.3 cm. No deviations from the Planck function have been firmly established.

Primordial nucleosynthesis is presently our deepest quantitative probe of the past history of the universe. With the cross sections of essentially all the relevant weak, electromagnetic, and strong reactions having been measured in the laboratory, the resulting nuclear abundances provide a relic record of conditions from a redshift of $z \sim 4 \times 10^9$, when the neutron–proton ratio froze out of equilibrium at $kT \sim 1$ MeV, to $z \sim 1 \times 10^8$, when the nuclear reactions ceased at $kT \sim 0.03$ MeV. The resulting abundances are sensitive to (1) the local expansion rate, which can be affected by such factors as additional particles, the theory of gravitation, and anisotropy; (2) the ratio of baryons to photons; and (3) the phase-space distribution of electron neutrinos, determined by the electron lepton number and the degree of anisotropy and inhomogeneity (after they drop out of equilibrium at a few MeV). For a recent calculation, see Yang et al. (1984). We discuss the results of these calculations below, but the essential conclusion is that the observed abundances remain consistent (within their uncertainties) with the predictions of the standard Big-Bang model.

Since the large-scale structure and evolution of the universe is controlled by the gravitational interaction, the believability of any statement we make about the universe depends on the degree of confidence that has been established in the present working theory, general relativity. A thorough review of the comparison with experimental and observational data has just been completed by Will (1984). The upshot is that general relativity passes all tests with flying colors. Especially impressive is the agreement with the observed orbital decay of the binary pulsar due to the emission of gravitational radiation, since this probes the structure of the gravitational interaction at a deeper level (higher powers of $v^2/c^2 \sim GM/Rc^2$) than the other tests.

B. Structure of the Standard Model

Although simplicity is a somewhat subjective concept, most cosmologists agree that the "standard model" is the simplest one consistent with the preceding evidence, although precise definitions of this model differ. It is presently the generally accepted working description of the background geometry and material content of the universe, the space–time canvas on which the evolution of all finite structures proceeds. However, one must resist the temptation to incorporate in the model any assumptions not supported by experimental or observational evidence.

If one accepts only the Cosmological Principle and the Equivalence Principle (that gravitation couples to matter only through a metric field), then the metric can be described by the Robertson–Walker form of the line element:

$$ds^2 = -dt^2 + R^2(t)[(1 - ku^2)^{-1} du^2 + u^2 d\Omega^2] \tag{1}$$

Here u is the dimensionless comoving radial coordinate, the sign of the spatial curvature $k = 0, \pm 1$, and we choose units in which the speed of light $c = 1$. The quantity $R(t)$ is a scale factor proportional to the (time-varying) distance between nearby comoving particles. (For details, see, e.g., Weinberg 1972.) With t_0 the present time and $R(t_0)$ denoted by R_0, the Hubble constant H_0 and deceleration parameter q_0 are defined by the Taylor series expansion of the scale factor:

$$R(t) = R_0[1 + H_0(t - t_0) - \tfrac{1}{2}q_0H_0^2(t - t_0)^2 + \cdots] \tag{2}$$

Of particular interest to us will be the relation between the redshift $z = (R_0/R) - 1$ and the various measures of the distance D of a source. There are essentially three measures of distance that are presently useful in cosmology.

1. If one knows the intrinsic total luminosity L of a source whose detected flux is $f = \int_0^\infty f_\nu d\nu$, then its *luminosity distance* is

$$D_L \equiv \left(\frac{L}{4\pi f}\right)^{1/2} = R_0u(1 + z) \tag{3a}$$

2. If one knows an intrinsic transverse size Δ of an object whose observed angular size is θ, then its *angular-size distance* is

$$D_A \equiv \frac{\Delta}{\theta} = R_0u(1 + z)^{-1} \tag{3b}$$

3. If an object whose local velocity transverse to the line of sight is $v(t)$ is observed to move at an angular rate $\dot{\theta}(t)$, then its *proper-motion distance* is

$$D_M \equiv \frac{v}{\dot{\theta}} = R_0u \tag{3c}$$

Integrating the photon equation of motion $ds = 0$ from the position u of the source then gives the redshift–distance relations, all related by powers of $(1 + z)$ as seen in Eqs. (3). For nearby sources we may employ the expansion (2), obtaining

$$R_0 u = H_0^{-1} z[1 - \tfrac{1}{2}(1 + q_0)z + O(z^2)] \tag{4}$$

Another quantity of interest is the photon travel time

$$t_0 - t = H_0^{-1} z[1 - (1 + \tfrac{1}{2}q_0)z + O(z^2)] \tag{5}$$

We next endow the universe with dynamics by imposing the Einstein field equations. We shall incorporate any possible cosmological constant term within the effective stress-energy tensor, which the Cosmological Principle requires to have the form

$$T_{\mu\upsilon} = (\rho + p)U_\mu U_\nu + p g_{\mu\nu} \tag{6}$$

where the four-velocity of the matter $U_\alpha = -\delta_{\alpha 0}$ in our comoving frame. This expression defines the total mass-energy density $\rho(t)$ and the pressure $p(t)$. Thus a cosmological constant Λ contributes a term $\Lambda/8\pi G$ to ρ and $-\Lambda/8\pi G$ to p.

Given an "equation of state" $p(\rho)$, the evolution of the fundamental dynamical variables R and ρ is governed by the "energy conservation" equations

$$\left(\frac{dR}{dt}\right)^2 - \left(\frac{8\pi G}{3}\right)\rho R^2 = -k \tag{7}$$

$$d(\rho R^3) + 3pR^2\,dR = 0 \tag{8}$$

Evaluating the Einstein field equations today gives the relations

$$\rho_0 \left(\frac{3H_0^2}{8\pi G}\right)^{-1} \equiv \Omega_0 = 1 + \frac{k}{(H_0 R_0)^2} \tag{9}$$

$$q_0 = \tfrac{1}{2}\Omega_0(1 + 3\alpha_0) \tag{10}$$

where $\alpha_0 \equiv (p/\rho)_0$. For example, if the present universe were dominated by vacuum energy density, $\alpha_0 = -1$; by nonrelativistic particles, $\alpha_0 \cong 0$; by relativistic particles, $\alpha_0 = \frac{1}{3}$; by a free massless scalar field, $\alpha_0 = 1$.

In order to proceed, we must make some assumption about the equation of state. One that would appear to be sufficiently general during the recent

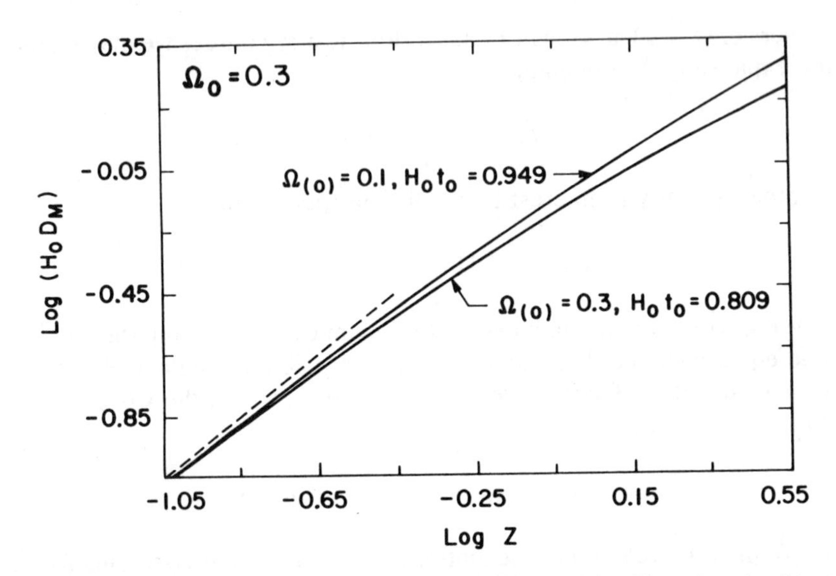

FIGURE 2a. Proper-motion distance versus redshift for models with various densities of nonrelativistic matter [$\Omega_{(0)}$] and vacuum energy density [$\Omega_{(-1)}$] for fixed total density parameter $\Omega_0 = 0.3$. The total age of the models is also indicated. The dashed line is the Hubble relation $D = H_0^{-1} cz$.

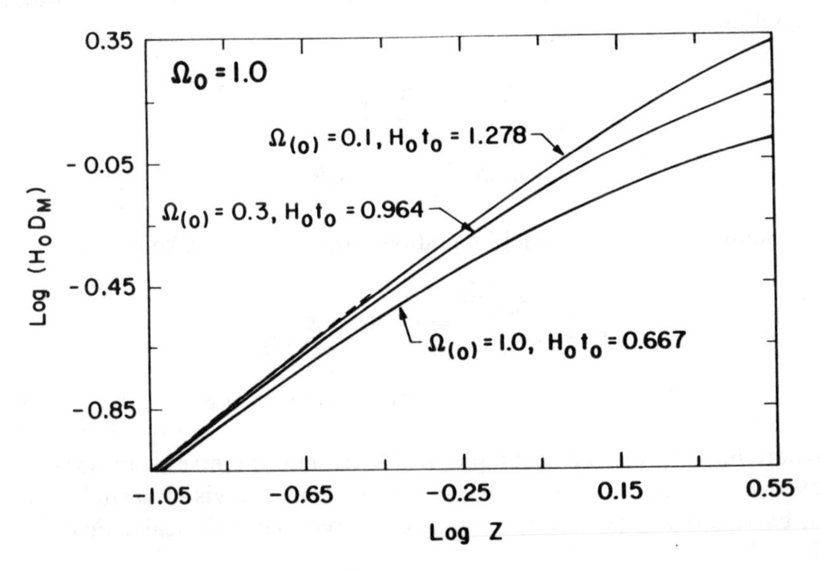

FIGURE 2b. Same as 2(a) for $\Omega_0 = 1.0$.

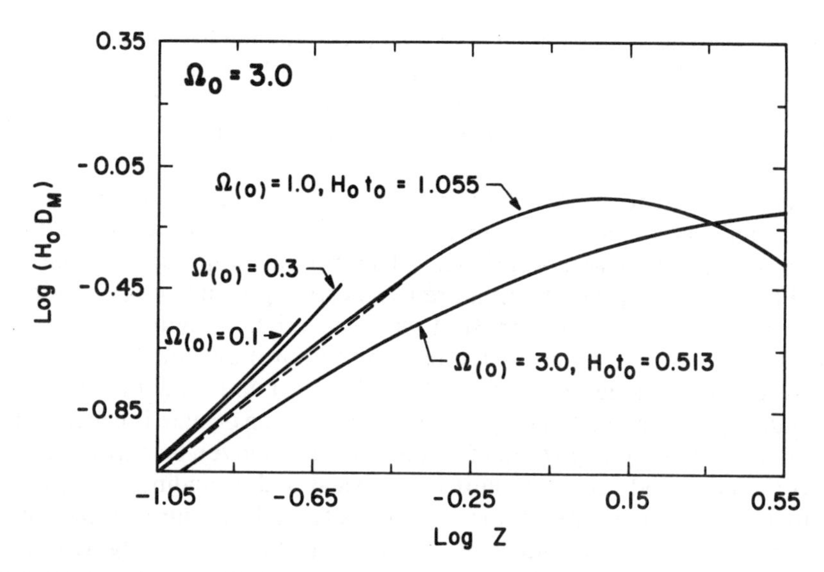

FIGURE 2c. Same as 2(a) for $\Omega_0 = 3.0$. Two of the models (without ages) had no high density epoch.

past is that the universe contained various distinct forms of noninteracting matter, each characterized by a particular constant value of $\alpha = p_{(\alpha)}/\rho_{(\alpha)}$. Since Eq. (8) applies to each form, we obtain $\rho_{(\alpha)} \propto (1 + z)^{3(1+\alpha)}$. From the above equations, we then obtain the differential relation between the (proper motion) distance and redshift:

$$\frac{H_0^2\,(dD_M)^2}{1 + (1 - \Omega_0)H_0^2D_M^2} = \frac{(1 + z)^{-2}\,(dz)^2}{1 - \Omega_0 + \sum_\alpha \Omega_{(\alpha)}(1 + z)^{3\alpha+1}} \tag{11}$$

with $\sum_\alpha \Omega_{(\alpha)} = \Omega_0$ and the density-weighted average $\langle\alpha\rangle = \alpha_0$.

Comparing an observed relationship $D_M(z)$ [or the corresponding relationships $D_L(z)$ or $D_A(z)$] with the integrated form of Eq. (11) then yields the basic set of parameters H_0 and $\Omega_{(\alpha)}$ that completely determine all properties of the standard model as long as conversion of one form of matter into another (by decays, annihilation, etc.) was negligible. As an example, in Fig. 2a, b, and c Eq. (11) is integrated out to redshifts $z \leqslant 3$ for models containing various amounts of nonrelativistic matter ($\Omega_{(0)}$) and vacuum energy density ($\Omega_{(-1)} = \Omega_0 - \Omega_{(0)}$). Also indicated is the approximate total age of those models that had a high-density past. Note that the two models with the largest values of $\Omega_{(-1)}$ can be ruled out because they have maximum redshifts (corresponding to minimum scales R) less than that of the observed quasars ($z \leq 3.78$). Note also that only the model with the largest value of $\Omega_{(0)}$ ($= \Omega_0 = 3$) is close to being ruled out from the limits on H_0 and the age of the globular clusters quoted above.

Within the standard model, the program of understanding the evolution of the universe as a whole can be simply viewed in terms of the two aspects of the initial-value problem. The present properties H_0 and Ω_0 (and thus also the present curvature k/R_0^2) can in principle be determined by methods involving the redshift–distance relations indicated above. Given these "initial values," the past (and future) evolution of the universe can be obtained by integration of Eqs. (7) and (8) *if* the equation of state $p(\rho)$ is known. Although the present equation of state is necessarily obtained in determining H_0 and $\Omega_{(\alpha)}$ (as indicated above), its extrapolation into the past (and future) requires a physical understanding of the constituents that contribute to the total density and pressure. Since only probes of the past can contribute to this physical understanding, it would seem premature to consider the future.

We know by direct observation of at least three constituents of the present universe. The microwave background radiation contributes an amount

$$\Omega_{(1/3)} = 2.3 \times 10^{-5}\left(\frac{100}{H_0}\right)^2 \qquad (T_0 = 2.7\ \mathrm{K}) \tag{12}$$

to the relativistic particle component. The ordinary stars and gas contribute an amount

$$\Omega_{(0)} \cong 0.01 - 0.02 \qquad \text{(visible)} \tag{13}$$

to the baryon component of nonrelativistic matter. Finally, the Newtonian dynamics of bound systems (galaxies and clusters of galaxies) provides a much larger lower limit to this component (Davis and Peebles 1983; Bean et al. 1983):

$$\Omega_{(0)} \gtrsim 0.1 \qquad \text{(clumped)} \tag{14}$$

which may or may not be baryons (more on this later). The *total* contribution of this "dark matter," like that of any other component, can only be obtained from its gravitational effects on the largest scales (i.e., the distance–redshift relation) unless one *knows* that it has all become bound within smaller systems or it can be detected directly. For example, gravitational lensing by a significant density of condensed objects (mass $\gtrsim 10^{-3}\ M_{\odot}$) can produce observable effects on the beams from quasars (Press and Gunn 1973; Gott 1981) or supernovae (Wagoner 1986).

Until recently, directly determined upper limits on the average density of gravitational radiation were too large ($\Omega_{(1/3)} \gg 1$) to be interesting. However, the remarkable regularity of the arrival times of pulses from some pulsars has allowed Romani and Taylor (1983) and Hellings and Downs (1983) to place the limit $\Omega_{(1/3)} \lesssim 10^{-4}$ on gravitational waves with periods of order the time interval over which the pulse arrival times have been measured. It is anticipated that this limit will be improved in the future, in particular from continued monitoring of the millisecond pulsar (Taylor 1986). Of course, limits on the large-scale anisotropy of the microwave background also imply limits $\Omega_{(1/3)} \lesssim 10^{-8}$ on the density of gravitational waves of cosmological wavelength, since $\Omega_{(1/3)} \sim$ (metric perturbation)$^2 \sim (\Delta T/T)^2$.

III. EXTRAPOLATION TO THE EARLY UNIVERSE

A. The Minimal Model

If we input only the known constituents of the present universe and use the laws of physics only throughout their experimentally verified range of validity to obtain the universal equation of state, we obtain what might be called the minimal model of the early universe. Although this allows one to consider only epochs corresponding to energies $\lesssim 10^2$ GeV, a richness of consequences one might not have anticipated emerges. This richness is indicated in Fig. 3, which depicts the evolution of the mass density of the various particle species predicted to have existed (or to still exist). Also indicated in Fig. 3 is an extrapolation to energies beyond the present "accelerator barrier" quoted above, which we shall consider in Section III.B. Let us take a brief tour of the past history of this model of the universe.

The first fundamental observational barrier that we reach is the photon barrier, corresponding to the epoch ($z \cong 10^3$) when atoms formed, allowing the photons to propagate freely thereafter. In the absence of subsequent reionization, this is the epoch that we view via the microwave background radiation.

Any structure present in the universe at that time will produce some anisotropy in the microwave background. This can occur via direct coupling to the photons (e.g., adiabatic fluctuations) and by their gravitational

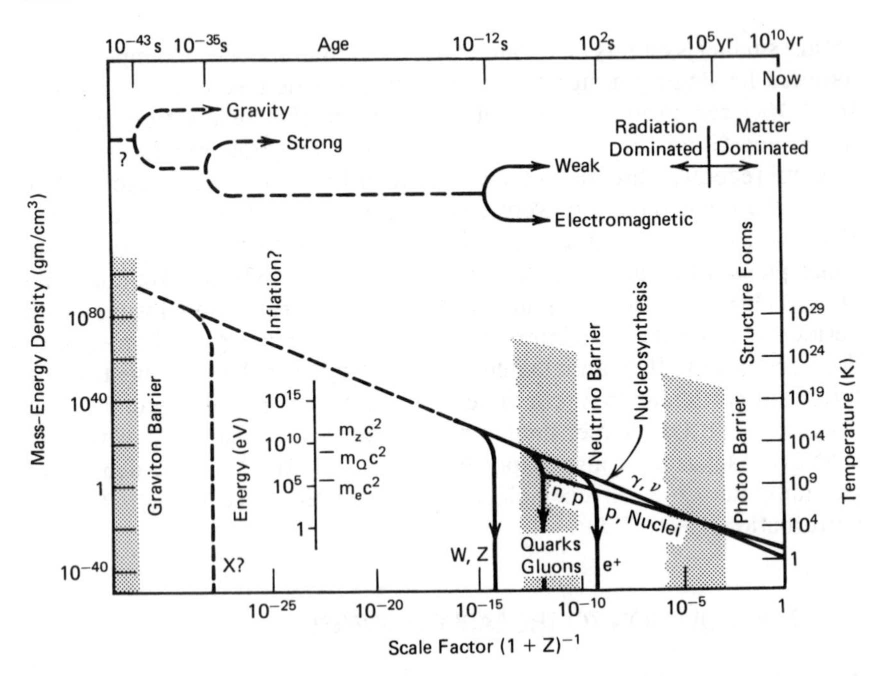

FIGURE 3. The past history of the universe, according to the standard model. The dashed portion of the curves represents the epochs for which the physics involved is not certain. The temperature refers to the photons. The unstable leptons (μ,τ) disappeared at about the same time as the free quarks and gluons combined into hadrons. The epochs of unification of the interactions are also indicated. [Adapted from Wagoner and Goldsmith (1983) and Wilkinson et al. (1985)]

effects (induced velocities, leading to Doppler shifts; effects on photon propagation; etc.). In principle, we can extrapolate the present amount of inhomogeneity on any large scale back to this epoch via the relevant gravitational growth equations and compare the resulting induced anisotropy with the observational limits indicated in Fig. 1. This technique has begun to put important constraints on some properties of the matter involved in the perturbations (e.g., Bond and Efstathiou 1984). However, this approach has not yet placed any new model-independent constraints on the nature of the background (unperturbed) cosmological model, our main concern in this article.

As we move back in time to a redshift $z = 10^7$–10^8, we reach the cosmic photosphere, the epoch at which the background photons were last created and destroyed (mainly by bremsstrahlung). Any energy input (via stars, particle decays, etc.) could have produced deviations from a blackbody spectrum only at subsequent times.

As we move to redshifts $z = 10^8$–10^{10} we reach the era of nucleosynthesis. At this epoch we pass the first pair-production threshold, as the photon temperature (increasing as $1 + z$) reached the equivalent electron rest-mass energy. At higher temperatures the densities of electrons and positrons were maintained comparable to that of photons by the fact that their interaction rate was much faster than the expansion rate of the universe. We say that the particles have "frozen out" of equilibrium after these rates became comparable. The stronger the interaction, the more delayed is the epoch of freeze-out, characterized by a temperature T_f.

If $kT_f \gtrsim mc^2$, then the relic abundance remains roughly comparable to that of photons. This is the case for all light neutrinos ($m \lesssim 1$ MeV), for which interactions such as pair production from $e^{\pm}$ via Z^0 exchange keep them in statistical equilibrium until they are freed at the neutrino barrier indicated. If we could detect the sea of neutrinos that fill the universe today (with a number density $\cong 100\ \mathrm{cm}^{-3}$ for each type, 3/11 that of photons, and temperature $\cong 2.0$ K), we would be viewing directly this second fundamental observational barrier, beyond which the universe was opaque to this probe. And as is well known, if any neutrino species had a mass $m \sim 30$ eV, they would contribute to the present density an amount $\Omega_{(0)} \sim 1$. It is therefore of great importance to verify the claimed determination of such a mass of the electron neutrino (Fackler 1985), and improve the mass limits for the other families, as well as to continue the search for neutrino oscillations and decays.

If $kT_f \ll mc^2$, then the final abundance is much reduced. A lower limit is provided by the assumption of equal numbers of particles and antiparticles. Detailed calculations of the annihilation process in the expanding and cooling environment (Steigman 1979) can be summarized by the reasonably accurate formula

$$\Omega_{(0)} \cong \frac{1.8 \times 10^{-12}}{g_f^{1/2} \langle \sigma v \rangle} \left(\frac{100}{H_0} \right)^2 \left(\frac{mc^2}{kT_f} \right) \tag{15}$$

for the present contribution to the mass density. Here $\langle \sigma v \rangle$ is the annihilation rate in units of $10^{-15}\ \mathrm{cm}^3\ \mathrm{sec}^{-1}$, g_f is the total number of degrees of freedom (statistical weight) of the particles present, and

$$\frac{mc^2}{kT_f} + \tfrac{1}{2} \ln\left(\frac{mc^2}{kT_f} \right) \cong 45.4 + \ln(g_i m \langle \sigma v \rangle g_f{}^{-1/2}) \tag{16}$$

In the last term, g_i refers to the particle in question and m is its mass in GeV. Note that $\Omega_{(0)}$ is most sensitive to $\langle \sigma v \rangle$, not m. As an example, for

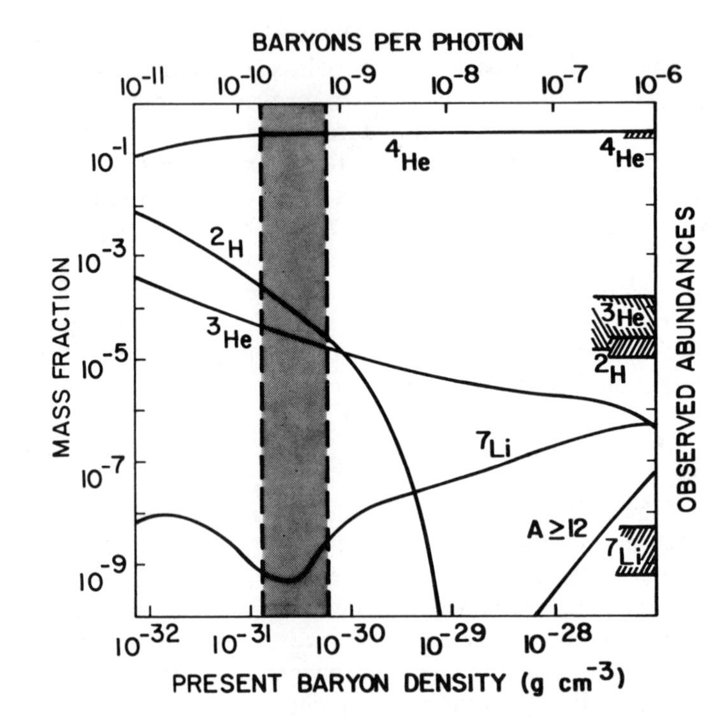

FIGURE 4. The abundances of nuclei produced within the standard model of the early universe compared with the range of observed abundances of those nuclei. Heavy elements ($A \geqslant 12$) are made mostly within stars. The range of models for which agreement is possible is also indicated. [Adapted from Yang et al. (1984) and Wilkinson et al. (1985).]

a stable massive weakly interacting particle, $\Omega_{(0)} \lesssim 3$ requires $m \gtrsim 2$ GeV (Lee and Weinberg 1977).

Primordial nucleosynthesis began with the nucleons produced from the primordial gas of quarks as the gluons confined them during the quark–hadron phase transition at $kT \sim 300$ MeV ($\sim 10^{-5}$ sec). As we have noted in Section II, their subsequent evolution can be confidently calculated within the context of this minimal model because essentially all the relevant cross sections are known from experiment. The latest results for the final abundances (Yang et al. 1984) are compared with the relevant ranges of observed abundances in Fig. 4. Let us briefly summarize the consequences of this comparison, using the latest available observational data (Steigman and Boesgaard 1985).

The most precise test of the minimal standard model comes from the observed mass fraction of ^{4}He. After reducing the effects of stellar contamination by observing systems with low abundances of heavy elements, this abundance now appears to lie in the range $X(^4\mathrm{He}) \cong 0.24 \pm 0.02$. The agreement with the predicted mass fraction for that range of baryon density that also produces approximate concordance for the other light elements provides strong evidence that deviations from the minimal standard model (due to inhomogeneity, anisotropy, neutrino degeneracy, other theories of gravitation, additional particles, etc.) which affected the expansion rate or the initial neutron–proton ratio must have been small at this early epoch (Wagoner 1973). Similarly, one can rule out the presence of significant numbers of antibaryons. In particular, if more than the three known families of light ($\lesssim 1$ MeV) neutrinos exist, the increased mass density [and thus the expansion rate via Eq. (7)] begins to lead to too much helium production (Yang et al. 1984). The decay width of the Z^0 will place constraints on the number of light neutrinos as well. More generally, this limit on the total mass–energy density at this epoch rules out a significant abundance of any component of the present universe that would have had a higher density than that of photons then. Thus unless production occurred between nucleosynthesis and the present (e.g., by decays), we conclude that

$$\Omega_{(\alpha)} \lesssim 10^{6-30\alpha} \qquad (\alpha \gtrsim \tfrac{1}{5}) \tag{17}$$

Although deuterium has been detected in molecular form in planets as well as interstellar clouds, and the radio hyperfine line still holds promise for future detection at greater distances, the displaced series of Lyman absorption lines from interstellar atoms presently provide the most reliable measures of abundance. However, as more lines of sight to nearby bright stars have been investigated, the range of derived abundances has increased. One problem appears to be possible misidentifications with hydrogen lines in the stellar winds. In addition, inhomogeneous astration would not be unexpected. Since no natural mechanism for producing deuterium has been found, but one can place no firm model-independent limits on the amount of astration, the best that can be done at present is to take (with a grain of salt) $X(^2\mathrm{H}) \gtrsim 2 \times 10^{-5}$ for the primordial abundance. From Fig. 4, this would then imply

$$\Omega_{(0)}\ (\text{baryons}) \lesssim 0.04\left(\frac{100}{H_0}\right)^2 \tag{18}$$

The interstellar hyperfine line of $^3\mathrm{He}^+$ has recently been detected by Rood et al. (1984) in three galactic HII regions with mass fractions in the range $10^{-4} \lesssim X(^3\text{He}) \lesssim 10^{-3}$, with upper limits in three other HII regions near the lower end of this range. Such values are much higher than those derived from meteorites, $X(^3\text{He}) \cong 2\text{–}6 \times 10^{-5}$, but are sensitive to assumptions about the physical conditions in the HII regions. However, since stars can both produce and destroy ^{3}He, such variation with location and age in the Galaxy are perhaps not too surprising. On the other hand, they further indicate that this data cannot put any reliable constraints on cosmology.

Important recent support for the minimal standard model has come from both observational and theoretical progress regarding ^{7}Li. In observations of more than two-dozen old (Pop. II) stars in our galaxy, Spite and Spite (1982a,b) have found that those with effective temperatures in the range 5500–6250 K have the same low level of abundance, $X(^7\text{Li}) \cong 5 \times 10^{-10}$. The lack of temperature dependence indicates that this lithium has not been depleted by convection, as might have been expected. In addition, its level is a factor of 10 below that normally attributed to Pop. I. Thus this could represent the primordial value. On the theoretical side, revised cross sections for some of the reactions involved in ^{7}Li production in the early universe have increased its predicted abundance (although the uncertainty is roughly a factor of 2). Both of these recent changes have gone in the direction to markedly improve the match between the observed and predicted values, as shown in Fig. 4.

In a wider context, another type of evidence remains one of the strongest reasons for believing in this primeval origin of these light nuclei (with the possible exception of ^{3}He). All other nuclei can be produced by either cosmic-ray bombardment of interstellar gas (^{6}Li, ^{9}Be, ^{10}B, ^{11}B) or stellar processes ($A \geqslant 12$).

To test more precisely and reliably these abundance predictions of the minimal standard model, we clearly need observations in a wider variety of locations in space and time. It is of crucial importance to identify systems in which stellar contamination has been negligible, as indicated by a much reduced level of heavy-element abundance. Prime candidates would appear to be the "Lyman-α forest" of absorption lines in the spectra of some quasars, usually interpreted to be produced by intergalactic clouds. However, there is recent evidence that heavy elements may be present at a low Pop. II level (Chaffee et al. 1985). The outer regions of extensive galactic halos, another candidate for some of the absorption line systems in quasars, might provide another relatively pristine sample of gas.

With its ultraviolet cutoff of ~1100 Å, the Hubble Space Telescope (HST; see Chapter 11) can observe the Ly-α through Ly-δ lines of deuterium from absorbers with redshifts $z \gtrsim 0.15$. The separation from the hydrogen lines will probably be resolvable with the high-resolution spectrograph. A much more difficult, but potentially as important, isotopic measurement would be that of $^3He/^4He$ from the same series of resonance lines, beginning at $\lambda = 584.3$ Å for HeI (requiring $z \gtrsim 0.9$) and 303.9 Å for HeII (requiring $z \gtrsim 2.6$). Although the predicted ratio $^3He/^4He$ is comparable to that of D/H, the isotope shift $\Delta\lambda/\lambda$ is six times less (4.6×10^{-5}). The resolution of the high-resolution spectrograph on the HST is $\Delta\lambda/\lambda \geqslant 5 \times 10^{-5}$, but the velocity broadening of the lines should be greater than this. In addition, the sources will be fainter. It is obviously also of great importance to determine $^4He/H$ from such systems but other elements will also have to be observed in order to establish the degrees of ionization. One must also remember that ground-based observations may be able to determine D/H at very large redshifts, while proposed space missions (such as FUSE) could determine D/H at very small redshifts.

As we move backward in time into the era when quarks and gluons existed as free particles, we encounter the epoch when the weak and electromagnetic interactions lost their separate identities, signaled by the appearance of the $W^{\pm}$ and Z^0 bosons, carriers of the weak interactions, prevented from annihilation and decay by pair production as indicated above. In the standard model of electro-weak unification, Higgs bosons or similar particles required to induce the masses of these particles should also have been present at energies above their rest mass. Unfortunately, no relics have yet been identified that could provide quantitative information about these two epochs, even though all the nucleons in the universe were presumably produced during the quark–hadron phase transition. Of course, the discovery of a new stable particle with a mass or abundance too large to be produced by cosmic rays or other galactic processes would provide a probe of the relevant epoch. One possibility is the axion, a pseudoscalar particle that could have been produced during the quark–hadron phase transitions, in amounts sufficient to dominate the present universe (Preskill, Wise, and Wilczek 1983). Although very light ($m \sim 10^{-5}$ eV), they would have always been nonrelativistic, and thus are one of the many candidates for the invisible matter. Sikivie (1983) has proposed an experiment that could detect such an abundance of axions via their interaction with electromagnetic fields.

B. Beyond the Physics Barrier

At energies above $\sim 10^2$ GeV, we presently have only theory to guide us in the continuation of this extrapolation back in time. However, the unification of the strong with the electro-weak interaction, predicted by many models to occur near energies $\sim 10^{16}$ GeV during this era, has a number of important potentially observable consequences. In many such "grand unified theories" (GUTs) the proton is predicted to decay at a rate (lifetime $\sim 10^{32}$ years) that should be observable with the very large detectors now operating. The observation of this decay would provide direct evidence for the existence of the X-boson responsible for the interconversion of quarks and leptons, but of course would not require that the universe had once been at such energies.

Another consequence of most such GUTs is the creation of magnetic monopoles at the epoch of grand unification. However, in the naive extension of our minimal model indicated in Fig. 3, generically far too many monopoles exist compared to the limits provided by present astrophysical constraints (Turner 1985). These monopoles are a consequence of singularities in the structure of scalar (Higgs) fields that break the grand unified symmetry, producing masses for the X-boson and other particles.

Both the baryon and lepton numbers (relative to photons) of the present universe may also be considered as "relics." It has been shown that these GUTs may contain the necessary ingredients (asymmetry of the laws of physics with respect to particles and antiparticles plus violation of baryon and lepton conservation) to produce a net baryon (or lepton) number in a universe that was initially symmetric. Although it has been shown to be plausible that this could have happened at this early epoch of symmetry breaking, experiments do not constrain these theories nearly enough to allow a definite prediction of the surviving numbers of leptons or baryons.

An ingenious way of avoiding the monopole problem, as well as explaining why the universe appears so uniform on the largest scales and was so nearly spatially flat [right-hand side of Eq. (7) negligible in the past] has been proposed by Guth (1981) and extended to avoid some of the initial problems by many others. The problem of homogeneity has been particularly vexing within the simple standard models, since some regions of the universe that we observe to have the same properties (as viewed via the microwave background radiation) have never been in causal contact unless the equation of state parameter $\alpha \leq -\frac{1}{3}$ in the remote past.

The inflationary model first proposed in detail by Guth is based on the fact that if these scalar fields exist, their vacuum expectation value could provide a significant contribution to the mass density that remains constant for some period of time, mimicking a cosmological constant. In order that such a contribution not dominate the density since primordial nucleosynthesis, a lower (zero) energy state must become accessible as the universe cools due to its expansion. But during the time following the

GUT era when this contribution might have dominated, the universe would have expanded much faster (exponentially) than in the minimal models, so that a small causally connected region could have expanded to at least the size of the presently observable universe. Besides greatly diluting the density of monopoles, this model allows structures such as galaxies to have originated from fluctuations over scales that were initially in causal contact.

At present, the study of this new class of standard models is one of the most exciting areas within the rapidly growing union of particle physics and cosmology. The discovery of Higgs particles in accelerators would place it on much firmer foundation, however. A much more serious problem is verification. Thus far, the only firm predictions of inflationary models are the absence of magnetic monopoles and the huge reduction of any initial curvature (giving $\Omega_0 = 1$). In principle, the origin and growth of structure can be computed within any particular model, but we have few guides to the correct physics at these remote epochs.

The final barrier indicated in Fig. 3 represents the epoch at which our theory of gravitation, general relativity, must be replaced by its quantum generalization, which does not yet exist. This onset of the Planck era is also the epoch at which the universe became opaque to gravitons, and where possibly the ultimate synthesis of all the fundamental interactions occurred. Any discussion of what occurred before this epoch is devoid of scientific meaning, yet many questions, such as the existence of a causal (particle) horizon mentioned above, depend on this unknown physics. In principle, relic gravitons (and possibly large-scale structure) bring us information from this theoretical boundary, but the gravitons would be undetectable if they resembled the neutrino and photon sea.

This extension of the standard model back in time clearly represents the greatest extrapolation in science. In the absence of terrestrial accelerators of such energies, it is tempting to use the very early universe as a "laboratory for particle physics." But it is very dangerous to employ a laboratory whose properties you do not know. The first priority should be the discovery and implementation of physically well-understood probes of the very early universe so that we can develop a cosmological model that has a much firmer foundation. Clearly, we need to know more about physics at very high energy before we make statements about cosmology at very high energy.

IV. PROBES OF THE RECENT UNIVERSE

The model of the past history of the universe that has just been described has the awesome potential to provide the context for our understanding of the evolution of both the laws of physics and the beautiful complexity of structure in which we are embedded. However, this fundamental importance of cosmology places a special burden on us to establish a much firmer foundation for this extrapolation. In this section we shall review some of the types of observations that we hope within the not too distant future can both help to check the standard model of the recent universe and determine its parameters. We mainly discuss, for each of the three measures of distance defined in Section II.B, what we believe are the most promising methods of determining the cosmological distance–redshift relation.

The history of the attempts of astronomers to determine the values of the Hubble constant H_0 and the deceleration parameter q_0 (Rowan-Robinson 1985) illustrates many of the dangers involved in the use of distance indicators that are not physically well-understood systems. Not surprisingly, unverified assumptions have often turned out to be invalid.

In Table 1 is listed the approaches we shall consider in determining the distances to objects in the Hubble flow [where random velocities contribute only a small amount to the redshift ($z \gtrsim 0.01$)]. As we shall see, only the first two require calibration via the classical construction of a "distance ladder" based on other measures of shorter distances. The uncertainty in the calibration of such methods reflects the uncertainties in each rung of this ladder. Crucial steps in this ladder have included (1) the determination of the distance to the Hyades cluster via the convergent point parallax method, (2) main sequence fitting, (3) the Cepheid period-luminosity relation(s), and (4) use of the brightest stars in galaxies. The problems associated with these and other measures of distance have been reviewed by Hodge (1981) and Rowan-Robinson (1985).

In assessing any method of distance determination, we should keep in mind the following questions:

1. What are the assumptions that define the model of the system being employed as a distance indicator?
2. How strongly are these assumptions tested by independent observations?
3. How accurately is distance determinable by this method? In particular, how precisely can the parameters of the model be determined, and how sensitively do these determine distance?

If one applies Newtonian dynamics [$V^2(r) \sim GM(r)/r$] to a spiral galaxy seen at inclination angle i, then one obtains the following equation for the

TABLE 1. Methods of Cosmological Distance Determination

Method	Basis	Tests	Problems
	A. *Luminosity Distance*		
Velocity-dispersion (Tully–Fisher)	$L \propto \mu^{-1}(M/L)^{-2}(\Delta V)^4$	L(IR) vs. ΔV Hubble diagram (for constant ΔV)	Calibration, selection effects, aperture corrections, evolution
Standard candle	Models of SNI	Hubble diagram, light curve, abundances	Calibration, selection effects, evolution
Dynamical	Hydrodynamical models of SNII	Light curve, colors, velocity	Uniqueness of models
	B. *Angular Size Distance*		
Absorption-emission	Emission $\propto \int n_e^2 F(T)\, dl$ Absorption $\propto \int n_e^q G(T)\, dl$	High-resolution observations	Spatial dependence of n_e and T
Gravitational lens	Time delay (differential bending and gravitational) $\propto H_0^{-1}$	Velocity dispersion within lens	Spatial dependence of mass density
	C. *Proper Motion Distance*		
Radio source expansion (two types)	Angular rate measured	High-resolution observations	Transverse velocity model dependent
Supernova expansion	Sharp, spherically symmetric photosphere ($\theta^2 \propto f_\nu/\mathcal{F}_\nu$)	Frequency dependence of flux, time independence of distance	Model atmosphere, galactic absorption

observed spread of velocities obtained from the rotationally induced Doppler shifts:

$$(\Delta V)^4 \propto \mu(M/L)^2(\sin^4 i)L \tag{19}$$

Here $M(r)$, $L(r)$, and μ are the total mass, luminosity, and average surface

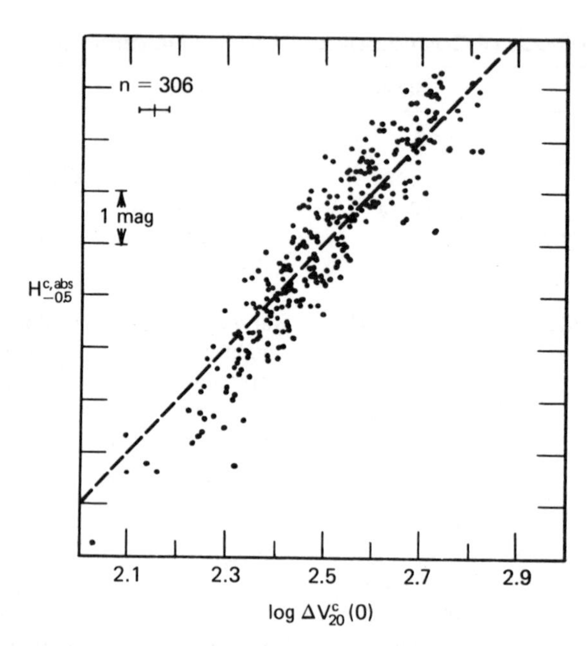

FIGURE 5. Relative absolute infrared magnitudes of 306 spiral galaxies versus total rotational velocity spread (Aaronson et al. 1982).

brightness within the radius corresponding to an observed range $\Delta V(r)$. It is found that V remains remarkably constant in the outer regions of the disk, providing strong evidence for the mass $M \propto r$ being dominated by a less-flattened invisible component, as summarized by Rubin in Chapter 9. Thus ΔV has a relatively sharp cutoff and so can be well determined. The luminosity L of a spiral galaxy converges fairly rapidly as r becomes large, but the use of standard isophotal sizes is still required.

Building on earlier work by Opik (1922), Tully and Fisher (1977) first proposed that Eq. (19) could be used to obtain the luminosity, and thus the distance, of spiral galaxies. Problems with internal absorption are reduced by using infrared luminosities (Aaronson, Huchra, and Mould 1979). Assuming that the inclination angle can be accurately determined, the key question is whether $\mu(M/L)^2$ depends only on L, even for spiral galaxies of a given class.

Observational evidence relevant to this question is presented in Fig. 5, where relative infrared absolute magnitude H is plotted versus the HI velocity width for a collection of nearby spiral galaxies whose relative distances were determined by assuming a uniform Hubble velocity flow (Aaronson et al. 1982). (Therefore some of the scatter should be due to the random component of the galaxies' redshifts as well as infall toward the Virgo supercluster.) The dashed line corresponds to constant $\mu(M/L)^2$. Although the scatter is not very small, this data does indicate a correlation of L with ΔV.

Using a carefully selected sample of 20 ScI galaxies in the velocity range 3000–13,500 km sec^{-1}, Bothun et al. (1984) obtain a Hubble constant $H_0 = 91 \pm 3$ km sec^{-1} Mpc^{-1}, consistent with earlier results of this group. Yet Tammann and Sandage (1986), using the same total data set, obtain $H_0 \cong 55$ km sec^{-1} Mpc^{-1}. Some of this discrepancy is due to different allowances for the dispersion in the relation at faint magnitudes (Malmquist bias) and some could be due to the use of different calibrators. In any case this difference reflects the problems that typically arise when complicated physical systems with significant dispersion in their properties are used as distance indicators.

Even in the absence of scatter, observational evidence such as that exhibited in Fig. 5 is only relevant to the determination of H_0, because the determination of $\Omega_{(\alpha)}$ from higher-redshift observations requires that the evolution of $\mu(M/L)^2$ be known. The crucial question is whether models of the structure and evolution of galaxies can be tested by independent observations strongly enough to produce confidence in any prediction for the dependence of $\mu(M/L)^2$ on the parameters that characterize galaxies. Certainly at present the number of (apparently) independent variables involved in any description of a galaxy is larger than is the case for some of the other distance indicators we shall consider.

What many now consider the most promising candidate for a cosmological "standard candle" is Type I supernovae. The major observational evidence is indicated in Fig. 6, which is a Hubble diagram relating the recessional velocity to peak blue apparent magnitude for a collection of SNI in E and SO galaxies, in which absorption should be minimum (Sandage and Tammann 1982). The line drawn corresponds to the relation

$$\langle M_B(\max)\rangle = (-19.73 \pm 0.24) + 5 \log\left(\frac{H_0}{50}\right) \tag{20}$$

which includes an estimate of the dispersion. In addition to the correction for the noncosmological component of the velocity of the Virgo cluster, selection effects will remain an additional source of uncertainty until it can be firmly demonstrated that at least a class of SNI are truly a standard set of objects. [Contrary to previous claims of some groups, Cadonau, Sandage, and Tammann (1985) now claim that the dispersion in the shape of the light curve for SNI is small.] Determining the calibration of the

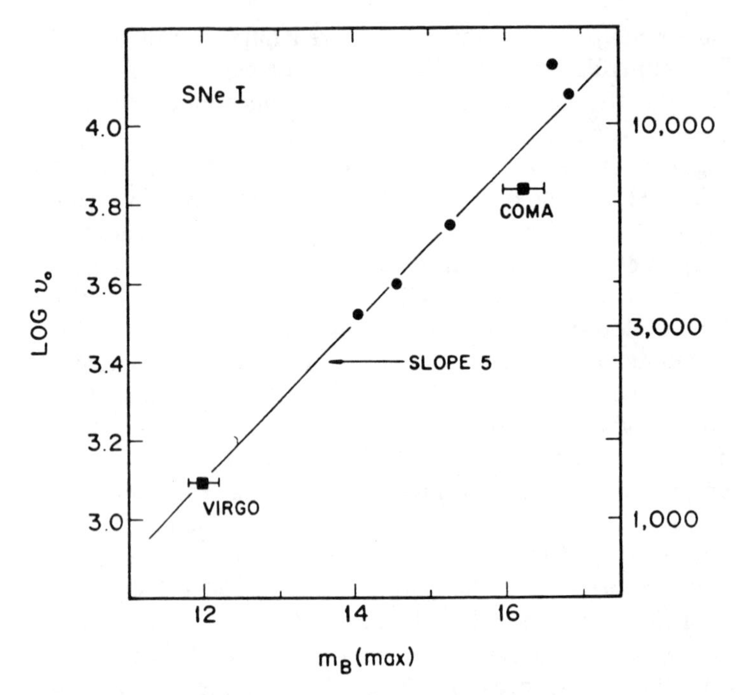

FIGURE 6. Recessional velocity (km sec^{-1}) versus apparent blue magnitude for Type I supernovae within E and SO galaxies, where absorption should be minimal. Supernovae within two clusters of galaxies are included, 6 in Virgo and 5 in Coma.

small number of SNI observed within elliptical galaxies is also a problem. And, of course, the unknown effects of evolution enter at larger redshifts.

However, there is now some hope that a physical basis may exist for believing that SNI are in some sense standard candles. The work of various groups on carbon-deflagration models of white-dwarf explosions induced by accretion of material from a binary companion indicates that the peak luminosity is related to the amount of ^{56}Ni produced (Arnett, Branch, and Wheeler 1985; Wheeler and Sutherland 1985). The decay of this isotope is thought to be the main energy input at late times, since it seems to explain the shape of the light curve and the decay of its daughter Co has apparently been identified spectroscopically (Woosley, Axelrod, and Weaver 1984). However, no accretion rate has yet been found to lead to the conditions required for the explosion. Again, what is needed are observational characteristics that can serve as independent tests of any proposed model. It is hoped that these will be forthcoming as the models are more fully developed.

A few years ago a different dynamical method for determining the peak luminosity of a supernova from distance-independent characteristics of the event was proposed. [For an updated summary, see Arnett (1982).] Basically the model only assumes an explosive release of energy, carried by a shock wave through the envelope of the presupernova star. The main parameters are the initial energy of the expolsion and the density profile $\rho(r)$ of the envelope, where the energy is diffusively released. The hope is that distance-independent observables such as the colors, velocity, and shape of the light curve are unique functions of the luminosity. Again, however, the models have not been developed fully enough to assess the power of this method. Note, however, that the assumption of a "standard candle" is not required. Similarly, the two methods for determining angular-size distance, to which we now turn, do not rely on the assumption of a "standard intrinsic size" (which has in the past also proved to be a dangerous assumption).

Consider a large region of optically thin ionized gas, in front of a different source of radiation. Then the emitted surface brightness (from bremsstrahlung, lines, etc.) from a given line-of-sight through the region will typically be of the form

$$I_{\nu 1} \propto \int n_e^2 F_{\nu 1}(T)\, dl \qquad (21)$$

while its scattering or absorption optical depth with respect to the background source will be of the form

$$\tau_{\nu 2} \propto \int n_e^q G_{\nu 2}(T)\, dl \qquad (22)$$

To avoid proliferation of assumptions, we shall consider a gas cloud that has been resolved at both frequencies, so that we have information on the variation of these path integrals across the region. Other possibilities have been summarized by Shaver (1984). In practice, one would only want to employ clouds that appeared spherically symmetric to further reduce the geometrical assumptions. If all quantities were spherically symmetric, and $T(r)$ could be determined from spectral information, then we see from Eqs. (21) and (22) that the normalization of n_e and the cloud scale size can be obtained if $q \neq 2$, giving

$$D_A \propto \tau_{\nu 2}^{2/(q-2)}\, I_{\nu 1}^{q/(2-q)} \qquad (23)$$

where the quantities on the right refer to the central values. Probably the most critical assumption involved in this method is the absence of inhomogeneities on scales that cannot be resolved observationally. For instance, the ratio $\langle n_e^2 \rangle / \langle n_e^q \rangle^{2/q}$ also appears in Eq. (23).

At present, the most promising application of this method is to the hot gas within clusters of galaxies, which both "cools" the microwave background via electron scattering (so $q = 1$) and produces X-rays via thermal bremsstrahlung. Silk and White (1978) have analyzed spherically symmetric models in detail, and shown that the method is most reliable if the scattering and emission are mostly confined to a well-defined core region, although the problem of possible smaller-scale clumping remains.

The major observational problem is that the predicted microwave cooling $[\Delta T_m / 3K \sim \int (kT/m_e c^2) n_e \sigma \, dl]$ is at the limit of detectability ($\Delta T_m \sim 10^{-4}$–10^{-3} K). Marginal detections have been claimed (Birkinshaw et al. 1981) but have not been strongly confirmed by independent observations (Andernach et al. 1983). Additional observational problems involve the contribution of other sources in the beam at both microwave and X-ray frequencies.

Another method that could in principle provide a measure of angular-size distances involves gravitational lenses, with at least five now known to exist. The most useful at present involves multiple images of the quasar 0957+561 (Young et al. 1981). (In general, there will be an odd number of images.) The redshift of the intervening cluster of galaxies has also been measured. If a time delay between variations in the brightness of the images can be measured, that provides a scale which in principle can be related to the distances of the lens and quasar (Refsdal 1966; Cooke and Kantowski 1975; Borgeest 1983).

However, there are two contributions to the time delay, which are typically of comparable magnitude. The first is the geometrical contribution (bending) to the time delay, which can be determined if the undeflected position of the quasar (two unknowns) can be obtained. Thus three images (with three time delays) would be sufficient to obtain this contribution. The second contribution is the gravitational time delay, proportional to the integral of the gravitational potential along the path of the ray through the lens. Evaluating this contribution requires knowledge of the distribution of gravitational mass throughout the lens. Thus the velocities of visible probes (which have relaxed to an equilibrium distribution) must be obtained throughout the lens in order to begin to obtain an accurate assessment of this contribution.

Another source of uncertainty has been pointed out by Alcock and Anderson (1985a). All measures of distance that involve propagation of a beam of photons over long distances in the universe are sensitive to the effects of inhomogeneities in the matter distribution along the beam because of their gravitational effect on the cross section of the beam. An especially lucid discussion of this process, involving the two optical scalars (rates of expansion and shear of the beam), has been given by Alcock and Anderson (1985b). A key point made by Alcock and Anderson (1985a) is that the rarity of gravitational lenses indicates that we are dealing with extreme conditions in the universe, so that this problem of inhomogeneities cannot as easily be reduced by averaging over many objects.

Thus it would appear that (at least for the forseeable future) time delays between gravitational lens images will be more useful for obtaining information about the distribution of mass within the lens as well as along the entire line-of-sight rather than as a measure of distance.

Finally, we turn to measures of proper-motion distance. Here, we shall discuss methods in which the source is resolved, so that the time rate of change of an angular size can be measured directly. In Section V, we discuss the method involving the early expansion phase of a supernova, in which the angular size is inferred from the ratio of observed to emitted flux.

The rapidly expanding jets seen at the core of many extragalactic radio sources are now believed to involve actual relativistic bulk motion of matter. The evidence has been reviewed by Begelman, Blandford, and Rees (1984) and Cohen and Unwin (1982) (see also Chapter 6). For a source moving with velocity $\boldsymbol{\beta}c$ (relative to another source) in some direction with respect to the line-of-sight $\mathbf{n}$ to us, the observed angular expansion rate is

$$\frac{d\theta}{dt} = \frac{\mathbf{n} \times (\boldsymbol{\beta} \times \mathbf{n})}{(1 - \mathbf{n}\cdot\boldsymbol{\beta})} \left(\frac{c}{D_M}\right) . \tag{24}$$

The apparent superluminal velocities observed in some sources is thought to be due to the factor $\gamma = (1 - \beta^2)^{-1/2}$ which appears in Eq. (24) as $\mathbf{n}\cdot\boldsymbol{\beta} \rightarrow 1$. The fundamental problem in obtaining D_M is obtaining accurately both the magnitude and direction of $\boldsymbol{\beta}$. Although the flux from superluminal sources is Doppler boosted, it is difficult to make use of observed fluxes in the absence of a tightly constrained physical model. Again, the short history of this method contains claims based on models that were subsequently disproven by observations. However, if one makes the reasonable assumption that Compton scattering is the primary X-ray emission mechanism, then one can calculate the minimum value of γ required to bring the calculated X-ray emission down to the observed value. Marscher and Broderick (1982) then claim that the observation of many such sources will allow one to perform the appropriate average over the directions of β, and the minimum values of γ will provide an upper limit to H_0. With our present knowledge of the geometry and physical conditions within these sources, it is however even difficult to determine the uncertainty in any such upper limit.

Recently, radio observations of the angular expansion velocities of supernovae a few years after the explosion have stimulated proposals that such systems be used to determine D_M (Bartel 1985). Here again, the major problem is the determination of the related velocity, although there are also large uncertainties in the geometry of the radio-emitting region (Marscher 1985). There have been proposals to employ optical emission lines from the supernova remnant. However, it is clear that large uncertainties will be present unless one can relate the velocity of the optically emitting material to that of the radio-emitting material. At present, the observational possibilities are not sufficient to constrain the geometry and physics of the system. Chevalier and Fransson (1985) have proposed the use of the X-ray flux to determine the velocity within the context of their model of the hydrodynamic interaction between the supernova and the circumstellar gas. At present, the model is not strongly tested by observations. In any case, the small size of the radio-emitting region ($\lesssim 10^{17}$ cm) and the weakness of the flux at all wavelengths makes it unlikely that such systems can be used to obtain distances wtihin the Hubble flow.

There are in principle other ways to determine the parameters that characterize the present universe other than the distance–redshift relations. But none of them are competitive at present.

Counts of a particular class of objects out to some limiting brightness or redshift can in principle provide information about the cosmological model (Weinberg 1972). But in practice they tell us more about the evolution of the objects employed.

A relatively clean way to measure q_0 would exist if a class of objects at cosmological distances were observed to emit coherently ("powerful pulsars") with some intrinsic period P_e over a time $\gg P_e$. The time of arrival t_N of pulse N would then be given by

$$t_N = (1 + z)P_eN - \tfrac{1}{2}H_0q_0(P_eN)^2[z + O(z^2)] + O(N^3). \qquad (25)$$

If one independently measures the redshift of the system, P_e can be obtained from the term linear in N, which then allows H_0q_0 to be obtained from the quadratic dependence upon N. However, any local gravitational acceleration will also contribute to this effect, so that in practice many objects would have to be used to average out this source.

If one instead attempts to measure the time dependence of the frequency of a spectral line, one employs the differential of Eq. (25), giving $dz/dt = -q_0H_0[z + O(z^2)]$. However, line broadening makes this approach far from feasible at present.

V. SUPERNOVA EXPANSION DISTANCES

In this section we discuss the last of the methods of cosmological distance determination referred to in Table 1. We choose to discuss this method in more detail becuase it involves a relatively simple, well-defined physical system, and therefore illustrates most clearly the approach to establishing the reliability of any distance indicator that we outlined by the questions posed at the beginning of the last section. This method, a generalization of the one first proposed by Baade (1926) as a means to determine the distance to variable stars, has been reviewed by Wagoner (1980). It was first applied to supernovae by Branch and Patchett (1973) and Kirshner and Kwan (1974). Details of the radiative transfer calculations and the most recent results are presented by Hershkowitz, Linder, and Wagoner (1986).

This method can in principle be applied to any expanding (or contracting) spherically symmetric object that emits continuum radiation from a well-defined (steep density gradient) photosphere surrounded by a lower-density atmosphere in which Doppler-shifted (''P Cygni'') line profiles are produced by resonant scattering. We will discuss the use of Type II supernovae (SNII) because the spectral analysis is simpler owing to the dominant abundance of hydrogen, and the models of the expansion (Weaver and Woosley 1980) are more secure. In Fig. 7a,b are shown optical and ultraviolet spectra of the SNII 1979c. Although the blue-shifted absorption and red-shifted emission characteristic of a P-Cygni profile is only readily apparent for the HeI line in this object, its well-defined optical continuum characterized by a color temperature that cools to $T_c \sim 5000$ K during the first month after maximum brightness is representative. A well-defined continuum also exists in the ultraviolet, with narrow features due to intervening galactic absorption and supernova atmospheric emission also present.

It can be shown analytically that there will be a discontinuity in the shape of a P-Cygni line profile (typically at its minimum) at a frequency shift corresponding to the velocity of the photosphere. Observations indicate that this velocity decreases slowly from $v \cong 0.03c$ during the first month after the shock wave from the explosion accelerated and heated

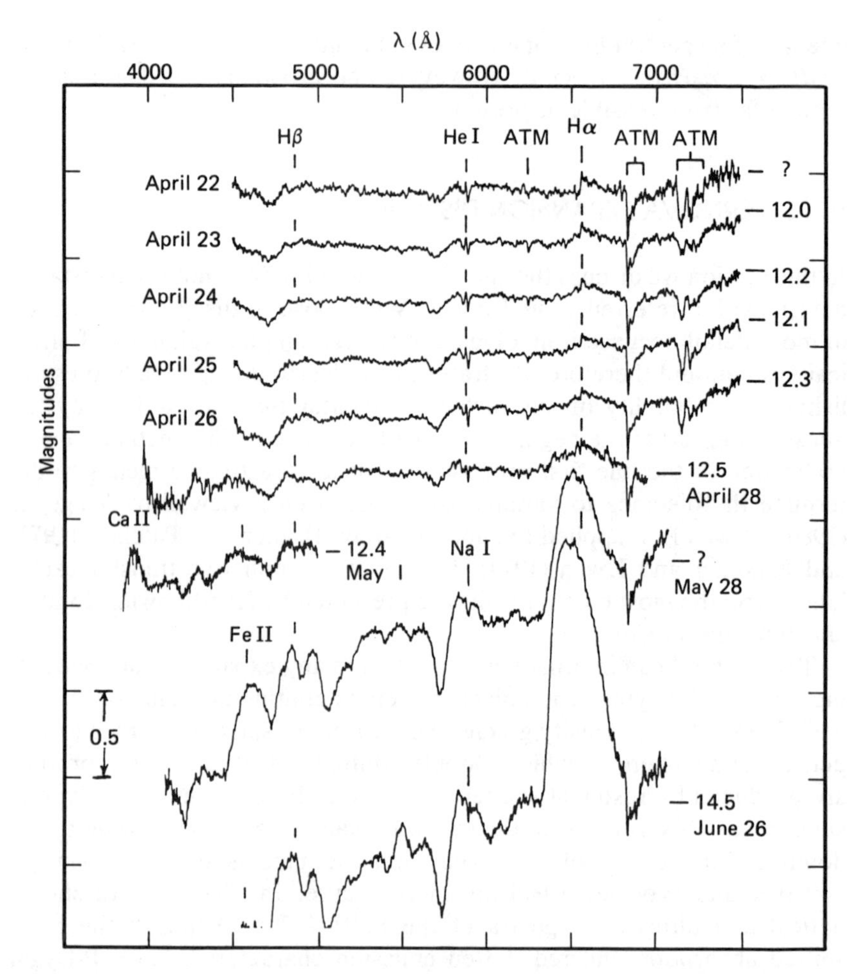

FIGURE 7a. Optical spectra of the Type II supernova 1979c during its early phase. Relative magnitudes are plotted, with individual values of m_v indicated (Branch et al. 1981).

the extended envelope of the star whose core collapsed to produce the explosion. The assumption of spherical symmetry thus allows us to determine the transverse velocity $v(t)$ that appears in the expression (3c) for the proper-motion distance D_M.

Determining the other required function, the angular expansion rate $d\theta/dt$, is not as straightforward. The angular size of the photosphere, $2\theta(t)$, is related to the received (f_ν) and emitted $(\mathcal{F}_\nu)$ flux per unit frequency (as measured in each frame) by the expression

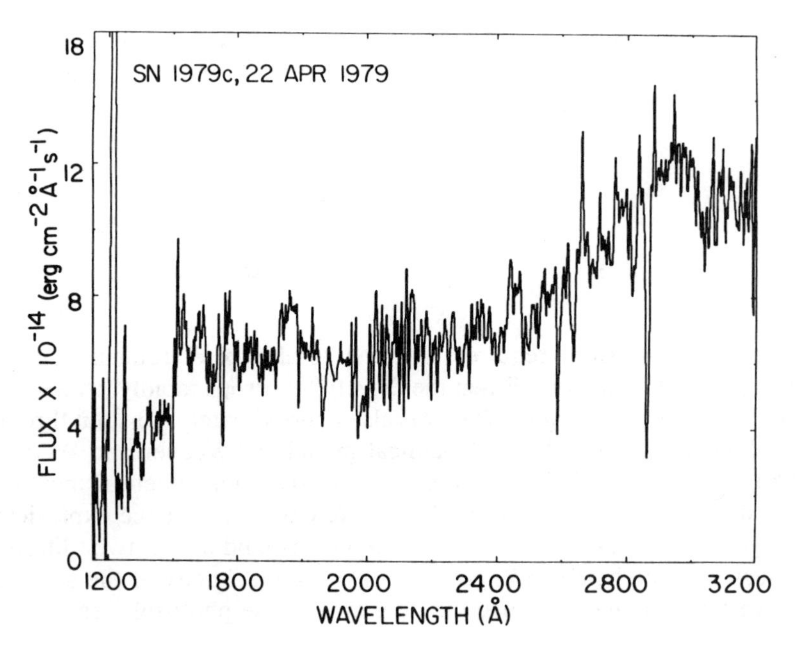

FIGURE 7b. IUE spectrum of 1979c on 22 April (Benvenuti et al. 1982). Note the linear flux scale.

$$\theta^2 = (1 + z)^3 \left(\frac{f_\nu}{\mathcal{F}_\nu}\right) \exp\,[\tau_I(\nu)] \tag{26}$$

where τ_I is the intervening optical depth due to absorption by dust in our Galaxy and the one containing the supernova (with redshift z). We shall discuss the main problems, the determination of $\mathcal{F}_\nu$ and τ_I, in turn.

It can be shown that the expanding photosphere of the supernova is quasistatic, in the sense that the escape time of a photon is much less than the hydrodynamical time scale r/v. Thus we can attempt to match theoretical spectra to the observed shape of f_ν at various times, yielding $\mathcal{F}_\nu(t)$. The catalog of emitted fluxes will be characterized by some set of parameters of a model photosphere. Programs for producing such catalogs of spectra are being developed at Stanford and by a collaboration with Giora Shaviv and Ranier Wehrse. The results that will be referred to below were computed from a preliminary model, in which the effects of lines on the continuum (due to their broadening by the velocity gradient within the photosphere) and the contribution of elements other than hydrogen to the (bound–free and free–free) absorptive opacity have been neglected.

Under these conditions the set of predicted fluxes $\mathcal{F}_\nu$ are characterized by two parameters: an effective temperature T_e and a scale height ΔR defined by the relations

$$\mathcal{F} = \int_0^\infty \mathcal{F}_\nu \, d\nu \equiv \sigma_R T_e^4 \tag{27}$$

$$\int_R^\infty \rho \, dr \equiv \rho(R) \, \Delta R \tag{28}$$

Radiative equilibrium requires that the total flux $\mathcal{F}$ remain constant through the photosphere. When the density $\rho(r)$ drops rapidly (as we have assumed), the scale height $\Delta R \ll R$ is also approximately constant through the photosphere. The hydrodynamical models of Weaver and Woosley (1980) give $\Delta R \cong 0.01R$ while the radius of the photosphere R increases from 0.3–1.5 $\times$ 10^{15} cm during the first few weeks after the explosion.

Figure 8 shows computed spectra corresponding to two different choices of scale height ΔR, compared with a blackbody at the same effective temperature. Because the density at these photospheres is much

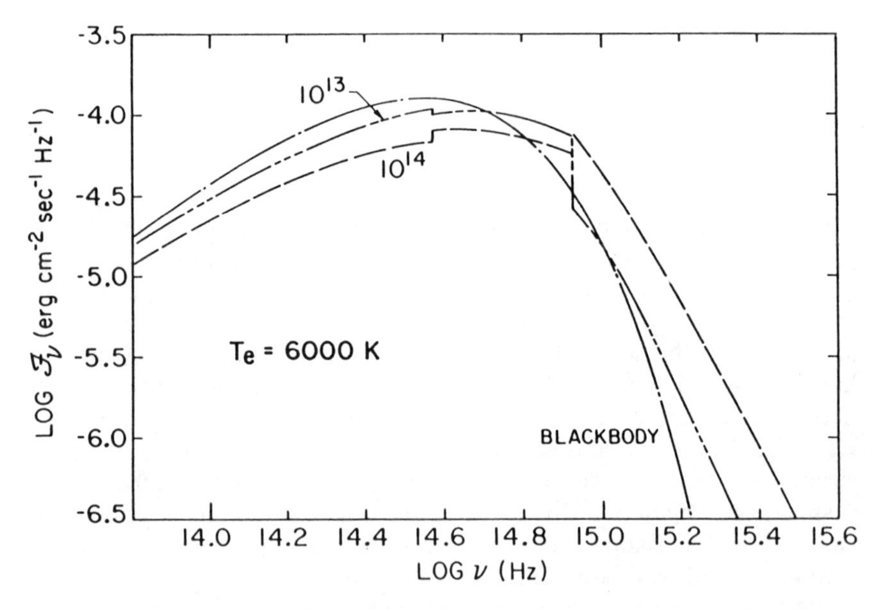

FIGURE 8. Example of predicted spectra for SNII, corresponding to two values of the scale height ΔR (cm) indicated. A Planck function for the same effective temperature T_e is also shown. (Hershkowitz, Linder, and Wagoner 1986).

less than those within ordinary stars at the same time effective temperature, the spectra are significantly different. For instance, scattering usually dominates absorption and the atomic level populations are controlled by the radiation field, as well as the matter temperature. Note the sensitivity of the Balmer jump (at $\log \nu = 14.91$) to ΔR, including its change in sign as ΔR increases (corresponding to density decreasing). As for blackbodies, the relative amount of flux in the ultraviolet depends strongly on the other parameter, T_e. Note, however, that the flux is diluted in the infrared and enhanced in the ultraviolet, compared to a blackbody with the same total flux.

The other unknown in Eq. (26), $\tau_I(\nu)$, could in principle be determined by standard methods if its frequency dependence was a universal function. It does seem that this function does have the same shape along various lines of sight through our galaxy (Schultz and Wiemer 1975; Nandy et al. 1975; Rieke and Lebofsky 1985). Then the strength of the interstellar absorption lines or the broad feature in τ_I at ~2200 Å can be used to obtain its normalization. However, although all SNII appear in spiral galaxies, their absorption functions may not be the same. In fact, within the Large Magellanic Cloud this function does begin to differ with decreasing wavelength at about 2400 Å (Clayton and Martin 1985). What is clearly needed is data for other galaxies.

In Fig. 9 we compare our best-fitting theoretical spectrum $\mathcal{F}_\nu$ with the observed flux f_ν from the SNII 1980k, after correction with the galactic absorption function normalized to a color excess $E(B - V) = 0.36$. The same normalization is obtained from the interstellar lines (in both galaxies) and the 2200-Å feature. Note that the value of ΔR characterizing the theoretical spectrum is consistent with that obtained from the hydrodynamical models mentioned above. For this supernova, as for all others, the observers have previously assumed that the spectrum was a blackbody. Fitting a Planck function to the spectrum shown in Fig. 9 yields a color temperature $T_c \cong 10{,}500$ K.

Now it follows from employing the integrated fluxes in Eq. (26) that using the correct flux $\mathcal{F}(T_e)$ rather than the blackbody flux $\mathcal{F}_{\mathrm{BB}} = \sigma_R T_c^4$ leads to a distance

$$D_M \propto \theta^{-1} \propto \mathcal{F}^{1/2}(T_e) = \left(\frac{T_e}{T_c}\right)^2 \mathcal{F}_{\mathrm{BB}} \tag{29}$$

Thus if this SNII is typical, distances to them obtained in the past should be decreased by this factor $(T_e/T_c)^2 \sim \frac{2}{3}$, and the values of $H_0 \cong (40\text{–}60)$ km sec^{-1} Mpc^{-1} previously obtained by this method should be corre-

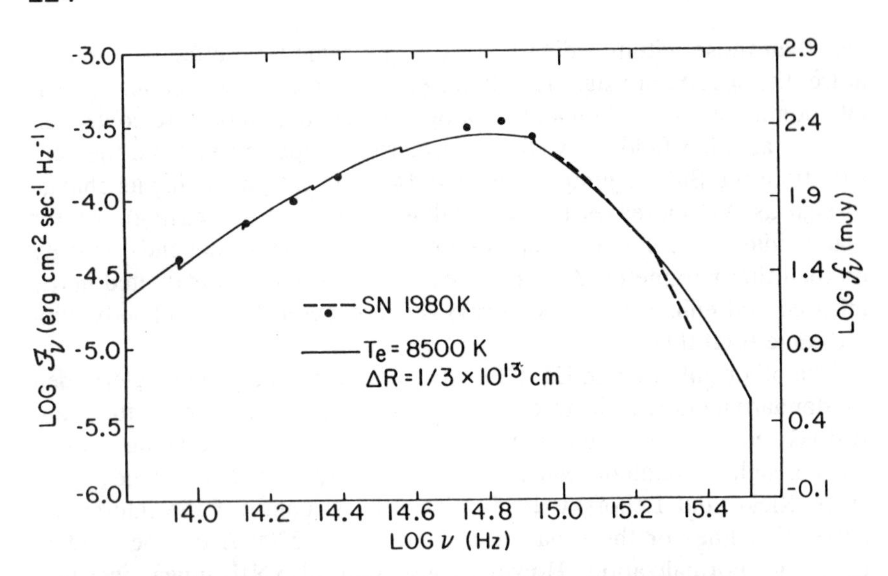

FIGURE 9. The best-fitting predicted emergent flux (left scale, solid line) is compared with the observed flux (right scale) from the Type II supernova 1980k. (Hershkowitz, Linder, and Wagoner 1986).

spondingly increased. Of course, this assumes that future refinements do not significantly alter the theoretical spectra.

The reliability of this method of distance determination is based on the fact that the model atmosphere assumptions involved in obtaining $\mathcal{F}_\nu$ can be tested in three ways, if we employ SNII whose intervening absorption is small or known:

1. The frequency dependence of the continuum must match the observed spectrum.
2. The distance

$$D_M = \frac{\int_{t_o}^{t} v(t')\,dt'}{\Delta\theta(t)} \tag{30}$$

   must be independent of time. It already seems that this ceases to be the case after a month or two, as the photosphere becomes diffuse.
3. The break in the line profiles (for moderate optical depths) must be present if the photosphere is sharp.

Once we have confidence in our model atmospheres, we can greatly reduce the uncertainties in distance determination due to intervening absorption in (at least) three ways:

1. The Balmer and other spectral jumps, which are sensitive to ΔR, are not affected by $\tau_I(\nu)$.
2. From Eq. (26) we see that comparing observations at two times gives

$$\frac{\mathcal{F}_\nu(t_1)}{\mathcal{F}_\nu(t_2)} = \frac{\theta^2(t_2) f_\nu(t_1)}{\theta^2(t_1) f_\nu(t_2)} \tag{31}$$

   Thus we can obtain the frequency dependence of the flux ratio (modulo normalization), which will be sensitive to both $T_e(t_1)$ and $T_e(t_2)$.
3. The actual value of θ is best obtained by employing Eq. (26) at the longest wavelength possible, where τ_I will be smallest. The theoretical spectra are also least uncertain at long wavelengths.

The observational program required for determining cosmological distances by this method is naturally divided into three parts:

1. Systematic searches for and spectrophotometry (IR + optical + IUE) of nearby SNII (apparent visual magnitudes $m_v \lesssim 13$) are needed to test the model atmosphere.

2. Obtaining H_0 from SNII at $0.01 < z < 0.1$ can be achieved with ground-based spectrophotometry in the IR + optical and the Hubble Space Telescope (HST) in the ultraviolet. On a Schmidt plate ($6° \times 6°$), $\sim 90\, z^3$ SNII per week could be discovered out to a redshift $z \lesssim 0.2$ from the ground. With CCDs (such as employed in John McGraw's transit survey at the University of Arizona) higher redshifts could presumably be reached because of improved background subtraction.

3. At redshifts $z \gtrsim 0.1$, the background light from the parent galaxy becomes comparable to that from the supernova under typical ground-based seeing conditions. This is illustrated in Fig. 10, which shows the redshift–magnitude relations for an average SNII a few days after maximum (T_c = 15,000 K) and a few weeks after maximum (T_c = 6000 K). [Observations over several weeks are required to obtain $\Delta\theta(t)$.] Also shown is the background light from a spiral galaxy within the supernova image size as seen from the ground and from HST. The corresponding night sky backgrounds are also indicated. We thus see that HST is re-

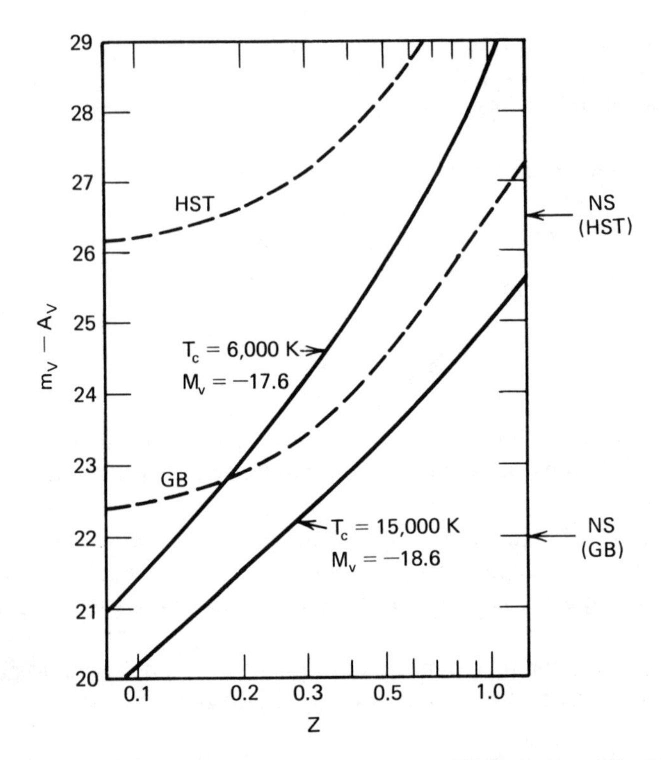

FIGURE 10. Redshift-magnitude relations for average SNII at two epochs indicated by the color temperature T_c and absolute visual magnitude M_v, for $H_0 = 50$. (A_v is the absorption in the V-band.) The amount of light from the parent galaxy (spiral arm region) is indicated by the dashed curves. A resolution area $\Delta\Omega = 1$ (arcsec)^2 was assumed for ground-based (GB) observations, while $\Delta\Omega = \pi \times 10^{-2}$ (arcsec)^2 was assumed for space (HST) observations. The night sky (NS) backgrounds are also indicated. A cosmological model with $\Lambda = p_0 = q_0 = 0$ was employed.

quired at all wavelengths. However, it may be possible to obtain the line profiles from the ground. For instance, with the proposed Keck 10-m telescope, it appears that with the required resolution of ~5 Å a signal-to-noise ratio of about 6 (at optical wavelengths) could be achieved for an SNII as faint as $m_v \cong 25$ after integration for a few hours. In that case, the low-resolution mode of the faint-object spectrograph (FOS) on HST could be used to obtain the continuum spectrum, allowing similar limiting magnitudes to be achieved after a few hours of observation. If the moderate resolution mode of the FOS (sufficient for the line profiles as well) were used, limiting magnitudes $m_v \sim 22$ could be achieved.

The information about deviations of the distance–redshift relation from linearity that can be obtained depends strongly on the redshifts that can be reached, since for small z the uncertainty in the deceleration parameter $\delta q_0 \propto z^{-1}$. For instance, determining θ and v to an accuracy of 5% at $z = 0.27$ gives $(\delta q_0)_{\text{rms}} = \frac{1}{2}$, which, however, is still much better than its present uncertainty.

However, it must be remembered that some SNII (such as 1979c) have been observed to be at least one magnitude brighter than the average indicated in Figure 10, comparable to SNI at maximum. In addition, the supernovae discovered at large redshift will tend to be the intrinsically brighter ones and have less intervening absorption. Thus a magnitude limit of $m_v \sim 25$ could make possible observations of SNII with redshifts $z \lesssim 0.5$.

In conclusion, we note that this method involves no evolutionary uncertainties, selection effects, aperture corrections, or need for many objects. It is fundamentally distinct from any method that involves "standard" objects.

VI. HORIZONS

Our generation has been the first to have the privilege of truly exploring the depths of the universe. This has been made possible in large part by the development of increasingly sensitive detectors across the spectrum of electromagnetic radiation. This is illustrated in Fig. 11, which shows how far back in time we have been able to look at each wavelength. From this figure it is also apparent how much more there is to be explored out to the photon barrier.

However, we have seen that there are other carriers of information about the universe. Neutrinos can probe redshifts $z \lesssim 10^{10}$, while gravitons can come to us from any epoch within the classical standard model. A different type of probe is relic nonrelativistic particles, which tell us about conditions on our past world line rather than our past light cone. The abundances of certain light nuclei bring us information about the nature of the universe just on this side of the neutrino barrier. And other relics (such as stable particles too massive or weakly interacting to have been discovered in accelerators) may bring us information from beyond the neutrino barrier. But the quality of this information depends on our physical understanding of the relic. Finally, it appears as if the large-scale distribution of matter in the present universe may reflect conditions in the very early universe, if the fluctuations that developed into this structure originated then.

This grand extrapolation into the past is based on our knowledge of

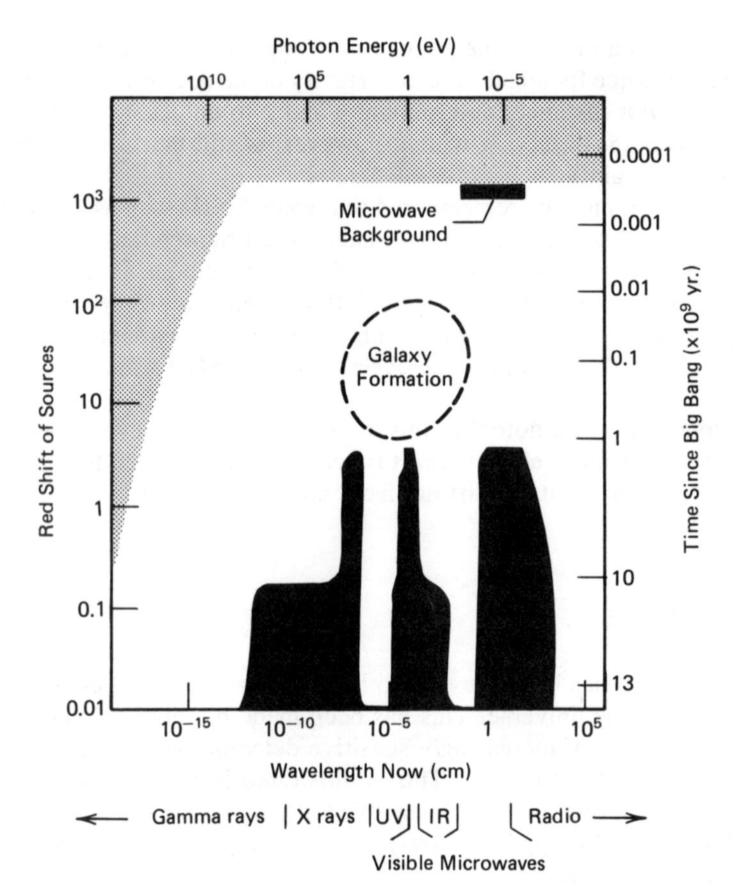

FIGURE 11. The extent to which we can explore the universe via photons. The dark areas indicate our present knowledge. The lightly shaded area is the region that we can never view directly because the photons are either scattered by electrons or (for gamma rays) collide with other photons, producing electron–positron pairs. [Adapted from Wagoner and Goldsmith (1983).]

the present universe as well. We have focused on the foundation of this knowledge: the determination of cosmological distances. As with all other probes, we have argued that in judging the reliability of any distance indicator, our physical understanding of the system employed should be given very high weight.

ACKNOWLEDGMENTS

I wish to thank David Branch, Nancy Evans, Stephen Hershkowitz, Eric Linder, Carl Pennypacker, and Giora Shaviv for their contributions to this article. This work was supported in part by NASA (grant NAGW-299) and NSF (grant PHY 81-18387).

REFERENCES

Aaronson, M., Huchra, J., and Mould, J., 1979, *Ap. J.*, **229**, 1.

Aaronson, M., Huchra, J., Mould, J. R., Tully, R. B., Fisher, J. R., Van Woerden, H., Goss, W. M., Chamaraux, P., Mebold, U., Siegman, B., Berriman, G., and Persson, S. E., 1982, *Ap. J. Suppl.*, **50**, 241.

Alcock, C., and Anderson, N., 1985a, *Ap. J.* (*Lett.*), **291**, L29.

Alcock, C., and Anderson, N., 1985b, preprint.

Andernach, H., Schallwich, D., Sholomitski, G. B., and Wielebinski, R., 1983, *Astr. Ap.*, **124**, 326.

Arnett, W. D., 1982, *Ap. J.*, **254**, 1.

Arnett, W. D., Branch, D., and Wheeler, J. C., 1985, *Nature*, **314**, 337.

Audouze, J., 1980, in *Physical Cosmology*, R. Balian, J. Audouze, and D. N. Schramm (eds.), North-Holland, Amsterdam.

Audouze, J., and Tran Thanh Van, J. (eds.), 1983, *Galaxies and the Early Universe: Proceedings of the 18th Rencontre de Moriond, Astrophysics Meeting*, Reidel, Dordrecht.

Baade, W., 1926, *Astr. Nachr.*, **228**, 359.

Bartel, N., 1985, in *Supernovae as Distance Indicators*, N. Bartel (ed.), Springer-Verlag, Berlin.

Bean, A. J., Efstathiou, G., Ellis, R. S., Peterson, B. A., and Shanks, T., 1983, *M.N.R.A.S.*, **205**, 605.

Begelman, M. C., Blandford, R. D., and Rees, M. J., 1984, *Rev. Mod. Phys.*, **56**, 255.

Benvenuti, P., Sanz Fernandez de Cordoba, L., Wansteker, W., Macchetto, F., Palumbo, G. C., and Panagia, N., 1982, *An Atlas of UV Spectra of Supernovae*, ESA Scientific and Technical Publications, Noordwijk.

Birkinshaw, M., Gull, S. F., and Moffet, A. T., 1981, *Ap. J.* (*Lett.*), **251**, L69.

Bond, J. R., and Efstathiou, G., 1984, *Ap. J.* (*Lett.*), **285**, L45.

Borgeest, U., 1983, *Astr. Ap.*, **128**, 162.

Bothun, G. D., Aaronson, M., Schommer, B., Huchra, J., and Mould, J., 1984, *Ap. J.*, **278**, 475.

Branch, D., and Patchett, B., 1973, *M.N.R.A.S.*, **161**, 71.

Branch, D., Falk, S. W., McCall, M. L., Rybski, P., Uomoto, A. K., and Wills, B. J., 1981, *Ap. J.*, **244**, 780.

Brandenberger, R. H., 1985, *Rev. Mod. Phys.*, **57**, 1.

Carr, B. J., Bond, J. R., and Arnett, W. D., 1984, *Ap. J.*, **277**, 445.

Cadonau, R., Sandage, A., and Tammann, G. A., 1985, in *Supernovae as Distance Indicators*, N. Bartel (ed.), Springer-Verlag, Berlin.

Chaffee, F. H., Jr., Foltz, C. B., Röser, H. J., Weymann, R. J., and Latham, D. W., 1985, *Ap. J.*, **292**, 362.

Chevalier, R. A., and Fransson, C., 1985, in *Supernovae as Distance Indicators*, N. Bartel (ed.), Springer-Verlag, Berlin.

Clayton, G. C., and Martin, P. G., 1985, *Ap. J.*, **288,** 558.

Cohen, M. H., and Unwin, S. C., 1982, in *Extragalactic Radio Sources*, D. S. Heeschen and C. M. Wade (eds.), Reidel, Dordrecht.

Cooke, J. H., and Kantowski, R., 1975, *Ap. J.* (*Lett.*), **195,** L11.

Davis, M., and Peebles, P. J. E., 1983, *Ap. J.*, **267,** 465.

Faber, S. M., and Gallagher, J. S., 1979, *Ann. Rev. Astron. Ap.*, **17,** 135.

Fackler, O., 1985, preprint.

Gibbons, G. W., Hawking, S. W., and Siklos, S. T. C. (eds.), 1982, *The Very Early Universe: Proceedings of the Nuffield Workshop, Cambridge*, Cambridge University Press.

Gott, J. R., III, 1981, *Ap. J.*, **243,** 140.

Guth, A., 1981, *Phys. Rev. D*, **23,** 347.

Hellings, R. W., and Downs, G. S., 1983, *Ap. J.* (*Lett.*), **265,** L39.

Hershkowitz, S., Linder, E., and Wagoner, R. V., 1986, *Ap. J.*, **301,** 220.

Hodge, P. W., 1981, *Ann. Rev. Astr. Ap.*, **19,** 357.

Iben, I., and Renzini, A., 1984, *Phys. Rep.*, **105,** 329.

Kirshner, R. P., and Kwan, J., 1974, *Ap. J.*, **193,** 27.

Kolb, E. W., and Turner, M. S., eds., 1985, *Proceedings of the Fermilab Inner Space/Outer Space Workshop*, to be published.

Lee, B. W., and Weinberg, S., 1977, *Phys. Rev. Lett.*, **39,** 165.

Marscher, A. P., and Broderick, J. J., 1982, in *Extragalactic Radio Sources*, D. S. Heeschen and C. M. Wade (eds.), Reidel, Dordrecht.

Marscher, A. P., 1985, in *Supernovae as Distance Indicators*, N. Bartel (ed.), Springer-Verlag, Berlin.

Nandy, K., Thompson, G. I., Jamar, C., Monfils, A., and Wilson, R., 1975, *Astr. Ap.*, **44,** 195.

Öpik, E., 1922, *Ap. J.*, **55,** 406.

Peebles, P. J. E., 1980a, *The Large-Scale Structure of the Universe*, Princeton University Press, Princeton, N.J.

Peebles, P. J. E., 1980b, in *Physical Cosmology*, R. Balian, J. Audouze, and D. N. Schramm (eds.), North-Holland, Amsterdam.

Preskill, J., Wise, M. B., and Wilczek, F., 1983, *Phys. Lett.*, **120B,** 127.

Press, W. H., and Gunn, J. E., 1973, *Ap. J.*, **185,** 397.

Refsdal, S., 1966, *M.N.R.A.S.*, **132,** 101.

Rieke, G. H., and Lebofsky, M. J., 1985, *Ap. J.*, **288,** 618.

Romani, R. W., and Taylor, J. H., 1983, *Ap. J.* (*Lett.*), **265,** L35.

Rood, R. T., Bania, T. M., and Wilson, T. L., 1984, *Ap. J.*, **280,** 629.

Rowan-Robinson, M., 1985, *The Cosmological Distance Ladder*, W. H. Freeman, New York.

Rubin, V., 1985, this volume.

Sandage, A., and Tammann, G. A., 1982, *Ap. J.*, **256,** 339.

Schultz, G. V., and Wiemer, W., 1975, *Astr. Ap.*, **43,** 133.

Shaver, P., 1984, in *Quasars and Gravitational Lenses, 24th Liege Astrophysical Symposium* (in press).

Silk, J., and White, S. M., 1978, *Ap. J.* (*Lett.*), **226,** L103.

Smoot, G. F., De Amici, G., Friedman, S. D., Witebsky, C., Sironi, G., Bonelli, G., Mandolesi, N., Cortiglioni, S., Morigi, G., Partridge, R. B., Danese, L., and De Zotti, G., 1985, *Ap. J.* (*Lett.*), **291,** L23.

Spite, F., and Spite, M., 1982a, *Astr. Ap.*, **115,** 357.

Spite, F., and Spite, M., 1982b, *Nature*, **297,** 483.

Sikivie, P., 1983, *Phys. Rev. Lett.*, **51,** 1415.

Steigman, G., 1979, *Ann. Rev. Nucl. Part. Sci.*, **29,** 313.

Steigman, G., and Boesgaard, A. M., 1985, *Ann. Rev. Astr. Ap.*, **23,** 319.

Tammann, G. A., and Sandage, A., 1986, *Ap. J.* (in press).

Taylor, J., 1986, *Ap. J.* (in press).

Tully, R. B., and Fisher, J. R., 1977, *Astr. Ap.*, **54,** 661.

Turner, M., 1985, in *Proceedings of First Aspen Winter Conference*, to be published.

Uson, J. M., and Wilkinson, D. T., 1984, *Nature*, **312,** 427.

Wagoner, R. V., 1973, *Ap. J.*, **179,** 343.

Wagoner, R. V., 1980, in *Physical Cosmology*, R. Balian, J. Audouze, and D. N. Schramm (eds.), North-Holland, Amsterdam.

Wagoner, R. V., 1986, in *Proceedings of the Conference on Theory and Observational Limits in Cosmology*, Vatican Observatory, to be published.

Wagoner, R. V., and Goldsmith, D. W., 1983, *Cosmic Horizons*, W. H. Freeman and Co., New York.

Weaver, T. A., and Woosley, S. E., 1980, in *Supernova Spectra*, R. Meyerott and G. H. Gillespie (eds.), American Institute of Physics, New York.

Weinberg, S., 1972, *Gravitation and Cosmology*, Wiley, New York.

Wheeler, J. C., and Sutherland, P. G., 1985, in *Supernovae as Distance Indicators*, N. Bartel (ed.), Springer-Verlag, Berlin.

Wilkinson, D. T., et al., 1985, *Report of Subcomittee for Gravitation and Cosmology, Physics Survey*, National Research Council, Washington, D.C.

Will, C. M., 1984, *Phys. Rep.*, **113,** 345.

Woosley, S. E., Axelrod, T. S., and Weaver, T. A., 1984, in *Proceedings of the Erice Workshop on Stellar Nucleosynthesis*, C. Chiosi and A. Renzini (eds.), D. Reidel, Dordrecht.

Yang, J., Turner, M. S., Steigman, G., Shramm, D. N., and Olive, K. A. 1984, *Ap. J.*, **281,** 493.

Young, P., Gunn, J. E., Kristian, J., Oke, J. B., and Westphal, J. A., 1981, *Ap. J.*, **244,** 736.

INTERSTELLAR MATTER

8.

Theory of the Interstellar Medium

GEORGE B. FIELD

George B. Field is the Robert Wheeler Willson Professor of Applied Astronomy at Harvard University and Senior Physicist at the Smithsonian Astrophysical Observatory. Under his directorship of the Harvard-Smithsonian Center for Astrophysics, many exciting and important discoveries in both observational and theoretical astrophysics were made. He is an expert in diverse areas of theoretical astrophysics, including the physics of interstellar matter, galaxy and star formation, and cosmology.

I. INTRODUCTION

The last research paper I wrote on the interstellar medium (ISM) was on the galactic fountain model with Paul Shapiro (Shapiro and Field 1976). I also wrote a review for the 1974 Les Houches Summer School (Field 1975). I feel like Rip Van Winkle, rubbing my eyes as I look around at what has happened to the science of the ISM since 1974. To better appreciate the situation, compare Spitzer's *Physical Processes in the Interstellar Medium* (1978) with his *Diffuse Matter in Space* (1968). What happened in between was the *Copernicus* ultraviolet spectroscopic satellite telescope. I'll refer often in what follows to the major impact that instrument has had on our understanding of the ISM.

II. THE SITUATION IN 1969

The year after *Diffuse Matter in Space* was published, Don Goldsmith, Härm Habing, and I proposed a theoretical model of the ISM, based on cosmic-ray heating and the consequent division of neutral (HI) regions into a cool cloud phase and a warm intercloud medium (Field, Goldsmith, and Habing 1969). As our proposal enjoyed a brief popularity, and remnants of it survive in the modern theory of McKee and Ostriker (1977), it is useful to look back and see what we were thinking in those days. This will give us a chance to see what survived and what was wrong in what we did.

At that time we were guided by diverse sources of information: optical interstellar absorption lines of calcium, sodium, and other elements; 21-cm measurements in both emission and absorption; the then recent discovery of a soft X-ray background that seemed plausible to attribute to the ISM; some free-free absorption measurements at low radio frequencies; a lot of information on extinction by interstellar dust; and a few determinations of free electron column densities from the plasma dispersion of selected pulsar radio signals. There was also a tremendous amount of spectroscopic information on photoionized hydrogen (HII) emission nebulae. Because the theory of such nebulae as localized regions near ultraviolet-emitting stars that photoionize and heat them to about 8000 K was well developed and tested even then, they appear to pose few problems of principle now, and I shall have little more to say about them today.

The many velocity components in optical interstellar lines and 21-cm data, and the patchiness of the dust extinction, are consistent with the idea that there is a large-scale distribution of neutral hydrogen (HI) gas in the galaxy, arranged in a layer several hundred parsecs thick with a mean atomic density $n_H = 0.7 \text{ cm}^{-3}$ at the plane, and that much of this matter is gathered into "clouds" having radii of about 5 pc and densities of 20 cm^{-3} or so, and occupying only a few percent of the volume.

Spitzer had argued that because the masses of such clouds would be too small to confine them gravitationally, and that because they would expand rapidly and dissipate if they were not confined somehow, there must be an intercloud medium of some sort having about the same pressure, in which they are embedded. What could this medium be? It had to have a high temperature, to compensate for its low density. Photoionized hydrogen (HII) came to mind because the theoretically predicted temperature would be about what was needed (8000 K), but from the pressure equilibrium condition we would require $n_H(\text{ICM}) = 0.12 \text{ cm}^{-3}$ and $n_e(\text{ICM}) = 0.14 \text{ cm}^{-3}$. This is about five times the electron density indicated by the pulsar dispersions, so thoughts turned to HI, which is heated by ionizing radiation. This seems to be a contradiction in terms, because Strömgren had shown that ultraviolet from a hot star creates a sharp ionization front between the HI outside and the HII inside; any partially ionized regions would be too thin to play much of a role. But I realized (Field 1962) that low-energy cosmic rays can do the job much better. They ionize hydrogen efficiently, releasing energetic electrons, and thus heat the gas. They are penetrating, and hence can heat the gas throughout a large volume. Pikel'ner (1967) proposed that an intercloud medium heated by cosmic rays would satisfy the observations. In my 1969 paper with Goldsmith and Habing we quoted the results of our detailed study of the heating and cooling of HI, emphasizing the role of cosmic rays (Goldsmith, Habing, and Field 1969). In this study we followed the pioneering work of Pikel'ner (1967) as well as the detailed calculations of Spitzer and Tomasko (1968); Spitzer and Scott (1969) published a study parallel with ours.

The result of this work is that cosmic-ray heating produces an unusual relationship between pressure and density (Fig. 1), with a stable high-temperature phase at low density and a stable low-temperature phase at high density. I had previously shown (Field 1965) that the intermediate-temperature phase is highly unstable toward a so-called thermal condensation mode, and so it is presumed not to occur in nature. We identified the low-temperature phase with clouds and the high-temperature phase with a pervasive warm intercloud medium in pressure equilibrium with the clouds.

For an assumed hydrogen ionization rate $\zeta_H = 4 \times 10^{-16} \text{ sec}^{-1}$ we found that the intercloud medium would have $n_H = 0.21 \text{ cm}^{-3}$, $n_e = 0.016 \text{ cm}^{-3}$, $T = 7800$ K, and $p/k = 1800 \text{ cm}^{-3}$ K. The values of n_H and

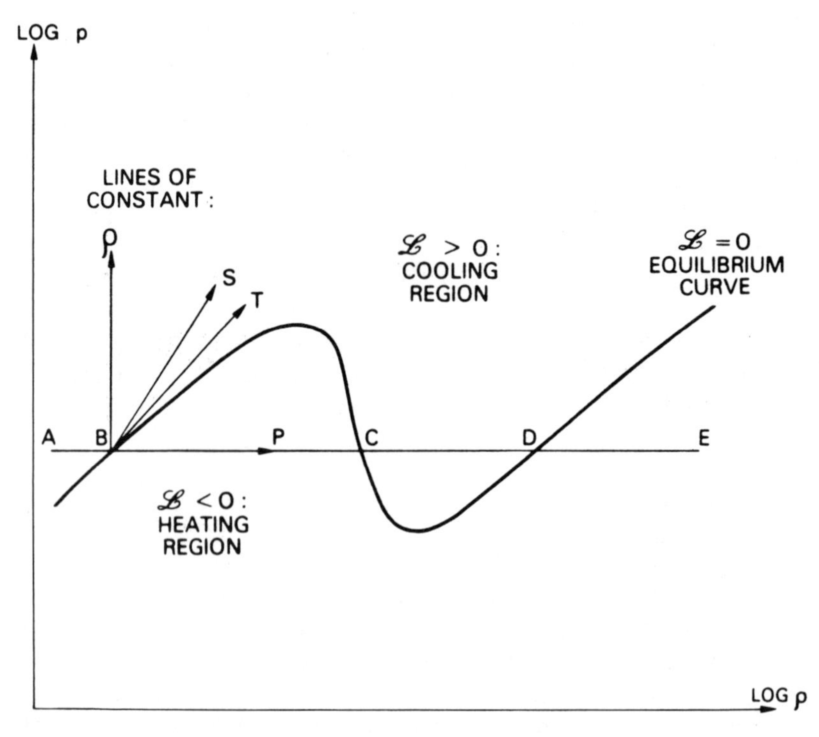

FIGURE 1. Sketch of the pressure–density relation of interstellar gas heated by cosmic rays or X-rays. The net heat loss (losses less gains) vanishes on the $\mathscr{L} = 0$ equilibrium curve. For gas placed at a point above the curve, $\mathscr{L} > 0$ and T decreases, while the opposite holds for gas at a point below the curve. For a given pressure, B and D are stable equilibria, while C is unstable. Lines of constant entropy S and temperature T are indicated.

T agreed with 21-cm observations, and n_e was in rough agreement with the value derived from pulsar dispersions, which has since been determined more precisely to be 0.025 cm^{-3} (Falgarone and Lequeux 1973).

On the other hand, the temperatures we predicted for clouds appeared to be too low—only 18 K, clearly at variance with 21-cm measurements indicating temperatures of the order of 100 K. In thinking about this discrepancy, we recalled the point made by Spitzer (1968), that atoms and ions in the gas would stick to the grains of interstellar dust, and further we noted that a particularly important coolant in the cloud phase, C^+, would be removed rapidly if the grains are negatively charged. At that time it was thought that grains would be negatively charged (Spitzer 1968). It is now believed that graphite grains are negatively charged, bare silicates are positively charged, and silicates with H_2O/NH_3 ice mantles are negatively charged (Watson and Salpeter 1972; Watson 1972). Hence C^+ would be depleted by sticking to graphite and core–mantle grains, with the result that the temperatures of diffuse clouds would be higher than otherwise. For example, if the depletion factor δ of C (defined as the gas-phase abundance relative to the cosmic abundance of C) is 0.03, cloud temperatures in the cosmic-ray heating model rise to 90 K, as required by observation.

Finally, we showed that the high value of n_e in clouds (due to cosmic-ray ionization), combined with the relatively low temperatures there, would make them strong absorbers of low-frequency radio waves. We found factor-of-2 agreement with the observations of Ellis and Hamilton (1966).

On the basis of our model we made several predictions:

1. A considerable fraction of the observed 21-cm emission originates in the intercloud medium. This seems generally consistent with the subsequent observations (Radhakrishnan et al. 1972; Falgarone and Lequeux 1973, Baker and Burton 1975; Davies and Cummings 1975; Radhakrishnan 1974; Heiles 1976, 1980), but we discuss later the implications of more recent work by the Cornell group.
2. Clouds at higher altitudes z (hence, statistically, higher latitudes b) should be warmer than those at lower latitudes. This is because clouds at higher altitudes experience lower ambient pressure, and therefore expand. Because cooling is proportional to n_H^2 and heating is proportional to n_H, T must increase to compensate. There is some indication that this effect exists; see below.
3. The temperatures of clouds should be inversely correlated with the gas-phase abundances of cooling elements. Subsequent calculations by Mészáros (1972; 1973ab) have confirmed that progressive depletion of elements in a cloud causes it to heat up and expand (Fig. 2). We return to comparison with the observations below.
4. The value of ζ_H in a given cloud should be directly obtainable by comparing its low-frequency absorption (proportional to $n_e^2 T^{-1.35}$) to its 21-cm absorption (proportional to $n_H T^{-1}$). Applying this relationship to the observations then available we could derive only an upper limit, $\zeta_H < 2.5 \times 10^{-15}$ sec^{-1}. A recent related experiment at Cornell is discussed further below.

Everything so far relates to diffuse clouds and the intercloud medium. In the same period people were working on the theory of what were then called dark clouds. These structures contain large amounts of dust, but

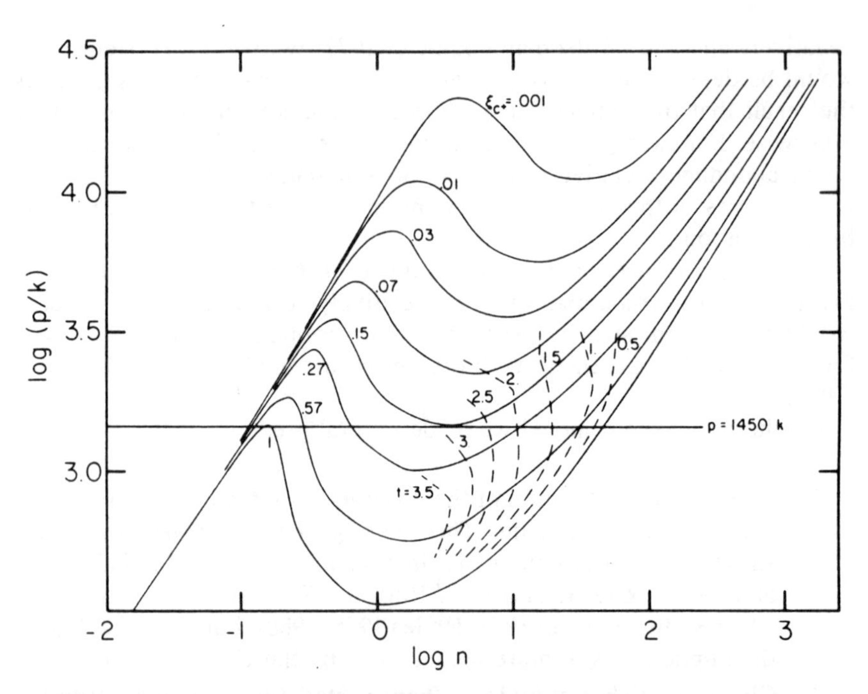

FIGURE 2. The effect of progressive depletion of cooling elements (primarily C^+) by sticking to grains, on the density of a cloud heated by cosmic rays with $\zeta_H = 1.2 \times 10^{-15}$ $\sec^{-1}$. If $p/k = 1450$ cm^{-3} K, a cloud moves from $n = 50$ cm^{-3} and $T = 30$ K to $n = 3$ cm^3 and $T = 500$ K in 3×10^7 years, at which time the depletion factor of carbon is 0.15. Elapsed times are given by the dashed lines in units of 10^7 years. (From Mészáros 1972, courtesy of the *Astrophysical Journal.*)

little HI is observed, so theorists speculated early that in such clouds HI is catalyzed on the surfaces of dust grains to form H_2 molecules. In 1963 Salpeter and his colleagues at Cornell showed that indeed this should be the case, and they predicted that large amounts of H_2 should be present in the Galaxy (Gould and Salpeter 1963; Gould, Gold, and Salpeter 1963). Later the Cornell group included the photodissociation of H_2 molecules by ultraviolet line radiation (Hollenbach, Werner, and Salpeter 1971), and showed that from 10 to 90% of the gas in the Galaxy should be molecular. This dramatic prediction explained the dominance of dark clouds and prepared the way for the explosive growth of interstellar chemistry based upon ion–molecule reactions initiated by cosmic-ray ionization of H_2.

People were also considering the role of supernova remnants as constituents of the ISM. A few remnants were known to optical astronomers, and radio astronomers kept finding new ones. By applying the Sedov–

Taylor adiabatic blast wave solution, the energies involved were estimated, and it became clear that they would have important effects on the ISM. Mansfield and Salpeter (1974) were among the first to study such effects. Spitzer thought that the energy of supernova remnants contributed to the acceleration of clouds, which were known to have speeds of 10 km sec^{-1} or more in spite of the collisional energy losses they must be experiencing. In his 1976 Russell Lecture Salpeter agreed, and applied the recent calculations to show that there is more than enough energy available.

Spitzer also probed what might be happening at distances quite far from the Galactic plane, where individual clouds had been inferred to be present from interstellar absorption-line observations. Again calling on the pressure-balance principle, he proposed that an intercloud medium must be present there as well (Spitzer 1956). But this time, to account for the large scale height above the plane, he thought in terms of temperatures of a million degrees or more for the intercloud medium. He estimated the pressure at the base of the corona to be $p/k = 1000\ cm^{-3}$ deg, not far from the modern value, simply from the requirement that it be in equilibrium with the pressure of diffuse clouds. As we shall see, Spitzer's "galactic corona" of hot gas was prophetic of the contemporary model of the ISM.

Thus the picture in 1969 was this. There is a warm intercloud medium of HI, widely distributed, in which diffuse HI clouds are embedded. Here and there are dark clouds, largely composed of H_2. Wherever hot stars flare up, the surrounding medium is ionized to HII, and wherever supernova explosions occur, blast waves propagate into the medium and heat it to very high temperatures. Hovering serenely above the plane is a crown of million-degree gas, standing by to confine any cool gas that should find its way up there.

III. THE SITUATION IN 1975

Let us jump ahead to about 1975. The results of *Copernicus* are now coming in, and the theorists are checking their scorecards. Here's how some of them came out:

1. Heavy elements were observed to be depleted, as required in the 1969 model, but $\delta(C) = 0.2$ in ζ Oph (not 0.03 as required).
2. H_2 was observed in reddened stars, in quantitative agreement with the predictions of Salpeter and his colleagues. The inferred pro-

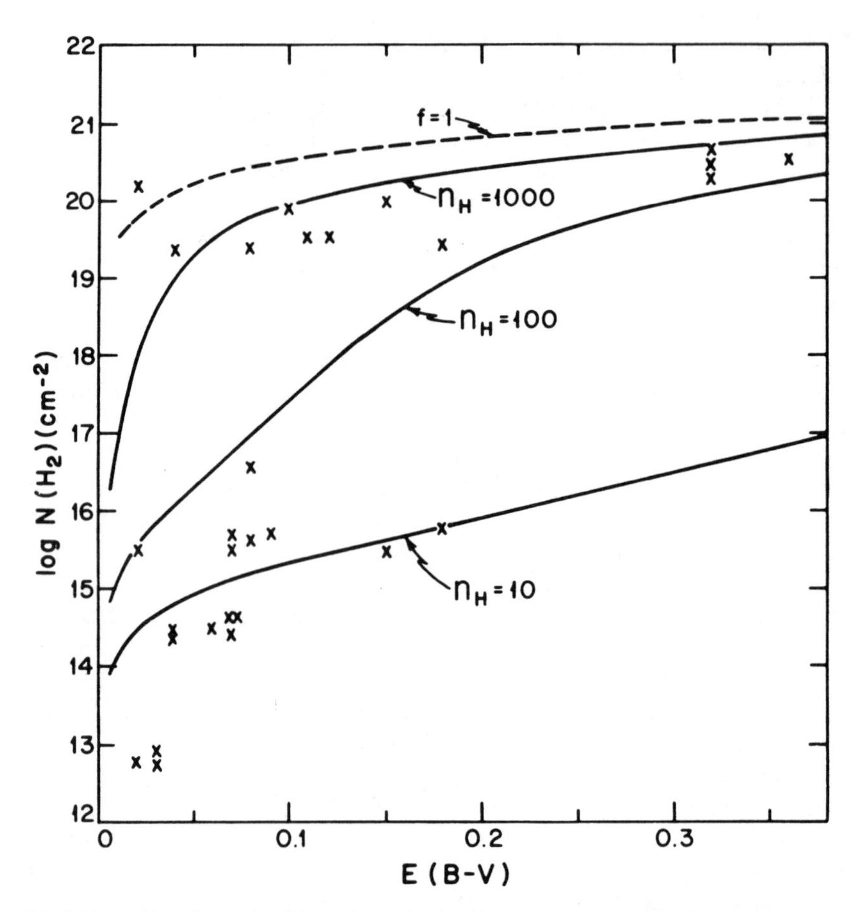

FIGURE 3. H_2 column densities $N(H_2)$ obtained by *Copernicus*, plotted against the color excess $E(B - V)$ of the star observed. As expected, $N(H_2)$ increases rapidly with $E(B - V)$ because of the shielding provided by grains against photodissociation of H_2. Since the density of the cloud where the H_2 is located is not known, the theoretical results of Hollenbach, Werner, and Salpeter (1971) are plotted for clouds of various particle densities n_H. One infers that column densities greater than 10^{20} cm^{-2} are associated with dense clouds having $n_H > 100$ cm^{-3}, while column densities less than 10^{15} cm^{-3} are associated with clouds having $n_H < 10$ cm^{-3}. (Courtesy of the *Astrophysical Journal*.)

duction rate of H confirmed the theoretical value for catalysis by grains (Fig. 3).

3. It was found that reddening correlates well with total column density $N(H) = N(HI) + 2N(H_2)$, confirming that the dust must be distributed like the gas as Spitzer always claimed (Fig. 4). This is

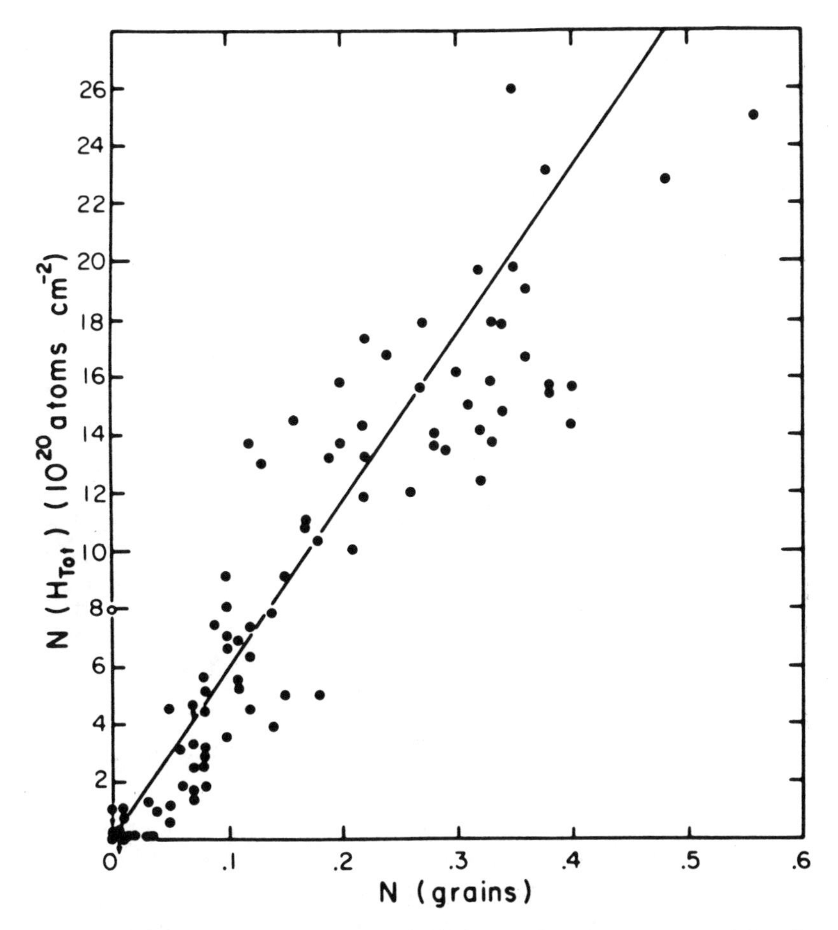

FIGURE 4. The correlation between the total column density of gas $N(H_{Tot})$ and the column density of dust grains N (grains) along the line of sight to various stars. N (grains) is measured by the degree to which the star is reddened, while $N(H_{Tot})$ is measured by the ultraviolet absorption lines of HI and H_2. (From Spitzer 1982a, courtesy of the Yale University Press.)

important because it makes it legitimate to infer that the gas is in clouds wherever the dust is.

4. OVI was discovered in the ISM and subsequently was found to be widely distributed. In his 1956 paper Spitzer predicted that the λ 1032/1038 doublet of OVI in the galactic corona could be detected. It is odd that no mention of that prediction appears in the discovery paper (Rogerson, et al. 1973); however, it is briefly referred to in

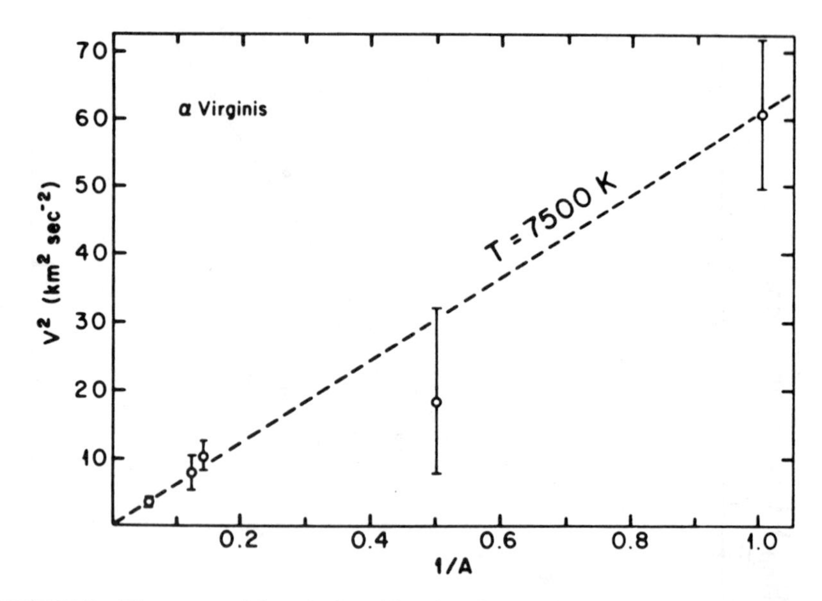

FIGURE 5. The square of the velocity width of the interstellar lines of elements of various atomic weights A in the star α Virginis. The dashed line corresponds to the results of thermal broadening for $T = 7500$ K. (From Spitzer 1982a, courtesy of the Yale University Press.)

later, more complete papers by Jenkins and Meloy (1974) and York (1974).

5. From the rotational temperatures of H_2 in diffuse clouds, kinetic temperatures of 60 to 120 K were inferred, agreeing with the estimates from saturated 21-cm profiles.
6. And now, quite recently, kinetic temperatures have been determined for gas toward unreddened stars, with 7500 K the result (York and Kinahan 1979); see Fig. 5. This hot gas appears to be HI, not HII, and tends to confirm the idea of a warm, neutral, intercloud medium.

It is astonishing the degree to which *Copernicus* seems to confirm many pre-*Copernican* theoretical ideas. Let me produce a few more agreements before I tell you about the disagreements and what they may mean.

While *Copernicus* was scanning the skies, 21-cm observers were not asleep. Using large new radio telescopes and multichannel receivers, they produced detailed surveys of the HI distribution with beamwidths of a few arcminutes and 1 km sec^{-1} resolution. They saw lots of detail but, among other things, "clouds" of HI not unlike those posited by theorists.

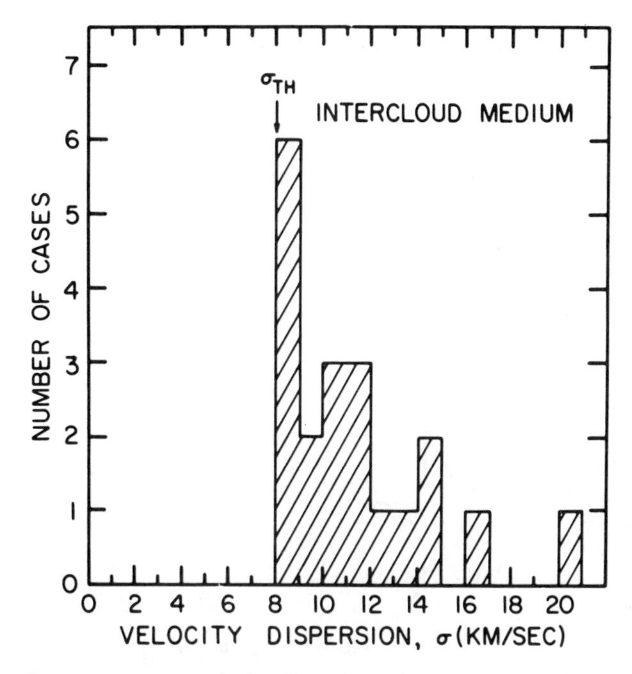

FIGURE 6. Root-mean-square velocity dispersions of the wide 21-cm emission component observed by Radhakrishnan et al. (1972) for 20 different directions. Note the absence of components with the dispersions less than 8.3 km sec^{-1}. (From Field 1973, courtesy John Wiley and Sons.)

Further, as I mentioned earlier, they found a ubiquitous intercloud medium of HI that is not absorbing, and on whose temperature they could reliably obtain only lower limits. Usually these lower limits are in the range of several hundred to 1000 K, although Davies and Cummings (1975) have determined temperatures as high as 10,000 K. In Fig. 6 is a histogram of 20 velocity dispersions of nonabsorbing (hence relatively warm) 21-cm components observed by Radhakrishnan et al. (1972). A number of components have dispersions of 8.3 km sec^{-1}; if this includes nonthermal as well as thermal motions, the temperature must be less than 8200 K. The fact that there is a sharp drop in the numbers of components narrower than 8.3 km sec^{-1} suggests that perhaps 8200 K is the actual temperature of the intercloud medium, not just an upper limit. In Fig. 7 I show the Mach numbers calculated for the nonthermal motions involved, assuming that T = 8200 K. This figure suggests that there is subsonic turbulence in the intercloud medium, a subject to which we return below.

The 21-cm observers also began to obtain evidence that the HI medium

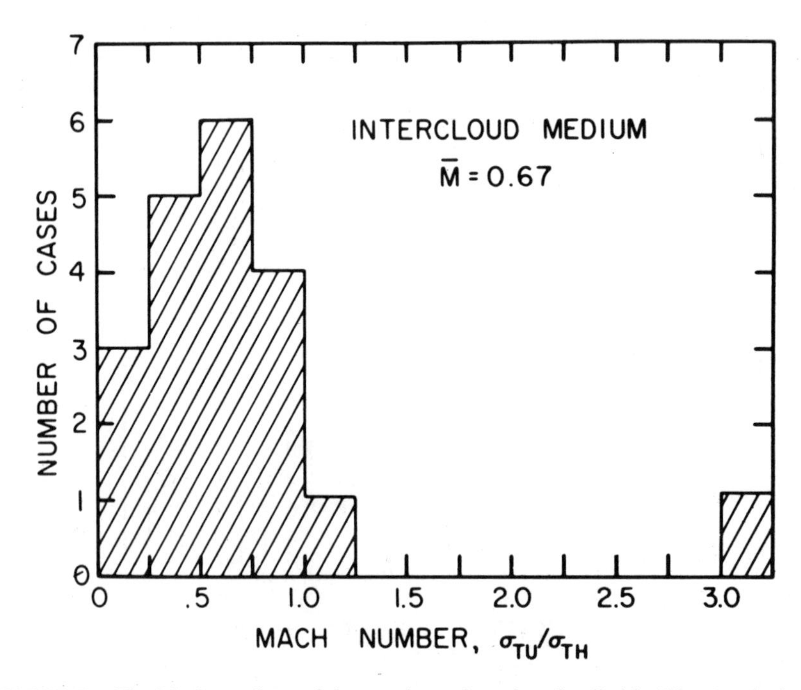

FIGURE 7. The Mach numbers of the nonthermal motions implied by Fig. 6, calculated on the assumption that $T = 8200$ K. Most of the motions are subsonic, with a mean Mach number $\overline{M} = 0.67$. (From Field 1973, courtesy John Wiley and Sons.)

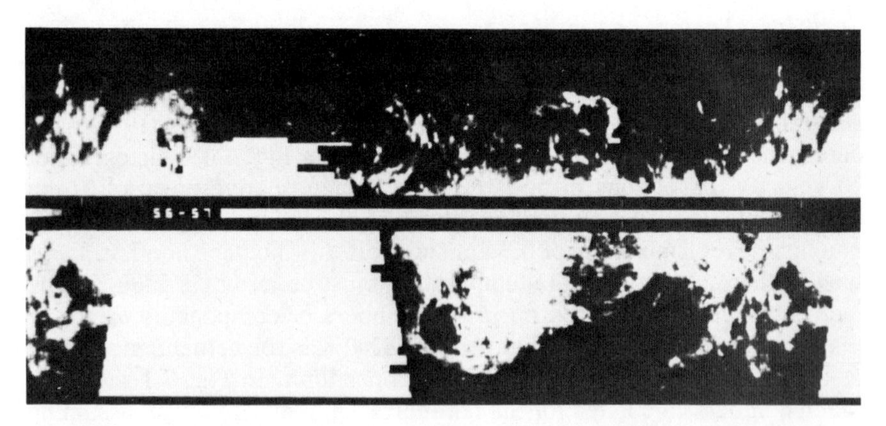

FIGURE 8. A photographic representation of the column densities of HI determined by 21-cm emission in the velocity range 3.2–6.3 km sec^{-1} (Heiles 1976). The region near the galactic plane is excluded, as is the region invisible from Hot Creek, California. (Courtesy of the *Astrophysical Journal.*).

(whether diffuse clouds or intercloud) is not distributed as simply as we postulated back in 1969. Figure 8 (Heiles 1976) is a photographic representation of 21-cm intensity in a certain velocity interval. As stated by Heiles, there are a number of HI shells visible in this figure; note particularly the one centered near $l = 198°$, $b = -40°$. Heiles gives evidence that this shell is expanding at 23 km $\sec^{-1}$ and has an energy of 4×10^{50} erg. Applying shock-wave theory, he infers an original explosive energy of 4×10^{51} erg, not unreasonable for a supernova explosion. From such observations it is evident that a quiescent steady-state model is not appropriate for the ISM.

A relatively new source of information is the survey of $H\alpha$ emission made with wide-angle Fabry Perot systems (see Reynolds 1984 for a recent summary). After taking account of the known HII regions, there remains a smooth background with an emission measure of about 3–5 cm^{-6} pc toward the galactic poles. This was not predicted in 1969. And Ellis (1982), exploiting a hole in the ionosphere above Tasmania to observe at frequencies down to 2.1 MHz, finds a smooth map of low-frequency absorption (Fig. 9), which can be explained by about the same emission measure as indicated by the $H\alpha$ measurements if one assumes $T = 8000$ K. Recall that in the 1969 model, low-frequency absorption was supposed to be due to free electrons in cool clouds; that is not necessary now that the $H\alpha$ shows that there is widespread HII. Taken together, the observations suggest that along with intercloud HI there is intercloud HII with

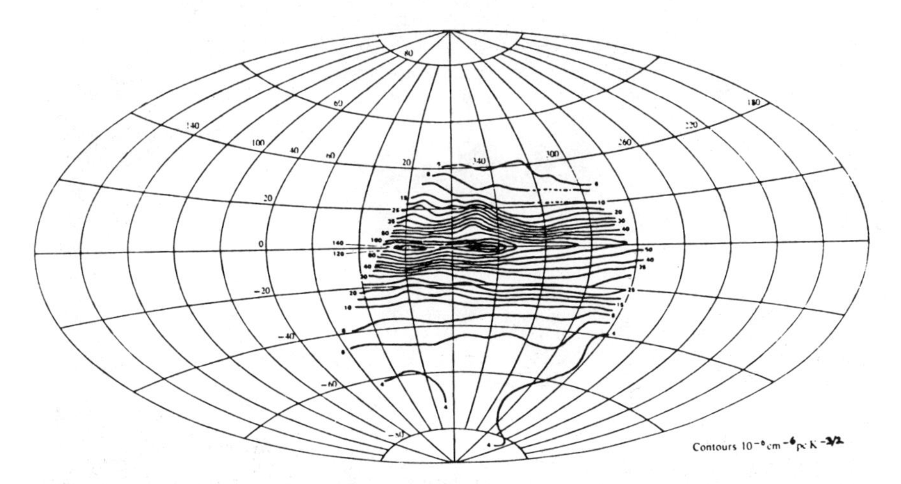

FIGURE 9. A map in galactic coordinates of the low-frequency radio absorption determined by Ellis (1982). The contours give $\int n_e^2 T^{-3/2} ds$ in units of 10^{-6} cm^{-6} $\mathrm{K}^{-3/2}$ pc. (Courtesy of the *Australian Journal of Physics*.)

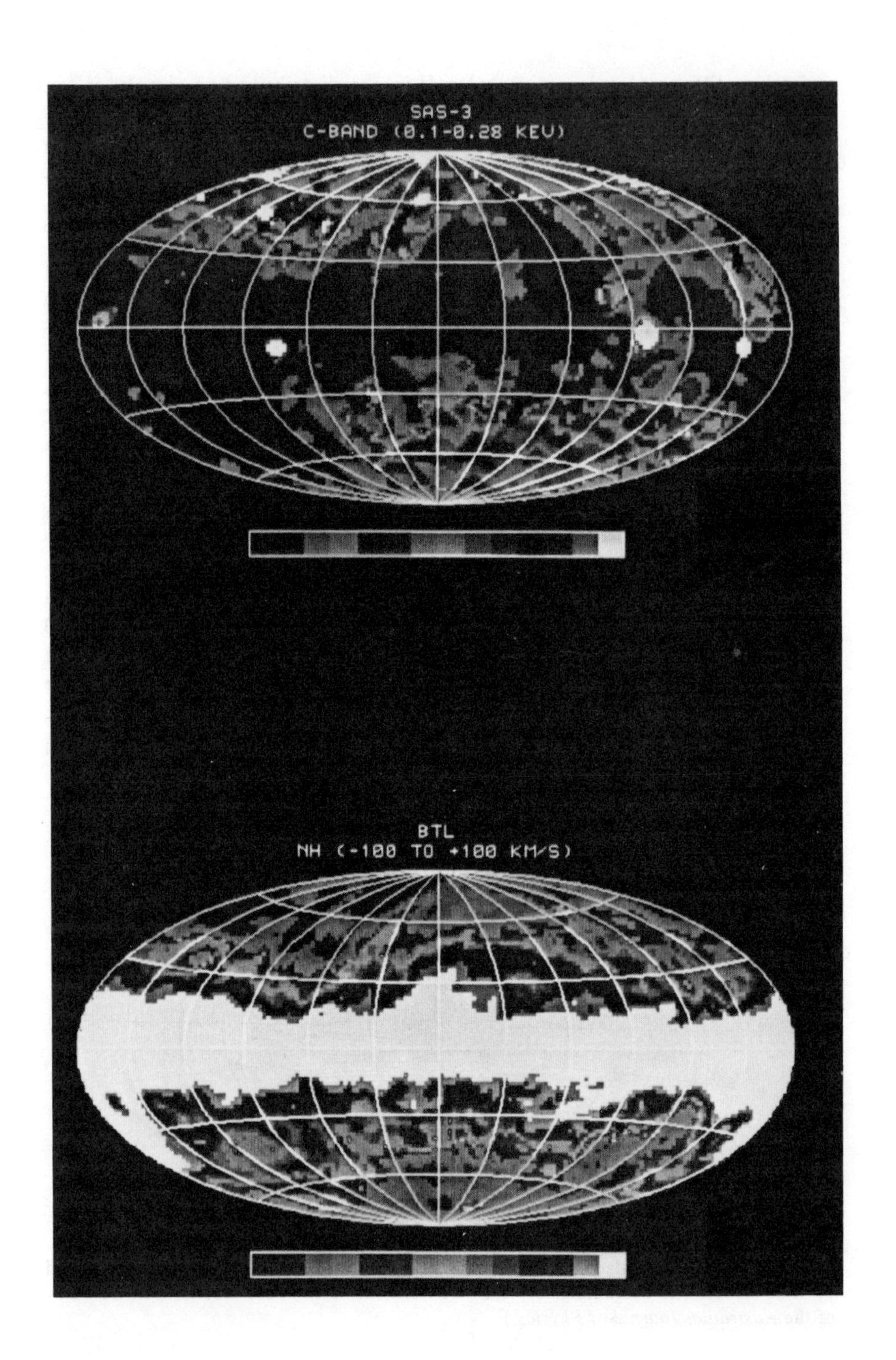
SAS-3
C-BAND (0.1-0.28 KEV)
BTL
NH (-100 TO +100 KM/S)

about the same temperature, and $n_e = 0.3\ \text{cm}^{-3}$. Only about 10% of the volume is filled by this HII, however, so this component can also explain the pulsar $\langle n_e \rangle = 0.03\ \text{cm}^{-3}$. The ionization of this material is probably due to large numbers of B stars (Spitzer 1978).

The cosmic-ray heating model ran into serious trouble when O'Donnell and Watson (1974) realized that it implies there should be free protons in diffuse clouds. Some free protons are required to charge exchange with H_2 to give H_2^+—a reactive ion-molecule that is needed to account for interstellar molecules, but too many protons violate the observations of the deuterated molecule HD, as follows. HD is formed via

$$D^+ + H_2 \rightarrow H^+ + HD + 0.039\ \text{eV} \tag{1}$$

It is destroyed by photodissociation. The charge exchange

$$D^+ + H \Leftrightarrow H^+ + D \tag{2}$$

insures that

$$\frac{n(D^+)}{n(D)} = \frac{n(H^+)}{n(H)} \tag{3}$$

Now $n(D)/n(H) = 1.8 \times 10^{-5}$ from direct *Copernicus* observations of the atomic lines. One can then calculate $n(H^+)$ from the observed ratio of $n(HD)$ to $n(H_2)$. In ζ Ophiuchi, a well-studied star, there is an HI cloud where one calculates that $n(H^+) = 3 \times 10^{-3}\ \text{cm}^{-3}$. One can then infer from the theory of ionization by cosmic rays or other penetrating radiation that the total rate at which HI is ionized is $\zeta_H = 1.6 \times 10^{-17}\ \text{sec}^{-1}$ in this cloud. This is about a factor of 25 less than that required in the original two-phase model for heating the warm ionized medium (WIM). Although the argument depends on uncertain rates for the mechanisms that neu-

←

FIGURE 10. The SAS-3 C-Band (0.1–0.28 keV) X-ray map obtained by Marshall and Clark (1984) (above), and the neutral hydrogen column density map obtained by Stark et al. (1985) and by Cleary, Heiles, and Haslam (1979) (below). Marshall and Clark interpret the apparent anticorrelation of the maps in terms of a two-component model. One component is due to hot ($3 \times 10^5 - 10^6$K) plasma located within about 200 parsecs, whose high pressure $p/k > 7000\ \text{cm}^{-3}$ K is probably due to an old supernova remnant. The other component is attributed to plasma in a galactic corona whose emission is partially absorbed by cool foreground gas. If the coronal gas is at 10^6 K, its density and pressure agree roughly with values predicted by McKee and Ostriker, and one may infer the existence of a galactic wind as explained in the text. (Courtesy of the *Astrophysical Journal.*)

tralize protons, the rates used in deriving this result are conservative. It has been concluded, therefore, that the cosmic-ray flux in clouds is lower than that required in 1969 and that something else must be responsible for heating them. (However, the final word on ζ_H may not be in; see below.)

Draine (1978) found that another mechanism, photoejection of electrons from grains, *can* heat the clouds, even if $\delta(C) > 0.03$, as observed, so cosmic rays are neither permitted nor required for heating clouds. This leaves open the heating mechanism for the intercloud medium, for it is believed that photoejection from grains is not adequate for that.

This brings us to the breakthrough in thinking that occurred in 1972. Recall that Bowyer, Field, and Mack (1968) had observed a diffuse soft X-ray background flux. Soon Kraushaar and his collaborators at Wisconsin were observing this phenomenon with a series of rocket flights sensitive to different energy ranges; a recent summary of their work is given by McCammon (1984). More recent results have been derived by Marshall and Clark (1984; Fig. 10). Cox tentatively attributed these X-rays to supernova remnants, and in a seminal paper with Barry Smith (1974) he pointed out that supernova remnants would be expected to percolate. Up to that time people had thought of individual explosions, one at a time, and had overlooked the fact that one remnant could overlap another before the latter gave out. Cox showed that the probabilities were such that this would happen, and that, consequently, channels of hot gas would form and remain open because it takes a long time for hot gas to cool.

Following Cox, most people have attributed the soft X-rays to a hot intercloud medium (HIM) created by random supernova explosions; according to Fig. 10, the sun is located within a supernova remnant. The connection with intercloud HI is as follows. Consider a diffuse cloud overrun by a supernova shock. Because the cloud density is so high, the hot postshock gas sees the cloud as an impenetrable wall, and simply reflects a shock back into the postshock flow. Now, however, the reflected shock is propagating in hot gas, and so is rather weak. Spitzer (1982b) shows that under certain conditions, one gets not a shock at all, but an acoustic wave. Hence one expects that hot tunnels are filled with a field of acoustic waves, rather like thunder after a lightning bolt, as Spitzer puts it. This acoustic field propagates into the intercloud gas, where it dissipates and suffices to keep it warm.

So at this stage we have diffuse clouds heated by electrons from grains, embedded in a neutral intercloud medium heated by SN-generated acoustic waves, intermingled with an ionized intercloud medium due to B stars and a HIM due to supernova shock heating.

IV. THE CURRENT SITUATION

Cox's realization that supernova remnants percolate focused attention on the role of supernovae in the interstellar medium. As we have already seen, Spitzer and Salpeter both concluded that they are the main source of cloud motions as well as of heating of the intercloud medium; they also appear to be the energy source for the HIM. They have other major effects as well, as developed in a series of papers by Cox, Chevalier, Cowie, McKee, Ostriker, and others. The idea is this: In its early phases, the supernova shock wave heats the ISM to X-ray temperatures. This phase can be described (with certain modifications) by a Taylor–Sedov blast wave solution. But as the shock slows down (as it must, because the energy of the supernova is being shared with a constantly increasing mass of material), there comes a point where one can no longer ignore nonadiabatic effects behind the shock front. This happens because the radiative cooling by bound–bound and free–free transitions over the time available begins to be a significant fraction of the (decreasing) internal energy of the gas. At this point the shock enters the radiative phase, and the gas immediately behind the shock cools to low temperatures, and even begins to recombine to HI. The interesting point is that because the postshock flow is subsonic with respect to the shock, there is time for approximate pressure equilibrium to be achieved behind the shock, and the cool gas is therefore compressed to high density. This provides a mechanism for regenerating diffuse clouds. In this oversimplified picture, diffuse interstellar clouds may originate as fragments of the shells of cool gas behind decaying supernova shock waves. One can readily see why the fragments are moving with considerable random velocities—it's because they were born that way. One can also picture how the hot gas behind the shock ultimately breaks free and streams around the decelerating clumps of cool gas to join up with the hot interiors of other remnants to form Cox's tunnels.

Draine and Salpeter (1979) pointed out that while the cores of interstellar dust grains are probably created in the cool gas flowing out of red giant stars, such grains can be destroyed when the clouds in which they reside are overrun by supernova shocks. This is a complicated problem because we are not sure of the chemical composition or physical state of the grains, nor do we yet have reliable temperature–time histories for shocks driven into clouds by passing supernova shocks. But the studies of Draine and Salpeter indicate that any shock front moving faster than 75 km sec^{-1} will heat the gas to the point that most materials which have been proposed for grain cores (including silicates and graphite) will be sputtered away by the postshock gas; evidence that this process occurs has been obtained in the Vela supernova remnant by Jenkins, Silk, and Wallerstein (1976). Draine and Salpeter estimate that this typically happens roughly every 10^8 years, so that interstellar dust grains must be regenerated constantly. This line of research has fascinating implications for interstellar chemistry—such as the patterns of gas-phase depletion factors in typical interstellar clouds, and the composition of the solid materials that went to form the planets and other bodies of our solar system—but there is not space here to pursue those topics.

Let me turn now to a detailed model of the ISM proposed by McKee and Ostriker (1977). This model, which takes into account the profound dynamical effects of supernovae, embraces both the hot tunnels of Cox and the cool clouds of previous authors, and seeks to derive the fractions of material to be found in the hot and cool states by imposing the constraints of mass and energy conservation, as well as approximate pressure equilibrium. It regards the ISM as in a state of statistical equilibrium, with matter constantly shifting from one state to another.

A qualitative description is as follows. It is assumed that the ISM consists of a low-density HIM in which are embedded diffuse clouds with a range of masses; each cloud has a cold core heated to the observed 80 K, presumably by the ejection of electrons from grains by near-ultraviolet radiation (Fig. 11). Each cloud core has an outer warm envelope that is largely ionized by the far UV photons available in general interstellar space from B stars; its temperature is about 8000 K, as expected for photoionized gas. Between this outer envelope of warm ionized medium (WIM) and the cold neutral medium (CNM) in the core is an intermediate layer composed of neutral gas (WNM), which has been partially ionized by soft X-rays, and thereby heated to about the same temperature (8000 K) as the WIM outside. Between the WIM and the HIM is a conductive boundary layer in which the temperature rises from about 8000 K in the WIM to about 8×10^5 K in the HIM (Fig. 12).

Supposing this to be the character of the ISM, the authors then inquire into what happens when a supernova explodes in it. As usual, the stellar ejecta travels out a few parsecs until the interstellar mass it has overtaken is comparable to its own mass; thereafter the propagation of the shock in the ISM is treated independently of all the parameters but the supernova kinetic energy E.

The shock propagates in the region of the lowest density, the HIM, which occupies most of the volume. But in contrast to previous blast wave solutions, explicit account is taken of the interaction with the clouds that are embedded in the shocked HIM. Because of the high temperature of the postshock HIM, there is a substantial thermal conduction flux into the much cooler clouds, with the result that they lose mass to the

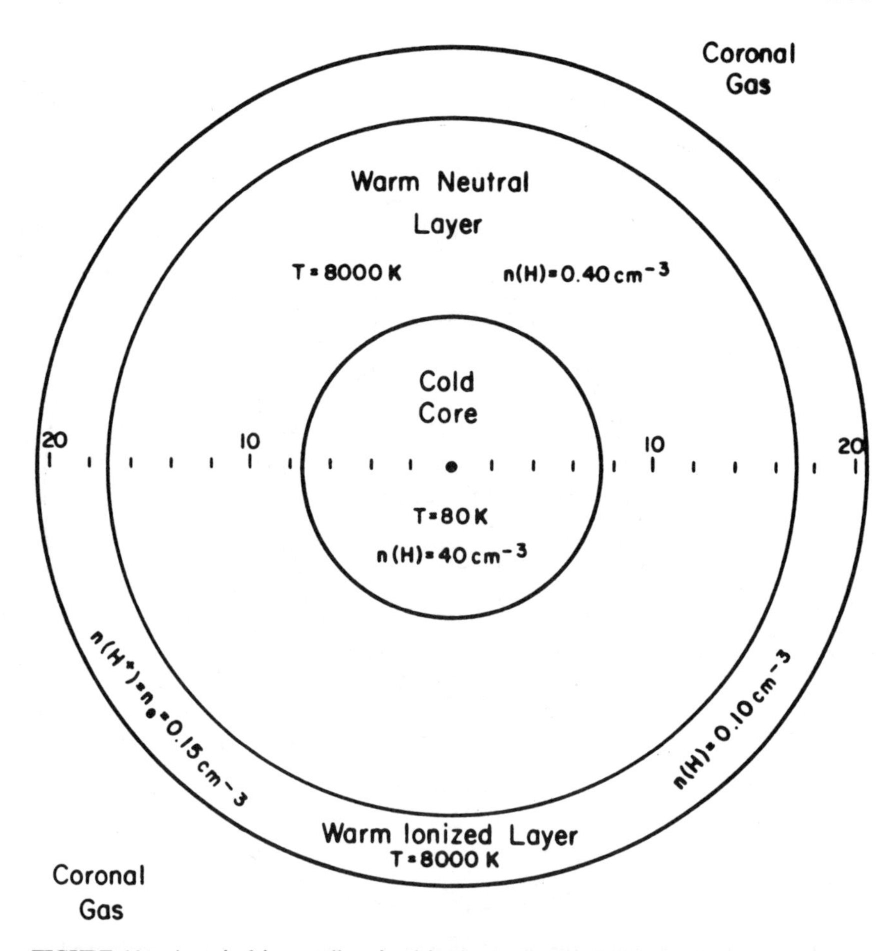

FIGURE 11. A typical interstellar cloud in the McKee–Ostriker (1977) model. The cold core (CNM) is HI, heated to 80 K by the ejection of electrons from grains. The warm neutral layer (WNM) is HI heated to 8000 K by penetrating X-rays. The warm ionized layer (WIM) is HII heated to 8000 K by the ultraviolet photons that are responsible for ionizing it. Surrounding the cloud is hot coronal gas (HIM). The density in each region adjusts so that the pressure is constant throughout. The distance scale is in light-years (about $\frac{1}{3}$ parsec), so that the core is about 2.5 parsecs in radius, and the whole cloud is about 7 parsecs in radius. (From Spitzer 1982a, courtesy of the Yale University Press.)

HIM by evaporation; this process is characterized by a parameter Σ, which depends on the properties of clouds. The mass from the clouds is distributed behind the shock when, as is inevitable, the shock slows down and the postshock gas begins to cool. According to the authors, the observed densities behind supernova shocks are in much better

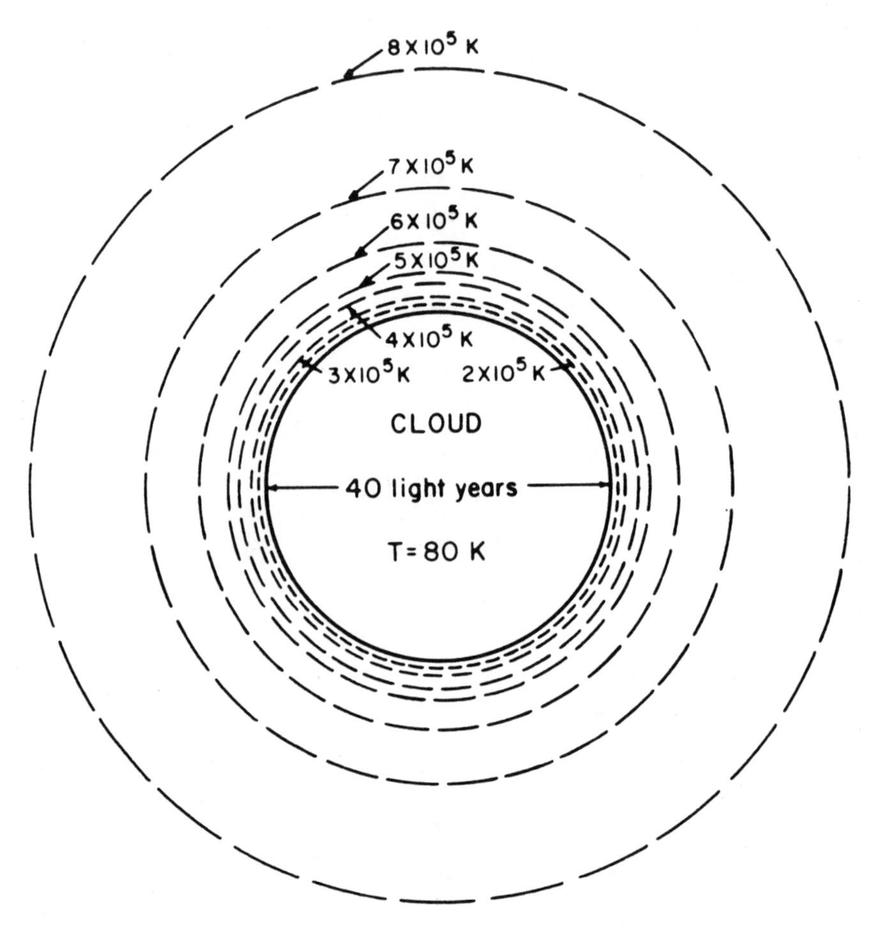

FIGURE 12. The same cloud as in Fig. 11, showing the conductive layer in the surrounding coronal gas that forms the transition from the assumed 5×10^5 K of the HIM to the 8000 K in the WIM at the outer edge of the cloud. In this figure, T (HIM) was taken to be 1×10^6 K rather than 5×10^5 K, but the idea is the same. (From Spitzer 1982a, courtesy of the Yale University Press.)

agreement with the cloud evaporation model than with adiabatic blast wave models.

After a time that depends on E and the mean density, the material begins to cool (largely at the interfaces between the hot postshock gas and the embedded cool clouds), and a thick shell forms as explained earlier; this occurs when $t \sim 10^6$ yr and $R \sim 200$ pc. The dense shell interacts with the clouds ahead of it as it sweeps over them, generally adding gas to each of them; in this way each cloud involved in this phase increases in mass, compensating as required for the loss of mass from clouds by evaporation during the earlier evolution of the shock. The existence of conductive interfaces has been confirmed observationally by Cowie et al. (1979).

The requirement of energy balance is simply that on the average the energy E has been radiated away at the moment when the shock waves from different supernovae begin to overlap. This requirement, it turns out, imposes a condition on the density of the HIM. (If the density were too high, the shocks would cool before they overlapped, and no tunnels—and hence no pervasive HIM—would form; if it were too low, the shocks would overlap before cooling, and the energy content of the HIM would increase with time, contrary to hypothesis.) A major achievement of the model is that the prediction for n(HIM) agrees with observations of the soft X-ray flux.

All of this depends on the parameter Σ, which determines the evaporation from clouds, and hence the moment when cooling begins. This is not known a priori, but depends on the properties of clouds. From the requirement that shocks overlap as they begin cooling, one finds a value for Σ, and hence a constraint on cloud parameters; in this way the model *requires* clouds and constrains their numbers and sizes.

An interesting point is that some supernova shocks will break out of the galactic disk, allowing some HIM to escape to higher altitudes, where it can form a corona of the type envisaged by Spitzer (1956). McKee and Ostriker discuss the energy balance of such a corona, or gaseous galactic halo as they call it. They estimate that the energy input E_i from supernovae is about 4×10^{40} erg sec^{-1}, while the radiative loss E_r for a corona of fixed base pressure is $10^{40}\ T_6^{-1.6}$ erg sec^{-1}; the negative temperature dependence reflects both the cooling law and the proportionality to the square of the density, which decreases with increasing temperature for fixed base pressure. Following Spitzer (1956), one may contemplate a static corona with $E_i = E_r$ in which case $T_6 = (4)^{-5/8} = 0.4$. However, such a solution is unstable, because an increase in temperature would make $E_r < E_i$, and the temperature would increase further. The instability may result in such a high temperature that the corona expands and escapes into intergalactic space, forming a galactic wind. Presumably the energy lost in such a wind, E_w, would increase with T, and hence provide a stable mechanism to dissipate E_i. McKee and Ostriker suggest that a wind could stabilize the corona with $T_6 \simeq 1$, but more detailed calculations seem necessary. In any case, the X-ray observations of Marshall and Clark (1984) are consistent with a corona; if $T_6 = 1$, the observations require that $n_e = 8 \times 10^{-4}$ cm^{-3} and $p/k = 1500$ cm^{-3} K at the base, close to

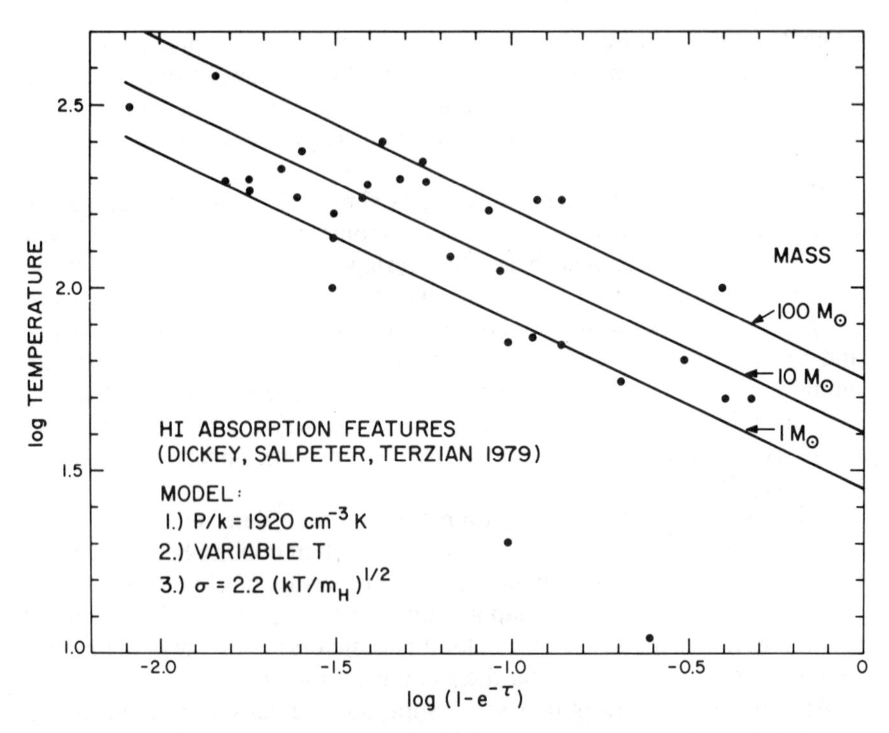

FIGURE 13. An interpretation of the correlation between the kinetic temperature of clouds thought to be the cause of 21-cm absorption features, and their optical depths, $\log(1 - e^{-\tau}) \simeq \tau$. We have plotted the observational data, together with the relation predicted by Eq. (10) for cloud masses 1, 10, and 100 $M_{\odot}$. It was assumed that p/k = constant = 1920 cm^{-3} K and that clouds of the same mass have different temperatures (hence optical depths) because of varying degrees of depletion of coolants and local heating rates.

the values predicted by McKee and Ostriker. The calculated radiative loss for the observed corona is 1.2×10^{40} erg sec^{-1}, considerably lower than E_i, so the corona may well be expanding to form a galactic wind.

The description of clouds already alluded to above assumes the existence of three phases: CNM, WNM, and WIM. McKee and Ostriker rely on observations (Hobbs 1974) to introduce a distribution of cloud masses, $f(M)\,dM \sim M^{-2}\,dM$. There is a large body of theory based on cloud–cloud collisions and agglomeration that yields $f(M)\,dM \sim M^{-1.5}\,dM$ as a first approximation (Field and Saslaw 1965; Cowie 1980). Below we show that $f(M) \sim M^{-1.8}$ follows from Fig. 13, consistent with the results of Cowie (1980). Once the temperatures of the various phases are assumed, the densities follow from the assumption of pressure equilibrium, and the filling factors follow from the parameter Σ referred to earlier.

TABLE 1. Properties of Neutral Cores and Ionized Envelopes of Clouds

Property	Cloud Cores (CNM)	Ionized Envelopes (WIM)
Temperature (K)	80 (assumed)	8000 (assumed)
Hydrogen density (cm^{-3})	42	0.25
Neutral hydrogen density (cm^{-3})	42	0.08
Electron density (cm^{-3})	Very small	0.17
Filling factor f	0.024	0.23
Number of intersections with line of sight (kpc^{-1})	11	83
Radii (pc)		
Largest	10	11
Mean	1.6	2.1
Smallest	0.4	2.1

The predictions of the model for the HIM are $p_{\rm HIM}$ (typical)$/k$ = 3700 cm^{-3} K, $n_{\rm HIM}$ (typical) = 3.5×10^{-3} cm^{-3}, $T_{\rm HIM}$ (typical) = 4.5×10^5 K, f(HIM) $\simeq$ 70%. The average values of n and T are about the same, but the average value of $p_{\rm HIM}/k$ is 7900 cm^{-3} K. Actual values of pressures in clouds agree reasonably with predictions of the model (Jenkins, Jura, and Loewenstein 1983).

The properties of the neutral cores and the ionized envelopes of the clouds are given in Table 1. The authors derived the radii of the smallest clouds from the rapid destruction of small clouds by various processes, and of the largest clouds by imposing the requirement of stability toward gravitational collapse.

In addition, if soft X-rays of about 60 eV (which are not observable at Earth because of interstellar absorption) are radiated by supernova shocks at the rate computed by Chevalier (1974), and absorbed in the outer layers of clouds, they form an intermediate envelope of warm neutral medium (WNM) at a temperature estimated by McKee and Ostriker to be 8000 K. The calculated flux of soft X-rays is sufficient to partially ionize an amount of gas equivalent to about $f_{\rm WNM} \sim 0.10 - 0.15$, with an atomic density of 0.32 cm^{-3} and an electron density of 0.06 cm^{-3}. From Fig. 1 we see that in the cosmic-ray heating case there is a maximum pressure that can be accommodated in the warm phase; the same phenomenon occurs in the X-ray heating case, where the maximum pressure is calculated to be p/k = 700 cm^{-3} K for the estimated ionization rate ($\zeta_H \sim 5 \times 10^{-16}$ sec^{-1}). As this is substantially below the ambient pressure in the McKee–Ostriker model (3700 cm^{-3} K), the question arises whether

X-ray heating can in fact sustain the WNM in the McKee–Ostriker model. The authors argue that depletion of cooling elements will raise the maximum pressure by a factor of 3, but even this falls short. They then conclude that WNM will be confined to regions of particularly low pressure, deriving a density of 0.16 cm^{-3} for it.

As we saw earlier, the heating of the WNM has been considered by Spitzer (1982b), who calculated the acoustic wave flux produced by reflection of supernova shocks from clouds and its absorption in a low-density partially ionized medium. Ikeuchi and Spitzer (1984) estimate that the energy dissipation rate in the WNM is $5 \times 10^{-27} f^{-1}$ erg cm^{-3} sec^{-1}, while Spitzer (1982b) gives the absorption length as $100x$ pc, where x is the ionization fraction in the WNM.

I wonder whether Spitzer's mechanism could be related to the velocity broadening of the intercloud HI referred to earlier. Suppose that we regard Fig. 7 as a measurement of the acoustic wave amplitude in the WNM, with the rms component of the wave velocity amplitude along the line of sight divided by the rms thermal speed of a hydrogen atom being $\overline{M} = 0.67$. Then from Spitzer (1982b) we find that the electron density in the WNM required to dissipate an acoustic wave field of this amplitude is

$$n_e = \left[\frac{43kT}{P^2 \langle v\sigma\rangle L_1}\right]^{1/2} \overline{M} = 0.025\overline{M} = 0.017 \text{ cm}^{-3} \tag{4}$$

where $T = 8000$ K is the temperature of the WNM, $P = 10^5$ years is a typical wave period, $\langle v\sigma\rangle = 4 \times 10^{-9}$ cm^3 sec^{-1} is the proton–H atom collision rate coefficient, and $L_1 = L/n_H n_e = 1.88 \times 10^{-24}$ erg cm^3 sec^{-1} is the cooling rate at 8000 K if the usual coolant ions, iron and silicon, are strongly depleted, as shown by *Copernicus*. (Note that even though the dissipation is due to "plasma slip," in which there is friction between the neutrals and the ions, which are stuck in the magnetic lines of force, this quantity is independent of the magnetic field strength.)

The value of n_e derived in this way enables us to calculate the value of the radiation loss L for any assumed value of the atomic density n_H:

$$L = L_1 n_H n_e = 3.2 \times 10^{-26} n_H \text{ erg cm}^{-3}\text{sec}^{-1} \tag{5}$$

If we set L equal to the available power (Ikeuchi and Spitzer 1984), we find that the WNM would have $f n_H = 0.15$ cm^{-3}.

There is a problem in the McKee–Ostriker model in accounting for the not-strongly-absorbing (NSA) HI observed by radio astronomers. For this component Falgarone and Lequeux (1973) obtain $\langle n_H\rangle = 0.16$ cm^{-3} at the plane and a gaussian distribution for the gas with a dispersion of 230 pc, with a column density $n_H = 1.3 \times 10^{20}$ cm^{-2} at the poles. They believe the NSA gas is pervasive; Heiles (1980) confirms this by an independent method. Payne, Salpeter, and Terzian (1983, 1984) have made great efforts to separate from their data not only absorbing gas, but also NSA gas which is possibly correlated with absorbing gas, and conclude that the column density of the remaining gas (called independent NSA, or INSA) is 1.1×10^{20} cm^{-2}. This is a lower limit to gas uncorrelated with clouds, and corresponds to a density of 0.13 cm^{-3} at the plane if the gas is distributed in a gaussian manner with $\sigma = 230$ pc. (Because of the possibility of side-lobe effects in the observations, Payne et al. note that some fraction less than half of the INSA gas could be spurious. We ignore this possibility in what follows.)

McKee and Ostriker would predict 0.16 cm^{-3} $\times$ (0.10 − 0.15) = 0.016 − 0.012 cm^{-3} for the mean density of HI in WNM envelopes, and 0.08 cm^{-3} $\times$ 0.23 = 0.018 cm^{-3} for the mean density of HI in WIM envelopes. The total, 0.03 cm^{-3}, is only 25% of the total NSA gas observed, and contrary to the observations of Payne et al. In their model all of the gas would be correlated with clouds (rather than <20%).

The observations seem to require that there is a warm intercloud medium of HI which occupies a substantial fraction of the volume; a medium with 0.3 cm^{-3} and a filling factor of ~0.5 (so that $fn_H = 0.15$ cm^{-3}; see above) would satisfy the observations and could reasonably be heated by acoustic waves as proposed by Spitzer if the amplitudes are those inferred from Fig. 7. The electron density in such a medium inferred from Eq. (4)—0.017 cm^{-3}—would correspond to $\zeta_H = \alpha n_e^2/n_H = 2.8 \times 10^{-16}$ sec^{-1} at $T = 8000$ K, and this is somewhat less than the value calculated by McKee and Ostriker from the soft X-rays expected from supernovae. The pressure in the medium would be $p/k = 2800$, higher than the earlier values (because of the smaller filling factor) but somewhat lower than the 3700 derived by McKee and Ostriker. If we include the pressure of an rms magnetic field of 3 μG (Spitzer 1978), the acoustic wave pressure indicated by Fig. 12 (which is 0.54 times the gas pressure), and the cosmic-ray pressure (which is 1.1 times the gas pressure), the predicted z distribution would be gaussian (as observed), with a dispersion of ~190 pc (compared to the observed 230 pc) if it is assumed that the various contributions to the pressure are proportional to one another at all altitudes.

Opposed to such a suggestion is the point made by McKee and Ostriker (p. 149): "The standard two-phase model . . . would rapidly self destruct ($t \sim 10^6$ years) as the relatively cool intercloud medium was swept up into dense shells and replaced by hot, low density shock-heated gas." I find this criticism convincing, but it seems to me that it would also apply to the WNM and WIM in their own model. Perhaps the envelopes of clouds are constantly swept away by shock waves, and are rapidly replaced as ionizing radiation heats and expands the outer layers of cold cloud cores. The same might be true in the alternative picture, with acoustic radiation replacing electromagnetic radiation. Note that a substantial ionization rate ($\zeta_H = 2.8 \times 10^{-16}$ sec^{-1}) is *required* in this model to provide the protons needed for ion-slip dissipation; the corresponding heating by the soft X-rays is not negligible, and will ease the situation somewhat.

V. REMARKS

As you know, Salpeter has been participating in the observation and analysis of HI absorption line experiments at Arecibo; I have already referred to some of the results of the Cornell group. I want now to return to the predictions made by Field et al. in 1969 and how the Cornell results bear on them.

I have already discussed the bearing of Cornell observations of NSA gas on the existence of a warm HI intercloud medium. In the course of the same observations, the Cornell group observed that the clouds at higher latitudes tend to have higher temperatures than those at low latitudes (Dickey, Salpeter, and Terzian 1977, 1978, 1979). We predicted this in 1969 on the grounds that at higher latitudes, the ambient pressure is reduced, and so cloud densities are lower, and, because of the quadratic dependence of cooling upon density, cloud temperatures are higher. The same effect would be expected in any theory in which the clouds are heated by a mechanism proportional to density. As this is true for heating by photoelectrons from grains as well as by cosmic-ray heating, this finding is in qualitative accord with the prediction of heating based on photoelectrons from grains.

We also predicted that clouds should have a range of temperatures if there are different depletion factors in different clouds as is expected in a statistical model based on depletion that increases with time. The calculations of Mészáros (1972, 1973ab) predict temperatures ranging from 20 to 370 K; most of the observations are in the range 50–500 K. Other contributors to variability in cloud temperatures are the relative numbers of small (50 Å) dust grains, as these are particularly effective as sources of photoelectric heating, and the numbers of nearby B stars, whose ultraviolet light is responsible for the ejection of electrons from grains (Jura 1976).

The Cornell group has discovered a relationship between the temperature of a cloud and its optical depth. Theory can explain this in a simple way as follows. We suppose that all clouds at the same altitude have about the same pressure, but there are different depletions and local heating conditions, and hence temperatures. There is also a range of cloud masses. Then, as pointed out by Liszt and Burton (1979), the optical depth τ to 21-cm H absorption varies as

$$\tau \sim \frac{na}{T\,\Delta v} \sim \left(\frac{pa}{\Delta v}\right) T^{-2} \tag{6}$$

where n is the HI number density in a cloud of radius a, T is the temperature, Δv is the velocity width, and P is the pressure. Hence if all clouds had the same radius and velocity width Δv, one would expect that $T \sim \tau^{-1/2}$; this is in fact the relationship found observationally by Dickey, Salpeter, and Terzian (1978). However, it is known that interstellar clouds have a wide range of masses (Cowie 1980). We may write the radius a in terms of the mass M and pressure p using

$$a \sim \left(\frac{M}{n}\right)^{1/3} \sim \left(\frac{MT}{p}\right)^{1/3} \tag{7}$$

and so (6) becomes

$$\tau \sim M^{1/3} p^{2/3} (\Delta v)^{-1} T^{-5/3} \tag{8}$$

If (Field 1973) we assume from the 21-cm H absorption observations that Δv is about 2.2 times the thermal width (proportional to $T^{1/2}$), we find that

$$\tau \sim M^{1/3} p^{2/3} T^{-13/6} \tag{9}$$

so that for fixed pressure

$$T \sim M^{2/13} \tau^{-6/13} = M^{0.15} \tau^{-0.46} \tag{10}$$

The dependence of T upon τ is nearly exactly equal to that observed by Payne et al. (1983) if no correction is made for INSA gas, although it is steeper than that found ($\sim\tau^{-0.34}$) after such a correction is made. However, (10) predicts a dependence on mass, albeit quite weak. In Fig. 13 we plot (10) for the (somewhat arbitrary) value $p = 1900$ k; for $p = 2800$ k, the mass labels in Fig. 13 would be reduced by a factor of 0.46. From the figure we deduce that if the model is correct, the observed negative correlation between T and τ can be explained by the fact that the dependence on mass is weak, together with the fact that most clouds have masses within the interval 0.1–10^3 $M_\odot$. (Is it significant that this includes the range of masses of stars?) We can also deduce from Fig. 13 that the volume density of clouds of mass M, per unit interval of mass M, is roughly proportional to $M^{-1.8}$, quite consistent with the observations discussed by Cowie (1980), which yield $M^{-1.8\pm0.2}$ for the same quantity. In summary, it appears that the Cornell data on 21-cm absorption by clouds can be interpreted as due to a power law distribution of cloud masses, each in equilibrium with an external medium at constant pressure. The temperatures that are observed (50–300 K) may be due to a variety of processes, including depletion of coolants or proximity to a source of energy such as low-energy cosmic rays or acoustic energy, and the fact that the model requires the cloud to respond to increased energy input by expanding and heating up so as to remain in pressure equilibrium.

Let us conclude with some comments about ζ_H. Earlier I described a method for determining ζ_H based on the chemistry of HD. Dalgarno (1984) notes that there is an alternative method which gives different results. Black and Dalgarno showed (1973) that OH is also produced at a rate which is proportional to ζ_H. Until recently, the photodissociation rate of OH, which determines its rate of removal, was very uncertain, but as a result of the work of van Dishoeck (1984), that is now no longer so. She has built detailed models of the clouds in front of ζ Oph and ζ Per, where OH has been observed, and finds from her models that the best fits are obtained when $\zeta_H = 4 \times 10^{-16}\ \text{sec}^{-1}$ and $2 \times 10^{-16}\ \text{sec}^{-1}$, respectively. As it is believed there may be dynamical phenomena in the ζ Oph cloud that have not been included in the model, she prefers the ζ Per result. What, then, of HD, which in 1974 gave a ζ_H an order of magnitude lower? Surprisingly, the HD observations also fit her model reasonably well, apparently because of revisions in the ultraviolet radiative transfer models that have occurred since O'Donnell and Watson's (1974) work. (What discrepancy there is can be resolved by using $D/H = 7.5 \times 10^{-6}$, or about 40% of the earlier values; this is not much outside the errors of observation according to Vidal-Madjar et al. 1977.)

We have said little about cosmic-ray heating here. It is clear, however, that if van Dishoeck is correct, there is substantial ionizing power in quite opaque interstellar clouds ($A_V = 0.99$ in ζ Per), which cannot be due to X-rays (as they cannot penetrate opaque clouds). Could it be due to low-energy cosmic rays, as originally proposed in 1969?

Salpeter has something to say on this also. In collaboration with Payne and Terzian (1984), he has employed a variant of the method we proposed to measure ζ_H: taking the ratio of an optical depth that depends on electron–ion collisions to that of HI. Instead of free–free absorption as we proposed in 1969 he uses free–bound transitions (high n), as proposed by Shaver (1976). No high-n recombination lines of C^+ or H^+ were seen in the spectrum of 3C 123, allowing an upper limit on ζ_H to be derived. Under one interpretation of the observations, $\zeta_H < 4 \times 10^{-17}$ sec^{-1}, far lower than van Dishoeck's result (albeit in a different region). On the other hand, another interpretation gives a much higher limit, $\zeta_H < 1.6 \times 10^{-16}$ sec, which is marginally consistent with her work. More work of this kind would be very interesting.

VI. CONCLUSIONS

There is no doubt that the two-phase model of 1969 is wrong; shock heating by supernovae is a dominant effect that has been confirmed observationally by soft X-ray emission from shock-heated gas and OVI absorption in interfaces between that gas and cooler gas. Furthermore, it seems likely that cool clouds are heated by photoelectrons from grains, not cosmic rays. And there is widespread ionization of hydrogen by B stars, which probably accounts for the pulsar dispersions, the $H\alpha$ emission, and the low-frequency free–free absorption. What survive are the tendency toward approximate pressure equilibrium, the need for substantial amounts of quite pervasive warm HI in approximate pressure balance with clouds, and the need for substantial depletion of cooling elements. Although the McKee–Ostriker model gives an account of the HIM, of the pressurization of the ISM, and of the life cycle of clouds, it has difficulty in accounting for the amount of warm intercloud HI. It appears that acoustic-wave heating, with smaller contributions from soft X-rays and perhaps even cosmic rays, can account for the temperature of the observed HI. What is needed is a better understanding of how such material gets to be widespread in a model like that of McKee and Ostriker.

REFERENCES

Baker, P. L., and Burton, W. B., 1975, *Astrophys. J.*, **198,** 281.

Black, J. H., and Dalgarno, A., 1973, *Astrophys. J.*, **184,** L101.

Bowyer, C. S., Field, G. B., and Mack, J. E., 1968, *Nature*, **217,** 32.

Chevalier, R. A., 1974, *Astrophys. J.*, **188,** 501.

Cleary, M. N., Heiles, C., and Haslam, C. G. T., 1979, *Astron. Astrophys. Suppl.*, **36,** 95.

Cowie, L. L., 1980, *Astrophys. J.*, **236,** 868.

Cowie, L. L., Jenkins, E. B., Songaila, A., and York, D. G., 1979, *Astrophys. J.*, **232,** 467.

Cox, D. P., and Smith, B. W., 1974, *Astrophys. J.*, **189,** L105.

Dalgarno, A., 1984, private communication.

Davies, R. D., and Cummings, E. R., 1975, *M.N.R.A.S.*, **179,** 95.

Dickey, J. M., Salpeter, E. E., and Terzian, Y., 1977, *Astrophys. J.*, **211,** L77.

Dickey, J. M., Salpeter, E. E., and Terzian, Y., 1978, *Astrophys. J. Suppl.*, **36,** 77.

Dickey, J. M., Salpeter E. E., and Terzian, Y., 1979, *Astrophys. J.*, **228,** 465.

Draine, B. T., 1978, *Astrophys. J. Suppl.*, **36,** 595.

Draine, B. T., and Salpeter, E. E., 1979, *Astrophys. J.*, **231,** 438.

Ellis, G. R. A., 1982, *Austr. J. Phys.*, **35,** 91.

Ellis, G. R. A., and Hamilton, P. A., 1966, *Astrophys. J.*, **146,** 78.

Falgarone, E., and Lequeux, J., 1973, *Astron. Astrophys.*, **25,** 253.

Field, G. B., 1962, "Thermal Instabilities in the Interstellar Medium," in *The Distribution and Motion of Interstellar Matter in Galaxies*, L. Woltjer (ed.), W. A. Benjamin, New York. p. 183.

Field, G. B., 1965, *Astrophys. J.*, **142,** 531.

Field, G. B., 1973, "Thermodynamic Structure," in *Molecules in the Galactic Environment*, M. A. Gordon and L.E. Synder (eds.), Wiley, New York.

Field, G. B., 1975, "Heating and Ionization of the Interstellar Medium: Star Formation," in Les Houches Session **XXVI**, *Atomic and Molecular Physics and the Interstellar Medium*, R. Balian, P. Encrenaz, and J. Lequeux (eds.), American Elsevier Publishing Co., New York, p. 469.

Field, G. B., Goldsmith, D. W., and Habing, H. J., 1969, *Astrophys. J.*, **155,** L149.

Field, G. B., and Saslaw, W. C., 1965, *Astrophys. J.*, **142,** 531.

Goldsmith, D. W., Habing, H. J., and Field, G. B., 1969, *Astrophys. J.*, **158,** 173.

Gould, R. J., Gold, T., and Salpeter, E. E., 1963, *Astrophys. J.*, **138,** 408.

Gould, R. J., and Salpeter, E. E., 1963, *Astrophys. J.*, **138,** 393.

Heiles, C., 1976, *Astrophys. J.*, **204,** 379.

Heiles, C., 1980, *Astrophys. J.*, **235,** 833.

Hobbs, L. M., 1974, *Astrophys. J.*, **191,** 395.

Hollenbach, D. J., Werner, M. W., and Salpeter, E. E., 1971, *Astrophys. J.*, **163,** 165.

Ikeuchi, S., and Spitzer, L., 1984, *Astrophys. J.*, **283,** 825.

Jenkins, E. B., and Meloy, D. A., 1974, *Astrophys. J.*, **193,** L121.

Jenkins, E. B., Jura, M., and Leowenstein, M., 1983, *Astrophys. J.*, **270,** 88.

Jenkins, E. B., Silk, J., and Wallerstein, G., 1976, *Astrophys. J. Suppl.*, **32,** 681.

Jura, M., *Astrophys. J.*, **204,** 12.

Liszt, H. S., and Burton, W. B., 1979, *Astrophys. J.*, **228,** 105.

Mansfield, V. N., and Salpeter, E. E., 1974, *Astrophys. J.*, **190,** 305.

Marshall, F. J., and Clark, G. W., 1984, *Astrophys. J.*, **287,** 633.

McCammon, D., *Proc. IAU Colloquium 81, The Local Interstellar Medium*, Madison, Wisconsin, June 4–6, 1984 (in press).

McKee, C. F., and Ostriker, J. P., 1977, *Astrophys. J.*, **218,** 148.

Mészáros, P., 1972, *Astrophys. J.*, **177,** 79.

Mészáros, P., 1973a, *Astrophys. J.*, **180,** 381.

Mészáros, P., 1973b, *Astrophys. J.*, **180,** 397.

O'Donnell, E. J., and Watson, W. D., 1974, *Astrophys. J.*, **191,** 89.

Payne, H. E., Salpeter, E. E., and Terzian, Y., 1983, *Astrophys. J.*, **272,** 540.

Payne, H. E., Salpeter, E. E., and Terzian, Y., 1984, *Astrophys. J.*, **89,** 668.

Pikel'ner, S. B., 1967, *Astron. Zh.*, **44,** 1915. (English translation *Sov. Astron. J.*, **11,** 737.)

Radhakrishnan, V., 1974, in IAU Symposium 60, *Galactic Radio Astronomy*, F. J. Kerr and S. C. Simonson III (eds.), Reidel, Dordrecht, p. 1.

Radhakrishnan, V., Murray, J. D., Lockhart, P., and Whittle, R. P. J., 1972, *Astrophys. J. Suppl.*, **24,** 49.

Reynolds, R. J., *Proc. IAU Colloquium 81, The Local Interstellar Medium*, Madison, Wisconsin, June 4–6, 1984 (in press).

Rogerson, J. B., York, D. G., Drake, J. F., Jenkins, E. B., Morton, D. C., and Spitzer, L., 1973, *Astrophys. J.*, **181,** L110.

Salpeter, E. E., 1976, *Astrophys. J.*, **206,** 673.

Shapiro, P. R., and Field, G. B., 1976, *Astrophys. J.*, **205,** 762.

Shaver, P. A., 1976, *Astron. Ap.*, **49,** 149.

Spitzer, L., 1956, *Astrophys. J.*, **124,** 20.

Spitzer, L., 1968, *Diffuse Matter in Space*, Wiley, New York.

Spitzer, L., 1978, *Physical Processes in the Interstellar Medium*, Wiley, New York.

Spitzer, L., 1982a, *Searching Between the Stars*, Yale University Press, New Haven.

Spitzer, L., 1982b, *Astrophys. J.*, **262,** 315.

Spitzer, L., and Scott, E. H., 1969, *Astrophys. J.*, **158,** 161.

Spitzer, L., and Tomasko, M. G., 1968, *Astrophys. J.*, **152,** 971.

Stark, A. A., Bally, J., Linke, R., and Heiles, C., 1985, in preparation.

van Dishoeck, E., 1984, *Photodissociation and Excitation of Interstellar Molecules: Calculations and Astrophysical Applications*, Ph.D. Thesis, University Of Leiden, Chap. **VI**.

Vidal-Madjar, A., Laurent, C., Bonnet, R. M., and York, D. G., 1977, *Astrophys. J.*, **211,** 91.

Watson, W. D., 1972, *Astrophys. J.*, **176,** 103.

Watson, W. D., and Salpeter, E. E., 1972, *Astrophys. J.*, **174,** 321.

York, D. G., 1974, *Astrophys. J.*, **193,** L127.

York, D. G., and Kinahan, B. F., 1979, *Astrophys. J.*, **228,** 127.

OBSERVATIONAL ASTRONOMY

9.

What's the Matter in Spiral Galaxies?

VERA C. RUBIN

Vera C. Rubin is a staff member of the Department of Terrestrial Magnetism at the Carnegie Institution in Washington, D.C. She has carried out leading observational work on the structure of galaxies. She was one of the first astronomers to recognize the importance of dark matter in galaxy halos.

I. INTRODUCTION

When we view the sky with our eyes, with a telescope, or with a photographic plate, we can see that the distribution of matter is clumpy. Atoms form into stars, stars into galaxies, and galaxies into clusters. Yet discoveries during the past 20 years have challenged this view of the universe and forced astronomers to examine more closely their underlying preconceptions. When we view the sky with our eyes, with a telescope, or with a photographic plate, what we actually see is that the distribution of *luminosity* is clumpy. Is there valid evidence that the distribution of optical luminosity describes the distribution of matter? The answer to this question is a resounding NO.

The answer remains NO even when we enlarge our wavelength base to include the entire range of the electromagnetic spectrum with which astronomers and physicists now routinely view the universe. Until the middle of this century, astronomical observations were made in the optical region of the spectrum, the region transmitted by the atmosphere and detected by the eye. This radiation, with a wavelength of 10^{-7} m, is characteristic of thermal radiation from stars like the sun with a surface temperature near 6000 K. In contrast, modern instrumentation is routinely used to detect radiation covering the range from 10^{-14} m to 100 m, which is emitted by a veritable zoo of astronomical objects: X-rays from quasars and diffuse interstellar gas; γ-rays from supernovae; infrared radiation from dusty galactic nuclei and the molecular cocoons that hide the births of stars; ultraviolet signatures from primordial gas clouds; radio radiation from pulsating stars and galactic gas. Yet the gravitational attraction of all the matter inferred from this radiation is not sufficient to predict the motions of stars and galaxies that we observe.

This then is the parodox I wish to discuss. We can employ Newton's law of gravitation to determine the mass of the sun from the motions of planets in the solar system, and the masses of more distant stars from the orbits of their binary companions. We can attribute masses to other stars that are consistent with the astrophysics of their radiation. But if we then infer the masses of galaxies and clusters of galaxies from the luminosities of their stellar populations, we find that the inferred mass is small; smaller by factors of order 10 or several 10s from that which we calculate by applying Newton's law directly to the observed motions of stars in galaxies and galaxies in clusters. But more significantly, the motions of stars and galaxies lead to a *distribution* of matter that differs in important respects from the distribution of luminosity. If luminosity were a fair indicator of mass, the mass of a galaxy would be concentrated near the center. Beyond the nuclear region of the galaxy, stellar orbital velocities would decrease inversely as the square root of the distance, in accord with Kepler's law, and as is observed for orbital velocities of bodies in the solar system. Instead, observations at both optical and radio wavelengths indicate that with increasing nuclear distance, orbital velocities in galaxies remain virtually constant at a high velocity. The most acceptable interpretation of this result is that a significant fraction of the mass of the galaxy consists of a dark component, distributed halolike, and extending to distances well beyond the optical galaxy. It is the gravitational response to this invisible matter that keeps the orbital velocities high.

In the following sections I briefly describe the early observations that taught astronomers about the dynamics of galaxies. Then I describe the recent results which convince us that as much as 90% of the mass of the universe is detectable only by its gravitational attraction. I conclude by listing the set of constraints that astronomical observations place on the properties of this dark matter.

II. A SELECTED HISTORY OF GALAXY DYNAMICS

Galaxies rotate. Even before astronomers knew what galaxies were, Slipher (1914a,b), working with the 24-in. reflector at Lowell Observatory, had detected inclined absorption lines in a spectrogram of the Sombrero galaxy (Fig. 1). Astronomers were familiar with inclined lines in planetary spectra. As the planet rotates, one side is carried toward the observer, one side is carried away from the observer, with a corresponding Doppler shift of the lines toward the blue and toward the red spectral regions, respectively. In the disk of a spiral galaxy, the gas, dust, and stars are all in orbit about a common center. On one side of the major axis, the stars and gas approach the observer, while on the other side, the stars recede from the observer. With two spectra of the Andromeda galaxy, Pease (1918) demonstrated convincingly that rotation was causing the observed inclined lines. The first spectrogram, exposed with the slit aligned along the major axis for 79 hours in August, September, and October 1917, revealed inclined lines; the second, exposed along the minor axis for an equivalent period in 1918, showed no velocity shift of the lines. This circumstance, a clear signature of a rotating galaxy, is illustrated in Fig. 2. Soon thereafter, Opik (1922) analyzed these observations, using Newton's law of gravitation in a procedure analogous to that which we discuss below, and determined the distance and the mass of M31. I would rank Opik's analysis as one of the major early achievements in extragalactic astronomy.

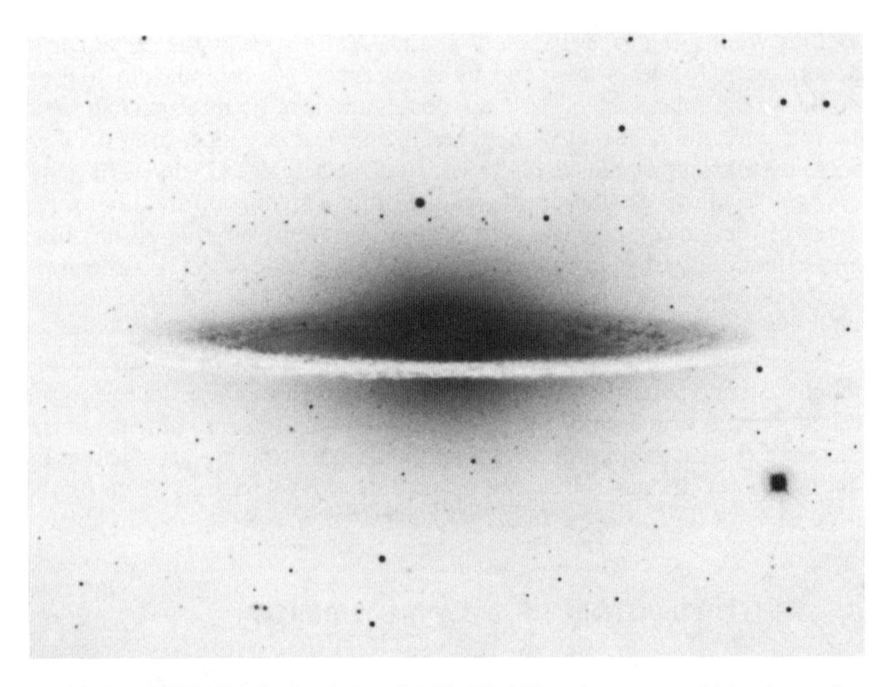

FIGURE 1. NGC 4594, the Sombrero, classified Sa^+/Sb^-. A spectrum of this galaxy, taken by V. M. Slipher (1914) at the Lowell Observatory, 1914, showed inclined lines that were correctly interpreted as a rotation of the stars about the center of the galaxy. Reproduction from a 30-min CTIO 4-m prime focus plate, baked 103a-O plate plus UG5 filter, taken by F. Schweizer.

Following this pioneering work, observations of rotational velocities of galaxies accumulated slowly. The contraints were instrumental. It is difficult to make spectroscopic observations of the velocities of individual stars, which are faint even in galaxies fairly close to our own galaxy. Progress was made during the 1960s, especially by Margaret and Geoffrey Burbidge (1975), who studied the spectra not of stars but of emission lines emitted by ionized gas clouds that orbit along with the hot young stars in the disks of spiral galaxies. However, exposure times were long, and most of the observations came from the brightest central regions of galaxies. Velocities were seen to rise rapidly from the nucleus, reach a maximum value, and then generally flatten out. Lacking observations to the contrary, astronomers assumed that just beyond the observed regions, the orbital velocities would start decreasing as is expected for objects orbiting a central mass.

With the 20/20 vision of hindsight, we know that this was an expec-

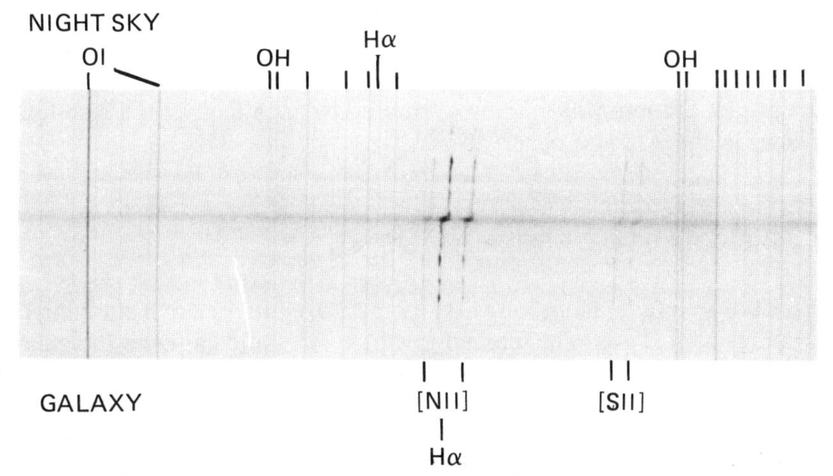

FIGURE 2. Spectrogram taken along the major axis of NGC 7541, an Sc spiral, with the KPNO 4-m spectrograph, which incorporates a Carnegie image tube. IIIa-J plate which had been baked and preflashed to enhance sensitivity; original dispersion 25A mm^{-1}, exposure 151 min. The straight vertical emission lines arise in the earth's atmosphere and are principally from OH; nightsky Hα is also present. The horizontal spectrum comes from the integrated starlight from the nuclear stars. The tilted emission lines are from the ionized gas clouds in the galaxy disk which are redshifted on the side where the rotation carries the stars away from the observer, and are blueshifted on the side where the rotation carries the stars toward the observer. The displacement of the Hα galaxy emission from the nightsky Hα emission is a measure of the redshift of the galaxy.

tation, not an observation. Yet astronomy has long been a science that is observationally limited, and important progress has often come from inferences based on observations. This time astronomers guessed wrong. In large measure this may have been because they were basing their expectations on the motions of stars in our own galaxy, which seemed to indicate that stellar orbital velocities decreased beyond the solar position. Although it is possible to trace in the literature statements questioning whether the inferred velocity decrease is actually observed, such doubts were not taken seriously until the early 1970s. By then, the voices of Schwarzschild (1954), Freeman (1970), and the 21-cm observations by Rogstad and Shostak (1972) and Roberts and Rots (1973) had to be heeded. Indeed, a rotation curve of M33, published by Louise Volders (1959) when 21-cm observations were in their infancy, shows that the velocities observed from the neutral hydrogen remain absolutely flat to large radii, in contrast to the then uncertain optical velocities which scattered enormously. The M33 rotation curve, startling for its time, seems to have made little impression on the astronomical community. Ed Salpeter and his colleagues (Krumm and Salpeter 1977) must be well aware of this phenomenon; when they started publishing in 1977 a series of flat rotation curves, the flatness was initially attributed to the effects of the side-lobes present in the Arecibo observations.

III. THE ROTATION OF SPIRAL GALAXIES

There are several reasons why astronomers are interested in studying the dynamics of galaxies. The manner in which stars and gas orbit the galaxy is a clue to the past history of the galaxy. Did the bulge and disk form from the collapse of a single gas cloud, or has an older bulge more recently acquired residual gas that has collapsed to a thin disk? Dynamics also plays a role in the chemical evolution of galaxies: as stars evolve, the heavier elements that they synthesize in their interiors are ultimately returned into the interstellar medium either by gradual mass loss or explosively in a supernova event. The degree to which this enriched material mixes throughout the galaxy is a function of the galactic orbits. Elongated orbits will enhance radial mixing. Orbital dynamics also enter into the interpretation of galaxy spiral patterns as a wave phenomenon. But at the present time the greatest interest in the dynamics of galaxies arises because orbiting stars serve as test particles in the gravitational potential of the galaxy, and hence reveal the mass distribution within the galaxy. It is the prospect of learning about the distribution of mass in galaxies that has led several groups of observers, both radio and optical, to undertake systematic studies of the rotation of galaxies.

The availability of large telescopes, of high-resolution long-slit spectrographs, and of efficient electronic imaging devices has recently made it possible to get high-resolution optical spectrograms of the faint outer regions of spiral galaxies. For the past eight years or so, my colleagues W. K. Ford, Jr., N. Thonnard, D. Burstein, B. Whitmore and I have been observing and studying the dynamics and mass distributions of spiral galaxies (Rubin, Ford, and Thonnard 1978, 1980; Rubin et al. 1982, 1985; Burstein et al. 1982). We have attempted to measure rotational velocities across the entire optical galaxy for galaxies of types Sa (disk galaxies with large central bulges, tightly wound arms, and few emission regions), Sb, and Sc (disk galaxies with small central bulges, open arms, and prominent emission knots). A sample of program Sc galaxies and spectra are shown in Fig. 3. Galaxies were chosen to be relatively isolated, to not be barred, to subtend an angular size such that the spectrograph slit could encompass the entire extent of the optical galaxy, and to be viewed at relatively high

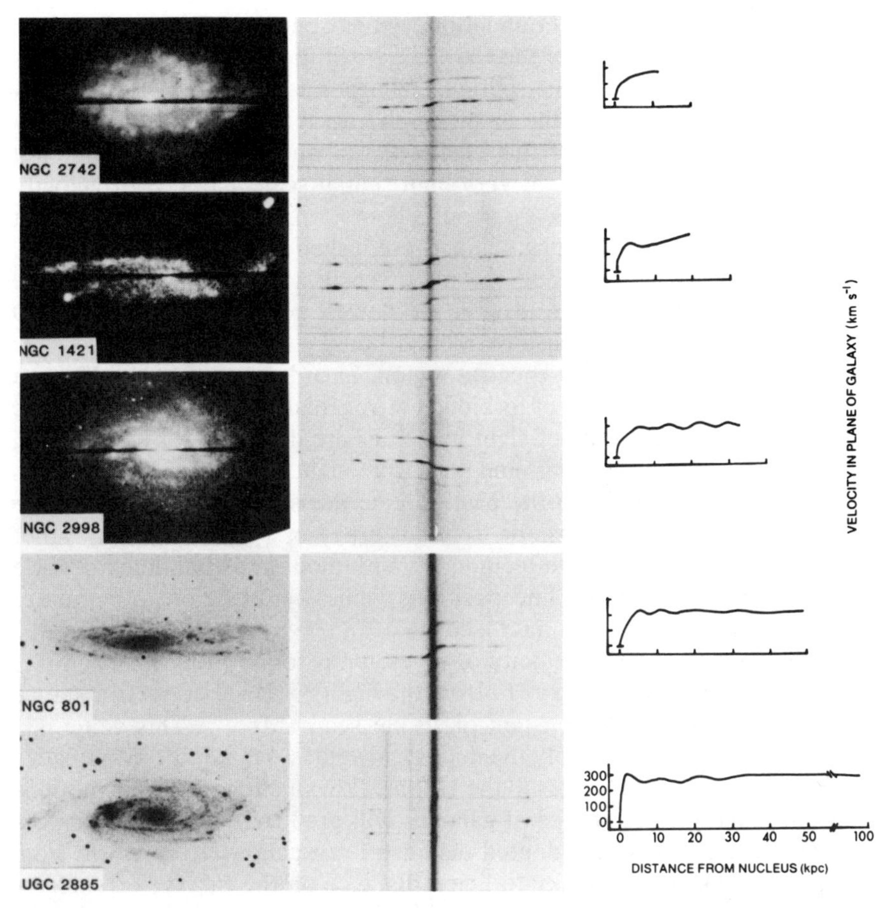

FIGURE 3. Spectrograms and rotation curves for five Sc galaxies, arranged according to increasing luminosity. The spectrograms are reproduced with wavelength increasing from bottom to top. The strong vertical line on each spectrum arises from the integrated emission from the stars in the nucleus. The strongest (horizontal) line in each spectrum is Hα, flanked by the [NII] lines. The undistorted emission lines crossing each spectrum come from OH in the earth's atmosphere. The right panel displays the rotational velocities as a function of nuclear distance.

inclination, so that a large component of the orbital velocity is along the line-of-sight to the observer.

The spectra are obtained generally with the 4-m telescopes at Kitt Peak National Observatory (near Tucson, Arizona), and Cerro Tololo Inter-American Observatory (near La Serena, Chile). Both observatories are operated by AURA, Inc., a consortium of U.S. universities, and funded by the National Science Foundation. A few spectra also come from the 2.5-m du Pont telescope of the Carnegie Institution of Washington at Las Campanas, Chile. The spectrographs incorporate an RCA C33063 ("Carnegie") image tube, and the image is photographed from the final phosphor of this tube. Use of this electronic enhancement device makes it possible to obtain spectra at very high spatial scale and at high velocity accuracy. Exposure times are about 3 hours, on Kodak IIIa-J plates that have been baked in forming gas and preflashed to enhance their speed. Without the image tube and the plate sensitizing, exposures would be about 10 times as long. And with the newest photon counting or CCD devices, exposures decrease to about 1 hour.

We observe in the red spectral region, in order to detect the Hα line (and its neighboring lines of forbidden nitrogen and sulfur) emitted by the ionized gas clouds. Line-of-sight velocities are determined by measuring the displacement of the emission lines from night sky (generally OH) lines of known wavelength. In the past, I have measured with a microscope that moves in two dimensions to an accuracy of 1 micrometer, a rather old-fashioned but accurate technique. With modern digital detectors, we now determine accurate line positions while we sit before a computer terminal, rather than at a microscope.

For each galaxy the velocity of recession is found from, for example, the separation of the galaxy Hα from the nightsky Hα. I derive a distance by adopting a Hubble constant of 50 km sec^{-1} Mpc^{-1}. This value is a scale factor and affects only the numerical values, but not the systematics of our results. Irregularities in the Hubble flow due to large-scale motions of our Galaxy or of groups of galaxies will produce only minor changes in our results. With an adopted distance for a galaxy, I transform from angular distance on the sky to linear distance on the galaxy; I transform from line-of-sight velocities to velocities in the plane of the galaxy by the geometrical factor [1/sin (inclination)]. A smooth rotation curve is formed by averaging together the velocities for the approaching and receding sides of the galaxies (Fig. 3). The photometric radius, R_{25}, is a measure of the optical galaxy; this is the radial distance where the surface brightness has fallen to 25 mag/arcsec^2, and encompasses all of the luminosity including the faint outer features generally detectable on a print of the Palomar Sky Survey.

Although an individual galaxy exhibits distinctive features in its rotational velocities, overall there are impressive systematic effects in the rotation of spiral galaxies. A few of the major conclusions follow.

1. Within a fixed Hubble type, there is a smooth progression of rotational properties from galaxies of lowest luminosity (and hence of lowest

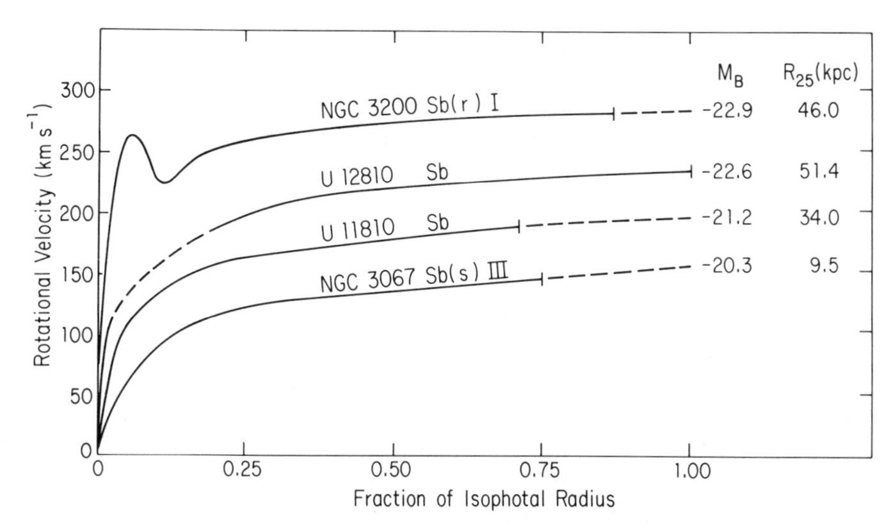

FIGURE 4. Rotation curves for four Sb galaxies plotted as a function of the fractional isophotal radius, R_{25}. This representation emphasizes the result that forms of rotation curves are similar, but amplitudes are not. With increasing luminosity, rotation curves rise rapidly at a smaller fraction of R_{25}, rise to a higher velocity, and have a longer (relative and absolute) nearly flat portion.

radius and lowest mass) to galaxies of highest luminosity. Rotation curves for galaxies of lowest luminosity have orbital velocities that increase slowly with distance from the nucleus, orbital velocities are low, and the maximum of the rotation curve is not reached until the outer radius of the galaxy. In contrast, rotation curves for galaxies of highest luminosity have steep velocity gradients near the nucleus, rotational velocities reach their maximum in a small fraction of a galaxy radius, and remain high across the entire galaxy. Hence the form and amplitude of a rotation curve are a good diagnostic of the luminosity of a galaxy, for a given Hubble type (Figs. 3, 4).

2. There is a marked similarity of form, but not of amplitude, for galaxies of different Hubble types. Maximum rotational velocities for Sa galaxies are higher by a factor of 1.7 than rotational velocities for Sc galaxies of the same (blue) luminosity, and higher by a factor of 1.3 than Sc galaxies of the same infrared luminosity. A high-luminosity Sa galaxy has a maximum rotational velocity of 375 km/sec; a low-luminosity Sa galaxy has a rotational velocity of 175 km/sec; corresponding values for Sc galaxies are 275 and 100 km/sec.

3. Virtually all the rotation curves are flat or even slightly rising out to the optical limits of the galaxies. Nowhere do we observe the Keplerian decrease in velocity with distance that is predicted if the distribution of galaxy mass follows the decrease in galaxy luminosity. The most direct interpretation of this observation is that the distribution of mass in spiral galaxies is very much less concentrated than is the light, and that this nonluminous mass extends to radial distances well beyond the optical galaxy. Some of the reasons behind these conclusions are detailed below.

IV. THE MASS DISTRIBUTION WITHIN SPIRAL GALAXIES

The determination of the mass of a disk galaxy from its observed rotation curve is a classic problem in dynamical astronomy, and valuable reviews are available (Perek 1962; Burbidge and Burbidge 1975). Analyses of the dynamics of disks have been made by various workers, and astronomers now speak of Mestel (1963) disks, Toomre (1963) disks, and Brandt (1960) disks. Unfortunately, it is not possible to determine a unique mass distribution from an observed rotation curve; only the mass distribution for an adopted model can be deduced. Generally, a disk or spheroidal model is adopted, and an integral equation is solved for the density distribution as a function of radius.

Several results follow from very simple assumptions. Let us assume that in a disk galaxy the gas and stars all rotate about the center in circular orbits, with velocities that result from the combined gravitational attraction of all matter in the galaxy. Although this is a poor assumption near the nucleus, the error in the total mass will be small. Then the equality of the gravitational and centrifugal forces on a mass m at distance r from the center is given by:

$$\frac{GM(r)m}{r^2} = k\,\frac{mV^2(r)}{r} \tag{1}$$

where $M(r)$ is the luminous plus nonluminous mass interior to r, $V(r)$ is the circular velocity at r, and G is the gravitational constant. The constant k is of order unity and depends on the geometry and mass distribution. For a sphere $k = 1$. If we adopt units such that $G = 1$ and ignore the dependence on k, we can write from (1),

$$V = \frac{[M(r)]^{1/2}}{r^{1/2}} \tag{2}$$

In several domains of interest, the solution of (2) is obvious.

1. *Central mass*. In the solar system, where essentially all the mass is in the sun, $M(r)$ is a constant beyond the sun. Hence the orbital velocities of the planets decrease as $1/r^{1/2}$. Mercury (r = 0.39 A.U.) orbits with V = 47.9 km/sec; Pluto, 100 times as distant (r = 39 A.U.) orbits with one-tenth the velocity, V = 4.7 km/sec. For spiral galaxies the flat rotation curves are convincing evidence that most of the mass is *not* restricted to the nuclear regions.

2. *Solid body*. For a body with uniform density, $M(r) \propto r^3$. Thus $V(r)$ increases linearly with r [Eq. (2)]; the object will exhibit solid body rotation. Some galaxies show approximately linear velocity curves near the nucleus, although the spatial resolution is often inadequate to define the curve accurately. Solid body rotation is rare, but not unknown, in the outer regions of disk galaxies. It may be important in barred spirals.

3. *Constant velocity*. When $V(r)$ is constant, $M(r)$ will increase with r [Eq. (2)]. This variation describes the pattern we have observed. For spiral galaxies the mass *density* $M(r)/r^3$ decreases with increasing r, but the mass interior to r increases linearly with r. The mass does not reach a limiting value at the limit of the optical luminosity. Such a mass distribution is very different from the light distribution in a galaxy.

We assume that the mass interior to r is the sum of a luminous and a nonluminous component, and that the mass is distributed in a sphere, an assumption we discuss below. The masses we determine for the galaxies in our study range from 6×10^9 to about 10^{12} solar masses. Although the local density falls as $1/r^2$, the differential volume increases as r^2, so the mass in any shell is a constant, $dM/dr \propto V^2$. In a disk galaxy the surface brightness of the luminous mass generally falls off exponentially, that is, more rapidly than $1/r^2$. Thus the ratio of dynamical mass M to luminous mass M_{lum} (i.e., to optical luminosity; both mass and luminosity counted in solar units) increases across a galaxy disk. Consequently, the ratio of the nonluminous mass to the luminous mass also increases with increasing radial distance. We present below ample evidence that the nonluminous mass extends well beyond the optical image. In such regions the value of M/M_{lum} is, of course, enormous.

We call the nonluminous matter a halo, because this is a convenient term, and because, lacking other evidence, we assume it to be spherically distributed. If the halo is not spherical, then the inferred halo mass from Eq. (1) will be smaller; for a disk $k = 2/\pi$ (Krumm and Salpeter 1977). The best evidence that the matter resides in the disk rather than in a halo comes from a study of the velocity dispersion in face-on galaxies, coupled with the observed HI scale height of edge-on galaxies (van der Kruit and Shostak 1982). Moreover, there are stability problems in placing too much mass in a disk configuration. Consideration of these stability constraints had led Ostriker and Peebles (1973) to postulate the existence of halos of relatively high mass, even before observations made such nonluminous matter mandatory.

Although the details of the dark matter remain elusive, it is remarkable that numerous constraints on its properties are imposed by the optical observations. We discuss a few of these constraints here. A few more will be added by the discussion later in this chapter.

1. *The nonluminous matter is not just part of an overall uniform background of mass distributed between the galaxies, but the matter is clumped around galaxies.* This conclusion follows because the density of nonluminous matter decreases, albeit slowly, with nuclear distance, but at large radial distances the density is still some 100 or 1000 times the mean density of the universe.

2. *The nonluminous matter is not as centrally concentrated as is the light,* as discussed earlier.

3. *The overall similarity of forms of the rotation curves for spirals of very different morphologies implies that both the dark halo matter and the luminous disk matter contribute to the total mass distribution at all radii within the optical galaxy.* Among galaxies of very different morphologies, the mass distributions must differ only by scaling factors. Figure 5 is a plot of the rotation curves for three galaxies with very different morphologies. NGC 2639, an Sa galaxy, has a bulge-to-disk luminosity ratio of 4.0; NGC 801, an Sc galaxy, has a bulge-to-disk luminosity ratio of 0.1. Yet even though these galaxies differ by a factor of 40 in the fraction of their total luminosity that arises from their bulges, the forms of their overall mass distributions are similar, as evidenced by the similar forms of their rotation curves.

4. *The fraction of dark matter is independent of galaxy luminosity and hence galaxy mass and radius.* The ratio of the dynamical mass M (i.e., luminous plus nonluminous) contained within the optical galaxy to the (blue) luminosity of the galaxy, M/L_B, is virtually constant within a Hubble type, even though the galaxies have luminosities that span a range of 100. From our observations we calculate mean values of M/L_B of 6.2 $\pm$ 0.6 for Sa's, 4.5 $\pm$ 0.4 for Sb's, and 2.6 $\pm$ 0.2 for Sc's, over a range of 100 in luminosity within each Hubble type. This result implies a close proportionality of luminous and nonluminous matter for all galaxies of a given morphology, regardless of whether the galaxy is a small, low-luminosity spiral, or a large, high-luminosity high-mass spiral. Somehow, the nonluminous matter knows how much luminous matter there is.

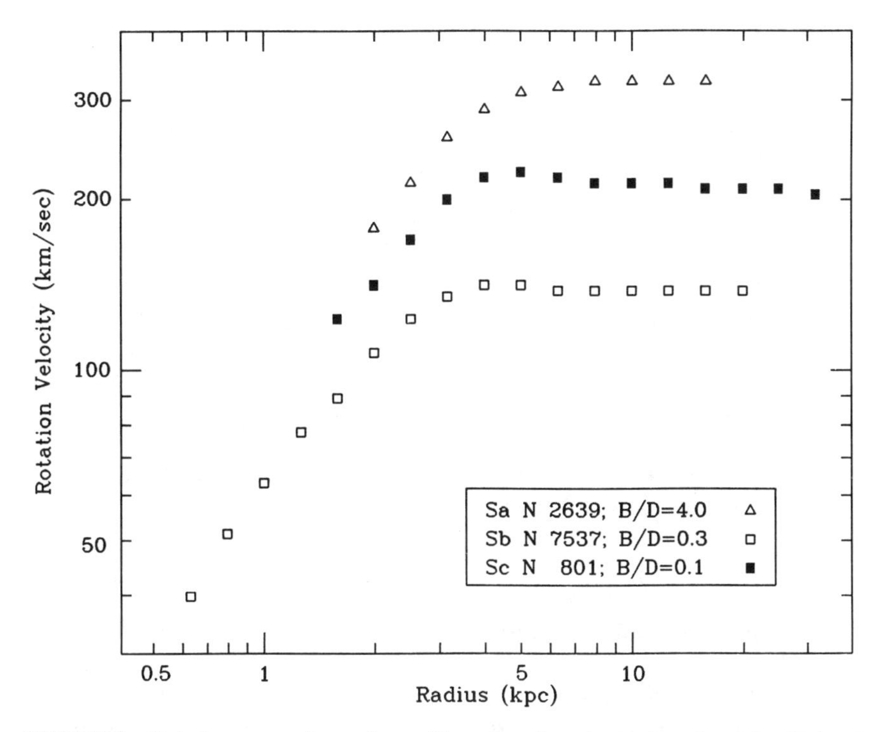

FIGURE 5. Rotation curves for an Sa, an Sb, and an Sc galaxy, plotted on a log V, log R scale to emphasize the similarity of form of the rotation curves, even for galaxies of very different morphology. For these three galaxies, the bulge-to-disk luminosity ratios differ by a factor of 40, yet the shapes of the rotation curves do not reflect these differences.

5. *The fraction of dark matter is independent of galaxy morphology.* From theoretical models of populations within galaxies, Larson and Tinsley (1978) *predict* values of M_{lum}/L_B of 3.1:2.0:1.0 for the stellar populations within galaxy Sa:Sb:Sc types. If allowance is made for the fact that Sc's contain, on average, more gas than do Sa's (e.g., Roberts 1975), then these ratios become 3.1:2.1:1.2 for the predicted luminous stars and gas in spirals. Thus the fairly close proportionality between the derived values of M/L_B (which mass contains both the luminous and the nonluminous portion) and the M_{lum}/L_B predicted (which mass contains only the luminous stars) implies that the proportion of dark matter is virtually independent of Hubble type.

6. *The dark matter contributes about half of the total mass interior to the optical radius, R_{25}.* The theoretical values of M_{lum}/L_B quoted above are *relative* values; we must exercise care in transforming them to absolute values. To do so we count up all the known matter in the solar vicinity. The most recent accounting of this "Oort limit" comes from Bahcall (1984a,b), who determines that M_{lum}/L_B in the galactic disk is about 1.5. Similar numbers come also from models of spiral galaxy stellar populations. This is an interesting but not surprising number. For a solar-type star, $M/L = 1$ (by definition, for we count in solar units); hence a disk population has an average stellar population that contains stars of mass slightly less than the sun. Thus the relative Tinsley–Larson ratios above are also good guesses of the absolute values for M/L. A comparison of these values of M_{lum}/L_B [3.1:2.1:1.2] with the values of M/L_B determined from the dynamical mass, [6.2:4.5:2.6], shows that M/M_{lum}, the ratio of total mass to luminous mass within the optical radius, is about 2 for spirals of all Hubble types and luminosities.

These six constraints on the properties of the dark matter are fairly tight, yet we can do even better in a few special cases. We discuss these below.

V. THE DISTRIBUTION OF MASS BEYOND THE OPTICAL IMAGE

How can we study the mass distribution beyond the optical galaxy? It seems tautological that it is not possible to carry out optical observations beyond the optical disk. However, occasionally nature offers us a rare chance to uncover one of its secrets. Such an opportunity arises with the polar ring galaxies, a class of galaxies which until now astronomers had known only as one of the oddities that from time-to-time show up on deep-sky exposures. Recently, François Schweizer, Bradley Whitmore, and I (1983) completed a dynamical study of one such object, A0136-0801, and it has afforded us the opportunity to draw conclusions concerning both the extent of the nonluminous matter and the form of the gravitational potential.

The spindle-shaped body of A0136-0801 is a low-luminosity disk of stars viewed nearly edge on (Fig. 6), with little or no gas and dust and no spiral structure. These characteristics, plus the distribution of its surface brightness, show that it is the type of galaxy classified S0; such galaxies represent a significant fraction of all disk galaxies. From the absorption lines of the stars, we have measured the rotation of the stars in the disk to be about 145 km sec^{-1} at an angular distance of 10″ (about 5 kpc). Along the minor axis we observe no rotational component, confirming evidence that we are viewing a rotating disk of stars. Moreover, the dispersion of velocity of stars near the center is very low, so the ratio of the rotational

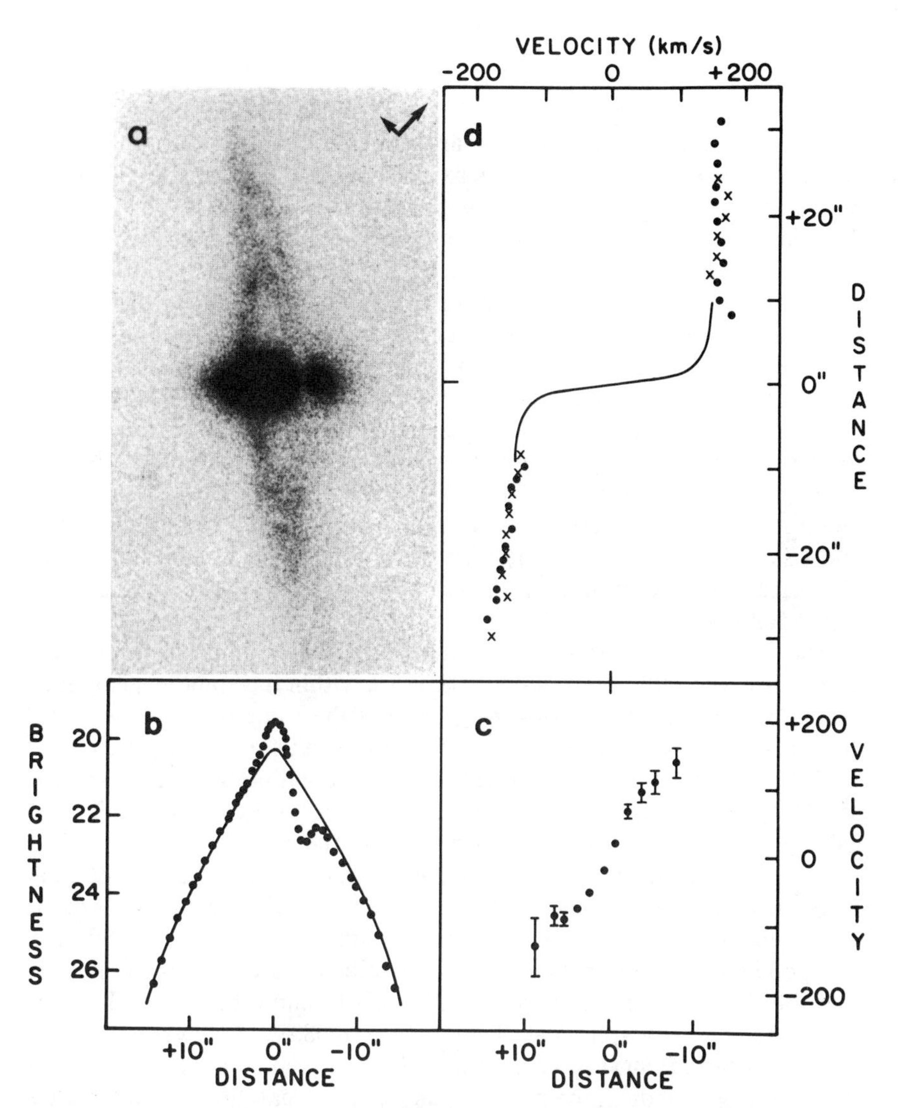

FIGURE 6. Collected data for A0136-0801. The four panels show, counterclockwise, (a) a reproduction of a 60 min 4-m CTIO blue IIIaJ + GG385 plate, taken by F. Schweizer; (b) the profile of B surface brightness (in mag arcsec^{-2}) along the major axis of the disk galaxy; (c) the rotational velocities (in km sec^{-1}) along the same axis; and (d) rotational velocities along the major axis of the polar ring (crosses and dots from two different spectra). The scales of the photograph and the distance scales of the plots are similar to facilitate intercomparison. In panel (a), the long arrow points north; the short arrow points east. The solid line in panel (d) represents schematically the rotational velocities in the disk of the main galaxy. The approximate match of these velocities at $r = \pm 10''$ with those in the polar ring suggests that the dark halo of the galaxy is more spherical than flat.

velocity to the velocity dispersion is high, another signature of a disk galaxy.

The outstanding feature of A0136-0801, and indeed all polar ring galaxies, is the large ring, also seen nearly edge on, that encircles the spindle and passes almost over the disk's axis of rotation. The ring is composed of gas, dust, and young stars. The gas reveals itself by its emission line spectrum, the dust by its absorbing effects as it passes in front of the spindle, and the stellar component by the knotty, bluish photographic appearance. The maximum extent of the ring is about 30″, or three times the extent of the disk. As a consequence, the motions of the particles in the ring offer a unique opportunity to probe the gravitational field perpendicular to the disk out to distances exceeding the radius of the disk.

Although the ring is roughly perpendicular to the main disk and extends to three times the latter's isophotal radius, it rotates just like a disk but at right angles to the plane of rotation of the spindle. The rotational velocities are approximately constant at $V = 159$ km sec^{-1} (on the NE side) and rise slowly from 135 to 170 km sec^{-1} (on the SW side) out to 31″. Nowhere does the rotation velocity show any sign of dropping off in a Keplerian fashion.

The mass of the spindle, calculated to the outer limits of its rotation curve, is 2.4×10^{10} solar masses; the mass within the spindle's isophotal radius, inferred from motions in the ring at the same distance, is 4.5×10^{10} solar masses; the mass within the outer radius of the spindle is 11.1×10^{10}. The mass to blue light ratios corresponding to these distances are $M/L_B = 9$, 14, and 25 and illustrate again the fact that the light in galaxies is more centrally concentrated than the mass.

What is the shape of this massive halo? If it is nearly spherical, circular velocities measured at a height z above the poles of the S0 galaxy will be nearly equal to those measured at $r = z$ within the plane of the S0 disk. The good match of the polar and disk velocities at around $r = 10''$ implies that the gravitational potential of A0136-0801 cannot be highly flattened and is more nearly spherical. Thus in A0136-0801, we have established that the rotation curve remains flat out to a distance about three times that of the optical disk; the mass continues to increase linearly with radial distance over this range. Moreover, the match of velocities in the disk with those in the ring permits us to establish that the dark matter is distributed in a form that is more nearly spherical than disklike.

Polar ring galaxies offer the additional opportunity to infer the past history of these curious objects. We suggest that the formation of polar rings around S0 galaxies is due to a second event such as the accretion of material from a companion during an encounter, or even a merger. Such an interaction disperses stellar material into the halo, while the gaseous material settles into a disk roughly aligned with the orbit of the former companion. Differential precession causes the disks with small tilts to settle into the plane within a few billion years. Disks with large tilts may survive for several Hubble times, only losing their central parts to settling. Although such a scenario is not certain, it predicts a frequency of polar ring galaxies that agrees with the detection frequency, and it also offers an understanding of the structure and dynamics of polar ring galaxies that is consistent with the current ideas of environmental effects on morphology of galaxies.

A second example of velocity measurements beyond the optical disk comes from the 21-cm neutral hydrogen observations of an Sc galaxy, NGC 3198. Although the optical and radio diameters of a galaxy are generally of the same order of magnitude (Bosma 1981), there exists a small group of galaxies for which the neutral hydrogen extent is several times that of the optical galaxy (Sancisi 1983). For one of these, NGC 3198, the Westerbork, Holland, synthesis radio telescope has been used by van Albada et al. (1985) to map its large-scale velocity field to a radial distance of 2.7 times its optical radius, R_{25}. In contrast to the curious polar ring galaxy described above, NGC 3198 is the prototype of a normal Sc spiral. Thus NGC 3198 is a prime candidate for an analysis of its mass distribution, within and beyond the optical galaxy.

The remarkable feature of NGC 3198 is that its rotational velocities are flat out to a radial distance of 2.7 R_{25}. Moreover, the velocity field is symmetric in all respects; velocities from the two sides differ by only a few kilometers per second, the inclination of the plane and the position angle of the major axis remain virtually constant with radial distance, and there is no evidence of a warped disk. van Albada et al. have modeled the mass distribution within NGC 3198 with two components; one component a normal disk with luminosity and mass decreasing exponentially with increasing radial distance (i.e., a disk with constant M/L); the second component a spherical halo with the density falloff consistent with a normal "de Vaucouleurs spheroid." Within these constraints, they find an infinity of models whose predicted velocities match the observations. These range from a "maximum-disk" model (shown in Fig. 7) to a "zero-disk" model in which all of the mass resides in the halo.

The characteristics of the nonluminous matter are impressive, even for that model which contains the minimum amount of dark matter. The luminous mass M_{lum} in the disk within R_{25} is 2.1×10^{11} solar masses; at this radius the amount of dark matter is equal to 0.8 M_{lum}. The mass M (luminous plus dark) interior to the final observed radius is 1×10^{11} solar masses. Here M/M_{lum} is about 5; the dark matter exceeds the luminous matter by at least a factor of 4. These figures represent the lower limit

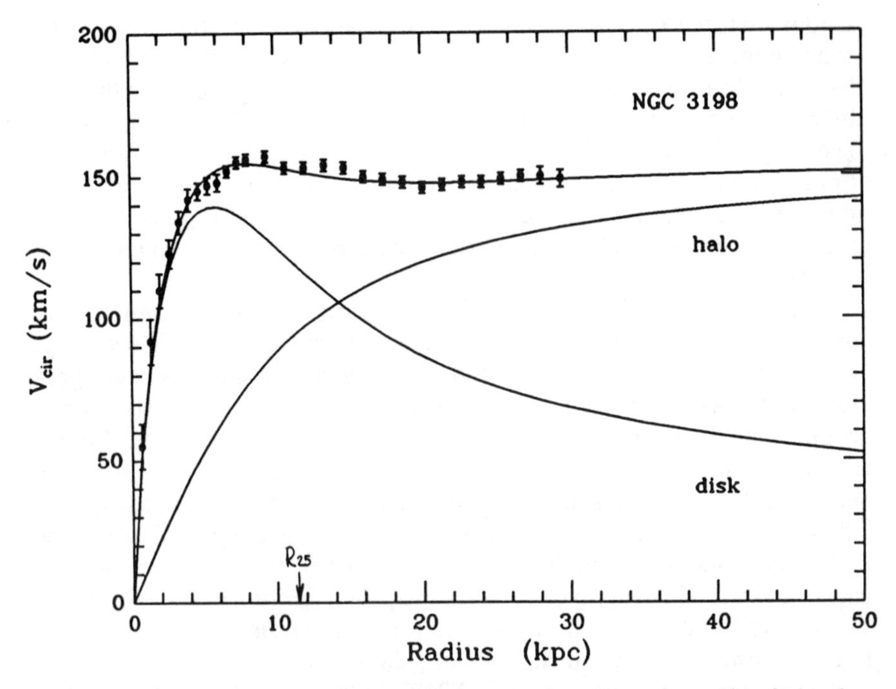

FIGURE 7. Rotational velocities for NGC 3198, an Sc galaxy, observed at 21 cm (van Albada et al. 1985). Velocities from neutral hydrogen are observed to 2.7 times the optical radius, R_{25}. A disk of constant M/L predicts the velocities shown by the "disk" curve; the velocities required to match the observations imply a halo of nonluminous matter which produces the velocities shown by the "halo" curve. The unlabeled curve plots the resultant total velocity. The model illustrated contains the maximum amount of matter permitted in the disk, a constraint imposed by the velocities at small radii. Equivalently good fits to the observations come from disks of smaller mass (and hence smaller M_{lum}/L), down to the limit of zero mass in the disk, that is, a "halo only" model.

to the dark mass which is consistent with the observations; the upper limit would place virtually all of the galaxy mass in the dark halo.

These studies of NGC 3198 and the polar ring galaxy support the following conclusions. In cases where it is possible to determine the mass distribution within a galaxy, of the order of half the mass within the optical galaxy is nonluminous and is not as centrally condensed as the luminous material. And in these few special cases the radial extent of the nonluminous mass is at least a few times the optical radius; the galaxy mass continues to increase linearly with radius out to this distance.

FIGURE 8. A view through the plane of our Galaxy, looking toward the galactic center, showing the distribution of star clouds and dust in the Milky Way. Viewed this way, our Galaxy shows a remarkable resemblance to external galaxies. Photo by D. L. Talent taken with the 0.6 Schmidt telescope of CTIO; Copyright Association of Universities for Research in Astronomy, Inc.

VI. THE DISTRIBUTION OF MASS WITHIN OUR GALAXY

For many research programs observations of our Galaxy are complementary to those of external galaxies. Some observations are easier to make in distant galaxies, as we look in directions away from the gas and dust layer in our Galaxy; it is difficult to study distant stars within our Galaxy if they are located in the plane (Fig. 8). However, some observations can be made most readily within our Galaxy. One such program is to deduce the mass of the galactic disk by its attraction on the stars at high galactic latitudes. From such a study, Oort (1960) pointed out that the acceleration of stars perpendicular to the plane was too high for the amount of material that could be counted in stars and estimated in atomic and molecular gas and dust.

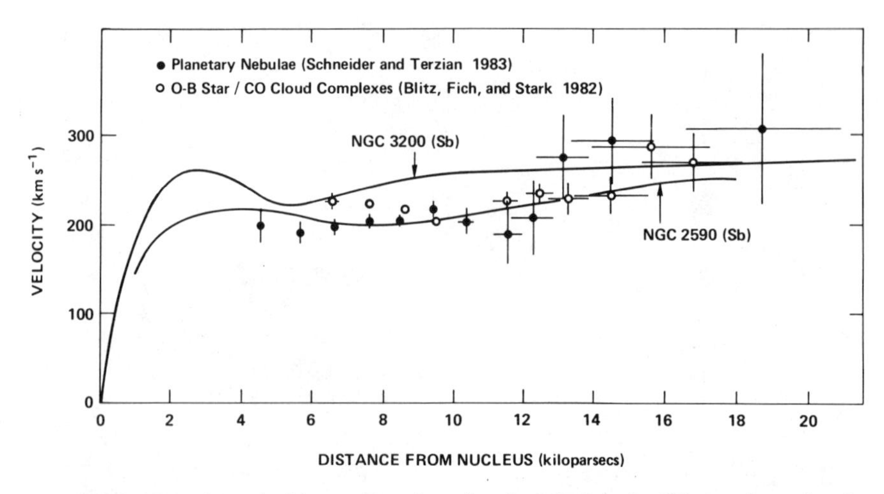

FIGURE 9. Rotation velocities as a function of nuclear distance for OB star plus molecular CO complexes in our Galaxy (Blitz et al. 1982), and for planetary nebulae (Schneider and Terzian 1983). Also shown are the rotation curves for two Sb galaxies.

This problem of the "Oort limit" remains with us today. A recent analysis by Bahcall (1984b) of the number density and motions of F and K stars above the galactic disk in the solar vicinity indicates that the gravitational response of the stars implies a disk mass equal to a mass about two times that which is counted in stars, gas, and dust. This, too, is nonluminous matter, but it resides in the disk rather than in a spherical distribution. There is yet no clear understanding of the relation of this dark disk matter to the dark halo matter.

The discovery of nonluminous matter in galaxies has forced us to revise our picture of our Galaxy (e.g., Bok 1981). While we once believed that we lived near the outside of a moderate-sized galaxy of moderate mass, recent observations by Blitz et al. (1982) of CO clouds and by Schneider and Terzian (1983) of planetary nebula have shown that the Galaxy extends well beyond the solar circle (Fig. 9), and that rotation velocities continue to rise for nuclear distances greater than the sun's. Analyses of the motions of the outlying globular clusters (Hartwick and Sargent 1978) and the dwarf satellites (Murai and Fujimoto 1980; Einasto et al. 1975; Lynden-Bell 1982) that orbit our Galaxy indicate that these tracers feel a gravitational potential equivalent to a mass distribution that continues rising beyond the solar circle. The "radius" of the galactic mass distribution is unknown, but models in which the mass extends to 10 times the nucleus-sun distance are not incompatible with the observations.

In astronomy, as in other sciences, bandwagon effects are hard to resist. Flat rotation curves and extended mass distributions are in such vogue that it may be difficult to notice discrepancies. But in at least two specific situations, falling rotation curves are seen. In spirals with extended HI gas disks, the disks are often warped (much like the flipped brim of a hat) beyond the optical galaxy. Sancisi (1983) has pointed out that velocities often decrease, perhaps 20 km sec^{-1}, at the start of the warp, and then flatten out at the lower level. The significance of this is not yet understood. And in tidally interacting galaxies, the complex gravitational field generally produces falling rotation curves. This effect, now established in a few well-studied cases, may make falling rotation curves a good diagnostic of a tidal interaction.

VII. NONLUMINOUS MATTER IN ELLIPTICAL GALAXIES

The rotational properties of elliptical galaxies are less well known than those of spiral galaxies, principally because the observations are more difficult. Elliptical galaxies lack significant quantities of ionized gas, or other relatively easy observational tracers of the gravitational potential. Therefore only absorption lines from the integrated stellar populations are available from which to deduce the line-of-sight velocities and the velocity dispersion. All of the stars along the line-of-sight through the elliptical galaxy contribute to the observed absorption line, and the viewing geometry is less certain than that in a disk galaxy. Until recently, it was believed that elliptical galaxies were rotationally supported. It was therefore surprising when observations by Bertola and Capaccioli (1975) and especially Illingworth (1977) showed that high-luminosity elliptical galaxies rotate only very slowly.

Rotational velocities are low, but rotation curves are flat. In general, however, velocities are known only for the bright inner regions. Hence knowledge of the large-scale distribution of nonluminous matter in elliptical galaxies has been indirect. A few very special ellipticals in cores of dense clusters of galaxies were bright enough so that velocity dispersions could be measured at large radial distances (Dressler 1979). For at least one elliptical, M87 (Fabricant et al. 1980), the radial variation of the X-ray gas temperature and density was used to determine the total gravitational mass. In these early studies, enormous masses and M/L values of order 100 resulted.

In a recent study, Forman, Jones, and Tucker (1984) have observed X-ray emission from a sample of 55 elliptical, S0, and Sa galaxies; they report the significant result that all have hot gaseous coronae. The mass of gas in a corona is of order 10^9 to 10^{10} solar masses, about 1% of the galaxy mass. Thus although this hot gas is not itself the nonluminous matter, it provides a unique tracer of the gravitational potential in the outer regions beyond the optical galaxy. Analysis shows that galaxy masses of up to 5×10^{12} solar masses are necessary to bind the hot gas to the galaxy; such masses are up to 100 times the mass that would be inferred from the luminous material. This result is an important one, and additional studies of nonluminous matter in elliptical galaxies and in dwarf galaxies may help to place even tighter constraints on the properties of nonluminous matter in spirals.

VIII. BINARY GALAXIES AND CLUSTERS OF GALAXIES

Galaxies congregate in clusters, and clusters cluster. Our Galaxy and its two orbiting satellites, the Magellanic clouds, reside in a loose group of about 20 or 30 galaxies, most of them small low luminosity dwarfs. This Local Group is an outlying member of the Virgo cluster, which, in turn, forms the bulk of the Virgo supercluster. All of our discussions above have related to the discovery over the past decade that nonluminous matter is associated with individual galaxies. Historically, however, the presence of nonluminous matter was inferred long ago by Zwicky (1933) and by Sinclair Smith (1936) from the motions of galaxies in clusters (Fig. 10).

As early as 1933, a sufficient number of radial velocities were available for galaxies in the Coma cluster to make an analysis of the cluster dynamics. Zwicky noted that the individual galaxies are moving so rapidly that their mutual gravitational attraction (calculated from their luminous mass) is insufficient to hold them together. If clusters are not flying apart (and the evidence is that they are not), then there must be enough dark matter present to hold them together. Zwicky deserves credit for being the first to uncover evidence for the existence of nonluminous matter. Astronomers were initially prone to think that this "missing mass" was an exotic property of clusters, unrelated to the more isolated galaxies. The importance of the recent observational work is that it demonstrates that nonluminous matter is a property of single galaxies as well.

Binary galaxies too can be used to probe the gravitational potential within their mutual orbits. However, the conventional analysis is complex because many of the dynamical parameters such as inclinations and shapes of the orbits can only be handled statistically for large numbers of pairs (White 1981; L. Schweizer 1984). Such studies typically find values of M/L in the 30–60 range, interior to the binary orbit. These high values offer supporting evidence that mass continues to increase beyond the optical galaxy.

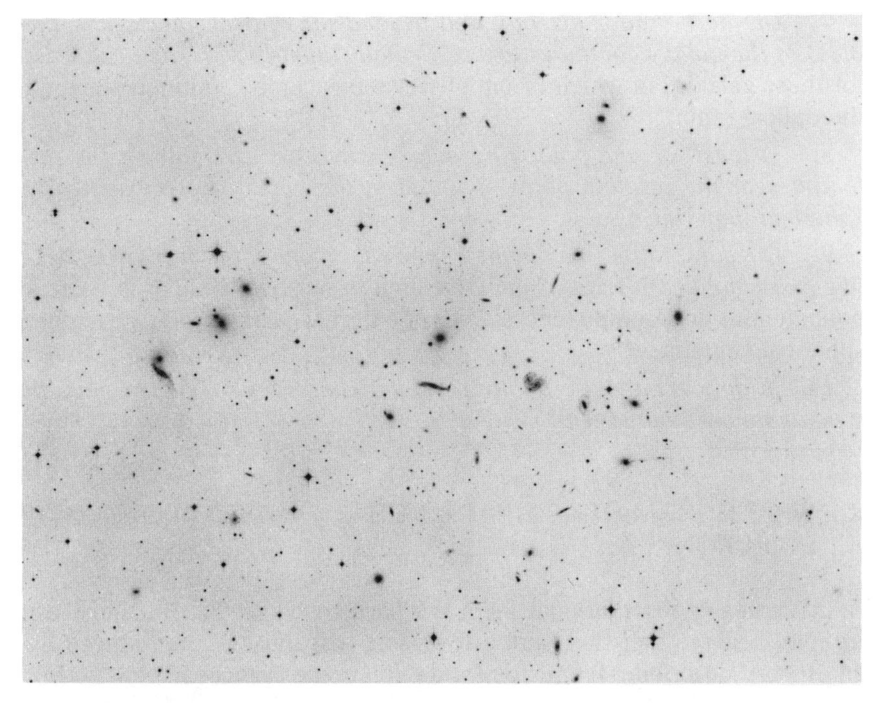

FIGURE 10. The Hercules cluster of galaxies, photographed with the du Pont 2.5 meter telescope at Las Campanas, Chile, by A. Dressler. Note the incredible variety of forms of galaxies within the cluster.

A novel approach to determining *M/L* for a single pair of galaxies has been carried out by Borne and Hoessel (1984). Using the abundant information contained in the photometry and the dynamics of an interacting pair, these authors are able to constrain the acceptable range of orbital and galaxy parameters. From that computer model that realistically reproduces the morphology and the observed velocities, they recover the viewing angle, the shape and velocity in the orbit, and ultimately the *M/L*, ratio of the binary system. The analysis of two pairs has produced values of mean *M/L* of about 10 ± 5 for the pairs.

IX. MORE CONSTRAINTS ON THE DARK MATTER

Earlier, I tabulated six constraints on the properties of the dark matter, which arise from optical observations. I now add four more which follow from the discussion above.

7. *The dark matter extends well beyond the optical galaxy and the mass of the galaxy continues to rise virtually linearly with radius* at least for those galaxies in which 21-cm observations can be continued beyond the optical radius.

8. *At least for one polar ring galaxy where we can sample the potential near the rotation poles of the disk, the potential is more nearly spherical than flat.*

9. *The dark matter may come in more than one form: it may be disklike and halolike.* Studies of the Oort limit in our own Galaxy indicate a nonluminous disk component; studies of external galaxies indicate a more spherical component.

10. *Elliptical galaxies too have a significant fraction of their mass in a nonluminous component.*

X. WHAT IS THE MATTER? WHAT ROLE HAS IT PLAYED IN GALAXY EVOLUTION?

In cosmology, observational facts are hard to come by. But most astronomers agree that there are a few facts that must be reproduced by any theory describing the history and evolution of galaxies.

1. The universe originated in a Big Bang, whose laws of physics, after the first few minutes, are the laws of physics we know today.
2. The 3 K black body radiation is the remnant of the primordial fireball; it has been expanding and cooling ever since. Because of the expansion of the universe, the distances between field galaxies is presently an increasing function with time.
3. After removal of the motion of the observer, the background radiation is isotropic on small scales to an impressively high degree, a few parts in 10^5 (Uson and Wilkinson 1984; Knoke et al. 1984).
4. The distribution of luminous galaxies is clumpy on enormous scales (Fig. 11). Regions of long strings of superclusters separate large regions void of any bright galaxies.

Also accepted by almost all cosmologists is the premise that Newton's law of gravitation holds for very large distances and for very small accelerations. All of the preceding discussion rests on the acceptance of Eq. (1). A few theorists do question this fundamental principle: efforts are underway by Milgrom (1983) and Bekenstein (1984) to modify Newton's law so that in a disk galaxy the distribution of mass is described by

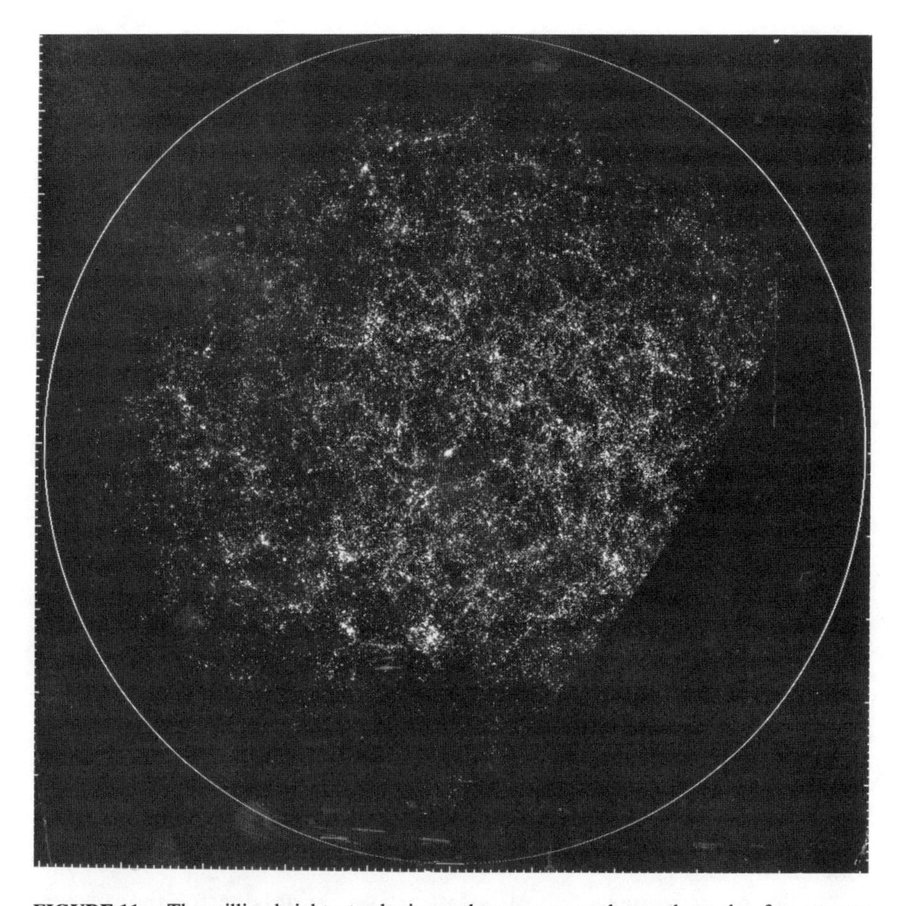

FIGURE 11. The million brightest galaxies as they appear on the northern sky, from counts by Shane and Wirtanen (1967) at Lick Observatory, newly reproduced by Seldner et al. (1977). The north galactic pole is at the center, the galactic equator is at the edge, and galactic latitude is a linear function of radius. Note the striking lacework pattern and the conspicuous voids.

the distribution of luminosity yet gives rise to a flat rotation curve. Most astronomers faced with the necessity of accepting the existence of dark matter or of modifying Newton's law of gravitation will opt to do the former.

Thus most astronomers unenthusiastically accept a universe in which much of the mass is nonluminous. Dark matter probably amounts to half of the galaxy mass within the optical image; it may amount to tens of galaxy masses beyond the optical image. Observations in a variety of spectral ranges have been unable to detect the dark matter. Massive halos do not radiate significantly in the ultraviolet, visible, infrared, X-ray or γ-ray region; they are not composed of gas or normal low-luminosity stars. We conclude with some current ideas as to what the dark matter can be. More details and extensive references are to be found in several recent interesting reviews (Blumenthal et al. 1984; Hut and White 1984; Peebles 1984).

The ordinary form of matter is baryonic, that is, protons and neutrons. If the dark matter is baryonic, it must be in objects of mass much less than a solar mass, so that its radiation remains undetected. Moreover, because deuterium, produced following the Big Bang, is destroyed in stars and in regions of high density, the presently observed deuterium abundance puts limits on the amount of baryonic matter that could have existed in the universe. If the universe is closed, that is, if there is a sufficient density of matter to ultimately halt the expansion, then the baryon number cannot be significantly higher than what we now observe in visible matter. If, however, the mass density in the universe is low, this limit on the baryonic number is invalid. However, galaxy formation in a baryon-dominated universe predicts fluctuations that would be observed today as fluctuation in the black body background radiation. The absence of such fluctuations has convinced many cosmologists that the dark matter cannot be baryonic.

Nonbaryonic forms of matter that have been suggested include neutrinos, gravitinos, photinos, axions, mini- and maxi-black holes, and even monopoles. The cosmology predicted by each of these particles is complex, but we can describe some of the major properties currently ascribed to such universes.

Neutrinos were one of the first attractive candidates for the missing mass. We know that neutrinos exist; they were originally assumed to have zero mass, but recent experiments have presented controversial evidence that their mass is near 30 eV. If their mass is in the range of a few 10s of electron volts, they contribute a mass on the order of that predicted by the galaxy dynamics. But neutrinos are hot, fast-moving particles, and a universe dominated by neutrinos would form enormous structures early in its history. Such a universe would produce a "top-down" cosmology, in which the largest structures form first, and substructures ultimately form clusters and then galaxies.

Such a cosmology has interesting properties, close to those observed and predicted. The Zel'dovich pancake model (1970, 1978) predicts that the earliest structures in the universe were sheets of gas that formed at the intersections of orbits of matter. Such sheets, with masses on the order of clusters and superclusters, fragment to form protogalaxies, from which the galaxies ultimately form. Regions away from the sheet would be empty of matter and would form the large voids that are presently being identified. These predictions match well the clusters and superclusters we see as long stings in the plots of the galaxy distribution (Fig. 11).

Yet two factors have dispelled the initial enthusiasm for a neutrino-dominated universe. First, it is still not established that the neutrino has mass; and second, the high temperature and hence high velocities of the neutrinos would form structures on scales as large or larger than those we observe. Fragmentation down to scales the sizes of galaxies would take an appreciable fraction of the age of the universe; galaxies would have only recently formed. Such a time scale seems unacceptable to our present ideas of galaxy evolution.

Many cosmologists have now turned to an alternative model, one in which the particles whose mass dominates the universe are cold rather than hot. Such particles, photino, graviton, axion (and there are more), have never been detected but are all allowed by the physics of the gauge theories from which they emerge. In such a universe the cold axions, for example, form clouds early on that withstand the expansion and clump on all sizes, from clusters of stars to clusters of galaxies. The smallest of these form first, and merge, in hierarchical fashion, to form galaxies and ultimately clusters of galaxies. Because fluctuations on all scales are condensing at the same time, many of the curious interactions between galaxies and their environments can be understood. The overwhelming drawback of such models is that they are based on particles whose existence is presently only postulated.

If we have been lucky, the real universe will incorporate some of the features of one or more of these models; if we have been unlucky, there is much to learn before we can describe galaxy formation. Theoretically, we have to learn how the formation of luminous galaxies proceeded in a universe dominated by dark matter. Observationally, we hope to detect young galaxies at a time close to their birth. But science progresses best when observations force us to alter our preconceptions. Astronomers will have to learn to carry out their research in a universe in which only a small fraction of the mass is luminous. We are surely in for some exciting surprises.

ACKNOWLEDGMENTS

I thank Drs. J. Bahcall, A. Dressler, P. J. E. Peebles, F. Schweizer, and Y. Terzian for permission to use photographs and figures from their publications.

REFERENCES

Bahcall, J. N., 1984a, *Ap. J.,* **276,** 156.

Bahcall, J. N., 1984b, *Ap. J.,* **287,** 926.

Bekenstein, J., 1984, Lecture at workshop on *Dynamics within Galaxies,* Rehovot, Israel, June.

Bertola, F., and Capaccioli, M., 1975, *Ap. J.,* **200,** 439.

Blitz, L., Fich, M., and Stark, A. A., 1982, in *Interstellar Molecules,* B. Andrews (ed.), Reidel, Dordrecht, p. 213.

Blumenthal, G. R., Faber, S. M., Primack, J. R., and Rees, M. J., 1984, *Nature,* **311,** 517.

Bok, B., 1981, *Sci. Amer.,* **244,** March, p. 92.

Borne, K., and Hoessel, J. G., 1984, preprint.

Bosma, A., 1981, *Astron. J.,* **86,** 1791.

Brandt, J., 1960, *Ap. J.,* **131,** 293.

Burbidge, E. M., and Burbidge, G. R., 1975, in *Galaxies and the Universe,* A. Sandage, M. Sandage, and K. Kristian (eds.), University of Chicago Press, Chicago, p. 81.

Burstein, D., Rubin, V. C., Thonnard, N., and Ford, W. K., Jr., 1982, *Ap. J.,* **253,** 70.

Dressler, A., 1979, *Ap. J.,* **231,** 659.

Einasto, J., Kaasik, A., Kalamees, P., and Vennik, J., 1975, *Astron. Ap.,* **40,** 161.

Fabricant, D., Lecar, M., and Gorenstein, P., 1980, *Ap. J.,* **241,** 552.

Forman, W., Forman-Jones, C., and Tucker, W., 1984, *Ap. J.,* **285,** 1.

Freeman, K., 1970, *Ap. J.,* **160,** 811.

Hartwick, F. D. A., and Sargent, W. L. W., 1978, *Ap. J.,* **221,** 512.

Hut, P., and White, S. D. M., 1984, *Nature,* **310,** 637.

Illingworth, G., 1977, *Ap. J.* (*Lett.*), **218,** L43.

Knoke, J. E., Partridge, R. B., Ratner, M. I., and Shapiro, I. I., 1984, *Ap. J.,* **284,** 479.

Krumm, N., and Salpeter, E. E., 1977, *Astron. Ap.,* **56,** 465.

Larson, R. B., and Tinsley, B. M., 1978, *Ap. J.,* **219,** 46.

Lynden-Bell, D., 1982, in *Astrophysical Cosmology,* H. A. Bruck, G. V. Coyne, and M. S. Longair (eds.), Pontifical Academy of Sciences, Vatican City State, p. 85.

Mestel, L., 1963, *M.N.R.A.S.,* **126,** 553.

Milgrom, M., 1983, *Ap. J.,* **270,** 365.

Murai, T., and Fujimoto, M., 1980, *P.A.S.J.,* **32,** 581.

Oort, J., 1960, *Bull. Astron. Inst. Neth.,* **15,** 45.

Opik, E., 1922, *Ap. J.,* **55,** 406.

Ostriker, J. P., and Peebles, P. J. E., 1973, *Ap. J.,* **186,** 467.

Pease, F. G., 1918, *Proc. Nat. Acad. Sci.,* **4,** 21.

Peebles, P. J. E., 1984, *Science,* **224,** 1385.

Perek, L., 1962, *Adv. Astron. Astrophys.,* **1,** 165.

Roberts, M. S., 1975, in *Galaxies and the Universe,* A. Sandage, M. Sandage, and J. Kristian (eds.) University of Chicago Press, Chicago, p. 309.

Roberts, M. S., and Rots, A. H., 1973, *Astron. Ap.,* **26,** 483.

Rogstad D. H., and Shostak, G. S., 1972, *Ap. J.,* **176,** 315.

Rubin, V. C., Burstein, D., Ford, W. K., Jr., and Thonnard, N., 1985, *Ap. J.*, **289,** 81.

Rubin, V. C., Ford, W. K., Jr., and Thonnard, N., 1978, *Ap. J.* (*Lett.*), **225,** L107.

Rubin, V. C., Ford, W. K., Jr., and Thonnard, N., 1980, *Ap. J.* **238,** 471.

Rubin, V. C., Ford, W. K., Jr., Thonnard, N., and Burstein, D., 1982, *Ap. J.*, **261,** 439.

Sancisi, R., 1983, in *Internal Kinematics and Dynamics of Galaxies,* E. Athanassoula (ed.), Reidel, Dordrecht, p. 55.

Schneider, S. E., and Terzian, Y., 1983, *Ap. J.* (*Lett.*), **274,** L61.

Schwarzschild, M., 1954, *Astron. J.*, **59,** 273.

Schweizer, F., Whitmore, B., and Rubin, V. C., 1983, *Astron. J.*, **88,** 909.

Schweizer, L., 1984, private communication.

Seldner, M., Siebers, B., Groth, E. J., and Peebles, P. J. E., 1977, *Astron. J.*, **82,** 249.

Shane, C. D., and Wirtanen, C. A., 1967, *Publ. Lick Obs.*, **22** (Part 1).

Slipher, V. M., 1914a, *Lowell Obs. Bull.*, No. 58.

Slipher, V. M., 1914b, *Lowell Obs. Bull.*, No. 62.

Smith, S., 1936, *Ap. J.*, **83,** 23.

Toomre, A., 1963, *Ap. J.*, **138,** 385.

Uson, J. M., and Wilkinson, D. T., 1984, *Ap. J.*, **283,** 471.

van Albada, T. S., Bahcall, J. N., Begeman, K., and Sancisi, R., 1985, *Ap. J.*, **295,** 305.

van der Kruit, P., and Shostak, G. S., 1982, *Ast. Ap.*, **105,** 351.

Volders, L., 1959, *Bull. Astr. Inst. Neth.*, **14,** 323.

White, S. D. M., 1981, *M.N.R.A.S.*, **195,** 1037.

Zel'dovich, Ya. B., 1970, *Astr. Astrophys.*, **5,** 84.

Zel'dovich, Ya. B., 1978, in *The Large Scale Structure of the Universe,* M. S. Longair and J. Einasto (eds.), Reidel, Dordrecht, p. 409.

Zwicky, F., 1933, *Helvet. Phys. Acta,* **6,** 110.

10.

Relativistic Motion in Quasars

MARSHALL H. COHEN

Marshall H. Cohen is Professor of Radioastronomy at Caltech. He has worked in a number of areas in radioastronomy, including observations of extragalactic radio sources, jets, and superluminal phenomena. He has been a world leader in the development of Very Long Baseline Interferometry.

I. INTRODUCTION

Quasars and other active galactic nuclei sometimes change so rapidly that extraordinary physical circumstances seem to be required. The most extreme examples are provided by "superluminal motion," where an elementary calculation of velocity suggests that components are moving faster than the speed of light! Many explanations for this phenomenon have been put forward, but the most favored one involves *relativistic* motion. This not only keeps the quasar safely within the boundary of known physical law, but also explains many other characteristics of these objects. This hypothesis—that quasars contain relativistic beams—unifies the otherwise disparate phenomena of superluminal motion, rapid flux variations, weak X-rays, power flow to the outer lobes, and strong bending. It may also explain the statistical relationships among several classes of sources. In this chapter we discuss some of these phenomena and introduce them as evidence favoring the relativistic beam hypothesis. The monograph by Begelman, Blandford, and Rees (1984) and Chapter 6 by Rees discusses the physics of these beams and their place in the general scheme of extragalactic radio sources.

Relativistic motion in quasars was first suggested as a possible solution to what was called the "Compton catastrophe" (see, e.g., Hoyle, Burbidge, and Sargent 1966; Scheuer and Williams 1968). The problem arose when rapid variations in flux density were discovered. If the time scale for variations is τ, then the size of the source cannot be bigger than $c\tau$, since changes cannot propagate faster than the speed of light, c. From this one can calculate a limit to the photon density in the source. Furthermore, one can also find the magnetic field density there. As τ becomes smaller the ratio of radiation energy density to magnetic energy density increases rapidly, and if it exceeds unity the source is catastrophically quenched by the inverse Compton process (radio or infrared photons of energy $h\nu$ are converted to X-rays of energy $\gamma_e^2 h\nu$ by scattering on electrons with energy $\gamma_e mc^2$). The problem was that the ratio was large, but the source appeared stable and the X-rays were weak. Rees (1967) suggested that the source was expanding relativistically so that, in the observer's coordinate system, the surface nearly kept up with its own radiation. The time scale for variations is thus drastically shrunk from what a comoving observer would measure, and the size is actually much bigger than is superficially calculated from $r \leq c\tau$.

This was only one of a number of possible explanations for the rapid variations. Another was that quasars were closer than the standard cosmological distance derived from the redshift and Hubble's law, and thus the photon density was reduced and the magnetic field increased, until the quenching effect became small. But no convincing alternative explanation for the redshift was ever suggested, and the more recent discovery of quasars surrounded by "fuzz" with stellar spectra is convincing evidence that they are indeed at their cosmological distances (Boroson and Oke 1984). Another series of suggestions involved changing the physical model of the source. The standard model used high-energy electrons in a magnetic field, with incoherent synchrotron radiation supplying the radio photons, which then scattered off the same electrons to make the X-rays. Various high-brightness coherent radiation processes were examined but none appears to have survived, and it is generally believed that the incoherent synchrotron process is at work in cosmic sources. Relativistic motion thus is the preferred explanation for the rapid variations. However, a more recent suggestion (Rickett, Coles, and Bourgois 1984) is that the variations are not intrinsic but are due to refractive scintillation effects in the local interstellar medium. This idea has merit, especially at low radio frequencies, and it may turn out that the variations do not make a strong case for relativistic motion.

The most powerful arguments for relativistic motion come from the phenomenon of superluminal motion, which was discovered in 1971. (Whitney et al. 1971; Cohen et al. 1971). The common model nowadays involves a narrow relativistic beam, aimed nearly at us (see Fig. 1). The density decreases outwards and the optical depth is unity near the base, where the beam is seen as a very bright spot (the "core"), possibly with a tail of decreasing intensity. The beam may also carry one or more luminous "blobs" or "components" that move away from the core. Because of the relativistic motion, the components display a large Doppler factor (blueshift) and, as with intensity variations, the time scale for transverse motion is shrunk for a terrestrial observer. When this is taken into account, the apparent transverse velocity is "superluminal," even though the true velocity is less than the speed of light.

The transformation between the rapidly moving coordinate systems also explains the weak X-rays, since in the moving source the magnetic field is stronger, and the radiation density less, than in a stationary source

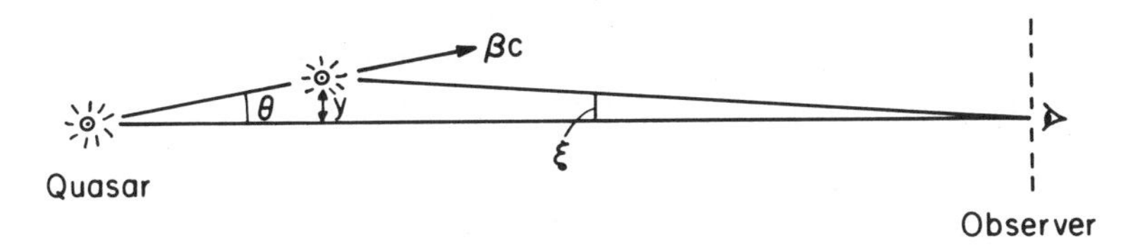

FIGURE 1. A luminous blob moves away from the nucleus of the quasar at speed βc.

with the same observed spectrum. In practice the argument is turned around, and the X-ray observations are used to derive a lower limit to the Doppler factor. Values between 5 and 10 are needed to explain the X-rays, and these values of the Doppler factor are roughly consistent with the values needed to explain the variability and the superluminal motion.

The plan for this chapter is as follows. Section II describes the relativistic beam model and gives some simple formulas that we will use. Section III is a summary of the observations of superluminal sources, which comprise more than half of the strong core-dominated radio sources. Section IV presents the X-ray limits and the geometric constraints that they imply, and Section V discusses some statistical questions associated with beamed sources.

II. RELATIVISTIC BEAM MODEL

The beam model is motivated by more than the need for relativistic effects. High-energy beams were postulated in the early 1970s (Rees 1971; Longair, Ryle, and Scheuer 1973) to provide energy to the outer lobes of double radio sources, and their use in explaining "hot spots" in the lobes dates from 1974 (Blandford and Rees 1974). Since then high-quality maps have directly outlined these beams (Bridle and Perley 1984). On scales of a thousand to a million parsecs they show up in great variety: one- and two-sided, straight and bent or twisted into various forms. The morphologies can perhaps be understood in terms of ram-pressure bending, combined in some cases with rotation or precession. The one-sidedness provides indirect evidence that some of these large-scale outer beams are relativistic, since in this case the Doppler effect makes the approaching beam much brighter than the receding beam. However, they are also discussed in terms of an intrinsic, perhaps episodic, one-sidedness (Bridle and Perley 1984; Perley 1984).

The superluminal effects arise close to the galactic nucleus, on scales of 1–100 parsecs from the center. Thus they are well beyond the resolving power of optical instruments or of conventional radio interferometers, and can be spatially resolved only with intercontinental very-long-baseline interferometry (VLBI). The VLBI maps usually show a linear structure that is aligned along the outer beam. Thus, radio data on all angular scales support the notion of a beam or jet that emerges from deep within the nucleus and ultimately powers the outer radio lobes.

In the simplest beam model a narrow relativistic jet comes out of the central region of the quasar. It carries a luminous blob moving with speed βc and Lorentz factor $\gamma = (1 - \beta^2)^{-1/2}$, and the blob moves at angle θ

to the line of sight (see Fig. 1). The Doppler shift seen by an observer in the reference frame of the quasar is

$$\delta = \frac{1}{\gamma} \frac{1}{1 - \beta \cos \theta} \tag{1}$$

and the net frequency shift seen by a terrestrial observer is $\delta/(1 + z)$, where z is the redshift of the quasar. The kinematics of such a beam have been discussed in detail by Blandford and Königl (1979).

If the observed variability time scale is defined as $\tau = S_\nu(dS_\nu/dt)^{-1}$, where S_ν is the spectral flux density (power per unit frequency per unit area), then the intrinsic time scale is $\tau^* = \tau\delta/(1 + z)$. When the intrinsic brightness temperature is estimated from the time scale and the Rayleigh–Jeans law, it varies as $T_b^* \sim S_\nu(D/c\tau^*)^2(1 + z)/\delta \sim (\delta)^{-3}$, where D is the distance. Values of δ near 10 are sufficient to reconcile apparent values of T_b near 10^{15} K with the "Compton limit" $T_b^* < 10^{12}$ K (Kellermann and Pauliny-Toth 1969).

Define v as the apparent transverse velocity of the blob, measured in the quasar frame:

$$v = \frac{\Delta y}{\Delta t_q} = \frac{\mu D_{\mathrm{lum}}}{1 + z} \tag{2}$$

where t_q is time measured in the quasar frame, μ is the apparent proper motion $\Delta\xi/\Delta t$, ξ is the angular separation between two components in the source, $y = \xi D_{\mathrm{lum}}(1 + z)^{-2}$ is the linear separation (Fig. 1), and D_{lum} is the luminosity distance. In this article we use a Friedmann cosmology with deceleration parameter $q_0 = 0.05$, and $h = H_0/100 =$ Hubble constant in units of 100 km sec^{-1} Mpc^{-1}.

From Fig. 1,

$$\frac{v}{c} = \frac{\beta \sin \theta}{1 - \beta \cos \theta} \tag{3}$$

When $v/c > 1$, the motion is called "superluminal."

Equation (3) follows immediately from Eq. (1) in Rees (1967), and is appropriate for a blast wave that ignites preexisting material as well as for a self-luminous blob. The "light echo" version of Eq. (3) is obtained for $\beta = 1$ and is appropriate when a shell of light is scattered from preexisting material (Lynden-Bell 1977).

The four quantities γ, θ, δ, and v/c are connected by Eqs. (1) and (3).

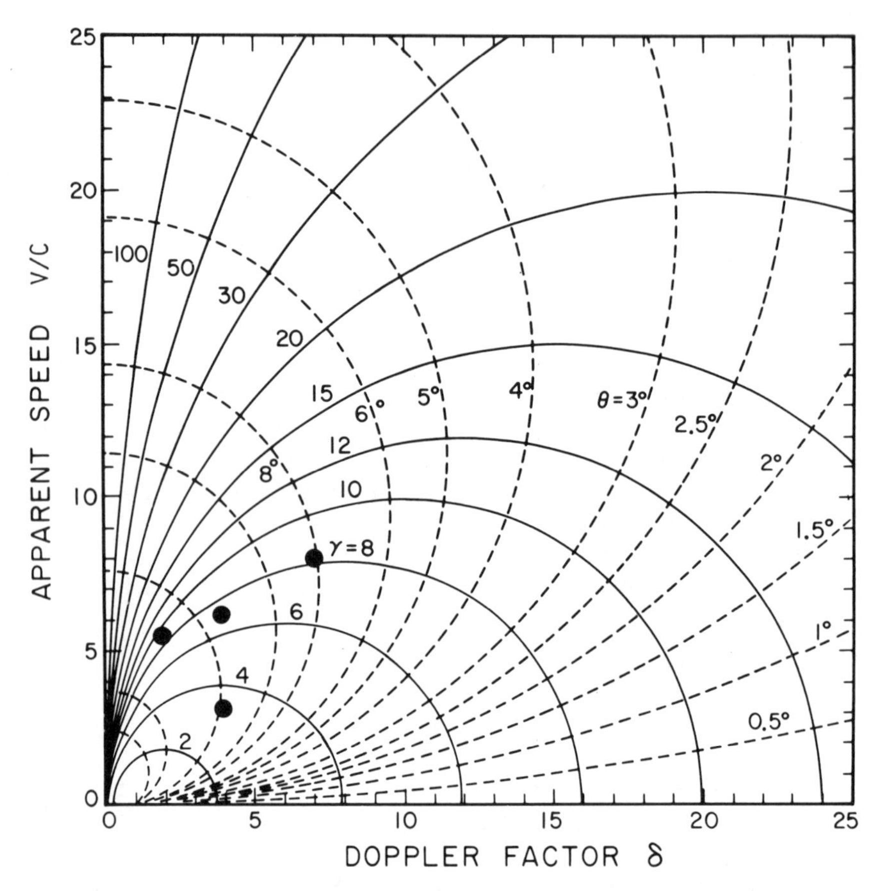

FIGURE 2. The apparent transverse speed v/c and the Doppler factor δ from Eqs. (1) and (3). The solid curves are lines of constant Lorentz factor γ, and the dashed curves are lines of constant θ, the angle to the line of sight. The unlabeled dashed curves correspond to θ = 10°, 15°, 30°, and 45°. The solid dots are measured values of $(v/c)h$ and δ_{min} from Table 2, for 0735+178, 1226+023, 0333+321, and 1641+399, in order of increasing $(v/c)h$. (Reproduced from *The Astrophysical Journal*.)

Various graphical versions of these equations have been published, and in Fig. 2 we reproduce the one used by Unwin et al. (1983). This is convenient since (as we show later) δ and v/c are the observable quantities, and from them γ and θ can be found.

The Doppler effect changes the apparent intensity of a moving source. At a fixed frequency the flux density S_ν of an optically thin source varies as $\delta^{3-\alpha}$ for an isolated point source, or as $\delta^{2-\alpha}$ for a uniform jet, where

α is the spectral index, $S_\nu \sim \nu^\alpha$ (Begelman, Blandford, and Rees 1984). Relativistic sources that are aimed at us will be much brighter than those that are sideways, and so will be preferentially selected in surveys of the sky. These sources also may display $v/c > 1$, so the fact that the majority of strong sources are observed to be superluminal (see the next section) supports the notion that we are looking at Doppler-boosted radiation, which requires relativistic motion.

III. SUPERLUMINAL SOURCES

Many compact radio sources show a "core-jet" structure at centimeter wavelengths, when observed at scales of one milliarcsecond (≡ 1 mas). In a typical case the "core" is a bright small component with a self-absorbed spectrum (i.e., with $\alpha > 0$) between 5 and 10 GHz. The "jet" is a narrow linear feature that may contain one or more bright spots or "components" with $\alpha < 0$. The core lies at one end of the jet, and there is no counterjet at a level of 50:1 or more. In the superluminal sources the jet components have proper motion, relative to the core, with $v/c > 1$.

A. 3C 345

Superluminal behavior is best illustrated with 3C 345, which has been studied extensively. Figure 3 shows a sequence of maps made with VLBI at 5.0 GHz. (Unwin et al. 1983). The eastern component D is the core, and C_2 and C_3 are two jet components. There is also an outer jet component C_1, which can just barely be seen at 5 Ghz but is prominent at lower frequencies, and an inner one C_4, which appeared recently and has been seen clearly only at higher frequencies. C_2, C_3, and C_4 all display superluminal motion relative to D.

VLBI maps do not contain absolute position information, and so the registration of maps made at different epochs or frequencies has been in question. However, Bartel et al. (1984) have made astrometric measurements showing that any proper motion of D, relative to the nearby quasar NRAO 512, is at least an order of magnitude less than the proper motions of C_2 and C_3 relative to D. It therefore may be assumed that D is stationary on the sky and that C_2, C_3, and C_4 are moving.

The following features can be seen in Fig. 3: (a) C_2 and C_3 separate from D. (b) The ridge is curved. (c) C_2 and C_3 evolve with a time scale of a few years. When C_3 was closest to D, they had roughly the same

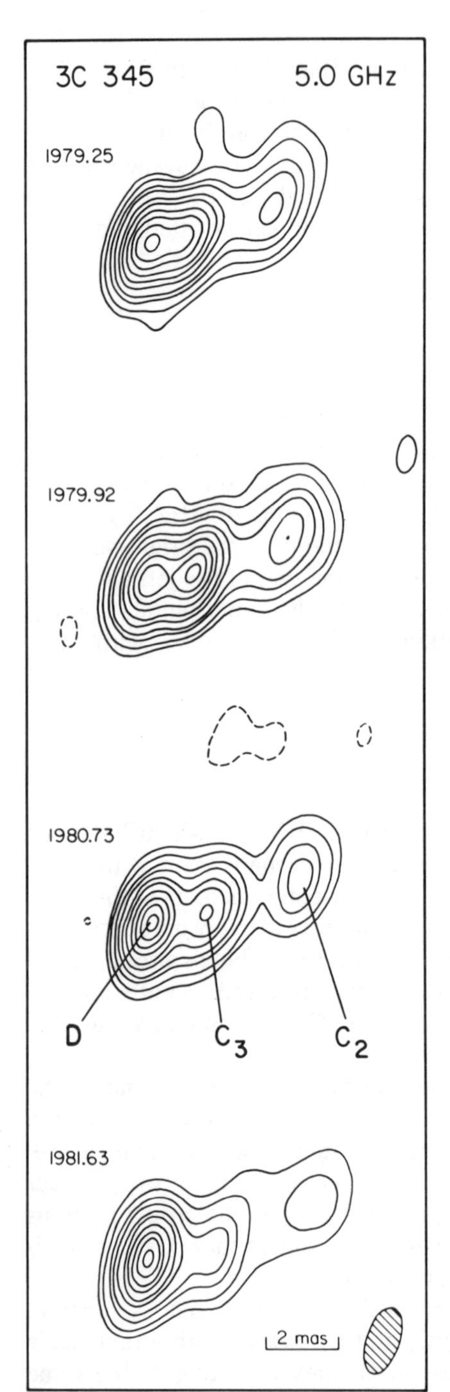

FIGURE 3. Maps of 3C 345 at 5.0 GHz at 4 epochs. The smoothing beam is shown as a shaded ellipse in the lower right corner. Contours are drawn at 2.5, 5, 10, 15, 25, 35, 50, 65, 80, and 95 of the peak brightness temperature, which is 8.4, 6.2, 9.9, and 18.4 × 10^{10} K at successive epochs. (Reproduced from *The Astrophysical Journal*.)

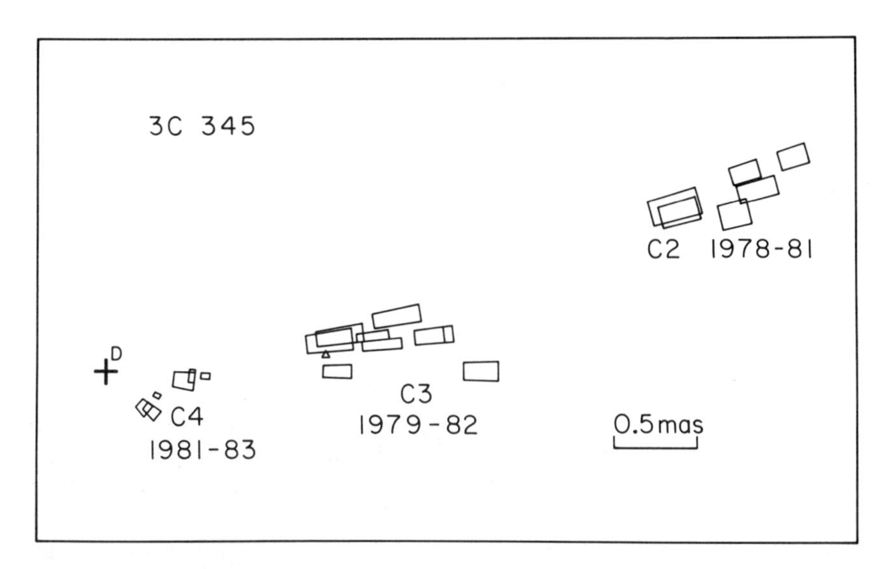

FIGURE 4. Locations of the peaks of the components in 3C 345. The core D is represented by + at the left. The rectangles roughly represent the errors. In every case the radius from D increases with time.

strength. The spectrum cannot be seen from Fig. 3 alone, but from other observations the spectral indexes were found to be $\alpha_D \sim 0.5$ and $\alpha_{C3} \sim -1$, from 5 to 10 GHz in 1980. The positions of the components may depend slightly on wavelength (Biretta 1985).

Figure 4 shows the locations of the peaks of the components since 1979. The new component C_4 was discovered in 1981 (Biretta et al. 1983; Moore, Readhead, and Baath, 1983), when it was 0.3 or 0.4 mas (1.5 h^{-1} pc, projected) from D. In 1981–1983 C_4 appeared to move along a straight track that was not radial from D, and not at the position angle of either C_2 or C_3. It also accelerated, from roughly $v/c = 4\ h^{-1}$ to $6\ h^{-1}$ in two years. Components C_2 and C_3 have $v/c = 9.5\ h^{-1}$ and $7.0\ h^{-1}$, respectively, but the errors are large enough that they both could have $v/c = 8.3\ h^{-1}$ (Unwin et al. 1983).

The origin of C_4 is unknown. One suggestion is that it emerged from D and traveled along a curved path close to the line of sight, with the apparent curvature strongly amplified by projection. There is no specific evidence supporting this; but it is simpler than models with multiple nuclei, and the track is curved at distances greater than 1 mas (Moore, Readhead, and Bååth 1983). The physics underlying the observed acceleration is unknown. It occurred while the component was on a straight portion of its track.

Components C_4, C_3, and C_2 form a jet that is about 5 mas long, corresponding to a projected length of 20 h^{-1} pc. C_1 lies at about 20 mas, or 85 h^{-1} pc, from the core. 3C 345 also has an outer one-sided jet that is 1000 times longer than the VLBI jet; and it has a large, amorphous, low-surface-brightness, outer lobe that surrounds the core and the outer jet. The inner VLBI jet curves toward the outer jet, which continues to curve in the same sense. The curvature is strongest close to the core (Readhead et al. 1983).

B. Other Sources

The discovery rate of superluminal sources has been low, because multi-epoch observations are needed, and VLBI observations are difficult. A number of new candidates have been found in the last few years, however, partly as a result of systematic VLBI surveys that now are in progress. (Pearson and Readhead 1984; Eckart and Witzel 1984) The surveys are especially important because they will give a solid basis for the statistical study of superluminal sources. Table 1 summarizes the knowledge of superluminal sources, as of late 1984.

Table 1 contains the following categories (column 8): all nine definite superluminal sources (SL), six of the sources that have been reported as possibly superluminal (Possible SL; there are another dozen candidates of this variety, generally with less supporting evidence), one source with subluminal internal motions (Sublum), three sources with well-determined VLBI structure that has been stable for several epochs, one (0538+498) with an unclassifiable structure change, and two others (0851+202 and 1730−130), which are included for completeness.

Table 1 is arranged in order of decreasing flux density. Nearly all these sources are variable (exceptions are the M87 nucleus and possibly several others), and column 4, $S_m(10.7)$, gives the peak flux density which has been reported at 10.7 GHz. The table is reasonably complete for core-dominated extragalactic sources with declination δ north of $-13°$ and $S_m(10.7) \geq 7.0$ Jy. In this context the core has dimensions 1 kpc or less and includes all the VLBI structure; its flux dominates that of any outer jet or halo. Three sources near the bottom of Table 1 are lobe dominated, and their tabulated core flux is only a small fraction of the total.

Column 5 gives the emitted luminosity at 15 GHz in a co-moving frame, in 10^{32} erg sec^{-1} Hz^{-1}, based on $S_m(10.7)$ with $\alpha = 0.2$ and assuming isotropic radiation (Schmidt 1972). If we assume that these sources are Doppler-boosted, then their radiation is anisotropic, and the luminosity is smaller than in column 5 by about $4\pi\gamma^2$.

Column 6 gives the typical values of internal proper motion as reported in the literature, and column 7 gives the corresponding apparent transverse velocity. Parentheses indicate uncertain values, and the negative signs indicate contraction rather than expansion (see the discussion below on 4C 39.25). The references (column 9) are to structure and flux density measurements. The Proceedings of IAU Symposium No. 110, "VLBI and Compact Radio Sources," includes a great deal of relevant recent work, and further references can be found there.

(i) Superluminal. The best cases, 3C 120, 3C 273, and 3C 345, all show multiple ejection with velocities (in the same source) differing by up to a factor of 2, although the accuracy is low and it is marginally possible that the velocity is constant in each source. In 3C 273 and 3C 345 the components lie on a strongly curved track. In 3C 120 new components have overtaken the position old components held at an earlier epoch, and now this is also happening with 3C 273 and 3C 345. There has been speculation that the C_2 and C_3 tracks in 3C 345 are not coincident, and this will be tested in a few years. Unfortunately, small offsets of this kind are masked in 3C 120 and 3C 273 by poor north–south resolution, caused by the low declination.

The jet components of 3C 120, 3C 273, and 3C 345 all evolve on a time scale of one to ten years, at centimeter wavelengths. They first appear close to the core with a flux density comparable to that of the core; as they move away their flux density at first changes little, but then the spectrum steepens and the component becomes weak. The core itself is stable to within a factor of 2, and it is suspected that some of its variation is due to blending with new components, before they can be identified separately.

Less is known about the other superluminal sources. 3C 279 went through a very active phase 15 years ago, before good maps could be made. Its study is seriously hampered by the low declination and poor north–south resolution, combined with the fact that the motion apparently has a strong north–south component. BL Lac has components that evolve with a time scale of months, and closely spaced observations are required to show the superluminal motion. 4C 39.25 is intriguing because for years it was a stable double, but in 1979 it began to change. The components have changed strength, they have elongated, and their peaks have moved closer together. One interpretation is that the source is contracting. Another is that the western component is the core, and it ejected a new component toward the eastern component, which has become weak. As yet the motions have been too small for these possibilities to be sorted out, but in both cases superluminal motion with $v/c \approx 2$ would be required.

TABLE 1. Superluminal Sources circa 1984[a]

Source Name		Redshift	$S_m(10.7)$	$L^*(15)$	μ	v/c		
IAU	Alt	z	Jy	(10^{32} cgs)	(mas/yr)	(H_0 = 100)	Structure	Ref.[d]
0316 + 413	3C 84	0.018	65	2.4	0.24	0.20	Sublum, complex	1–3
1226 + 023	3C 273	0.158	60	170	0.8	5.5	SL	4–7
2251 + 158	3C 454.3	0.859	27.4	2100	(0.3)	(9)	Possible SL	7, 8
1253 + 055	3C 279	0.538	20.3	630	.17	3.5	SL	4, 6, 7
0430 + 052	3C 120	0.033	16.8	2.1	2	3	SL	9–11
2200 + 420	BL Lac	0.070	16.4	9.1	(.75)	2.4	SL	7, 12
1641 + 399	3C 345	0.595	16.3	620	.35	8	SL	3, 13–18
0355 + 508	NRAO 150	—	15.1	—	—	—	Stable, double	11, 19, 20
2134 + 004	2134 + 004	1.936	13.0	5200	<.01	<0.6	Stable, double	7, 8
0923 + 392	4C 39.25	0.698	11.9	620	(−0.1)	(−2.6)	SL, contracting?	11, 21
2223 − 052	3C 446	1.404	9.7	2000	—	—	Possible SL	3, 22
0851 + 202	OJ 287	0.306[b]	9.2	94	—	—	Compact	11, 23
0235 + 164	0235 + 164	0.851[b]	7.4	570	—	—	Possible SL	11, 24
1730 − 130	NRAO 530	0.902	7.0	600	—	—	Extended	3, 25, 26

1928 + 738	1928 + 738	0.36	3.0	42	(0.6)	(9)	Possible SL	3, 27
0333 + 321	NRAO 140	1.258	2.9	490	.15	6.2	SL	7, 28
1807 + 698	3C 371	0.051	2.8	0.8	—	—	Possible SL	11, 29, 30
0735 + 178	0735 + 178	0.424[b]	2.7	52	0.18	3.0	SL	7, 24, 31
1228 + 126	M87 nucl	20 Mpc	2.5[c]	0.01	<0.5	<0.15	Stable, core–jet	32–34
0538 + 498	3C 147	0.545	1.5[c]	48	—	—	Variable, complex	35–37
0711 + 356	0711 + 356	1.620	0.7	200	(−.05)	(−2.5)	Possible SL	29, 38
0732 + 679	3C 179	0.846	0.3[c]	23	0.19	5.7	SL	39

[a] See text for definitions of column headings.

[b] Questionable redshift.

[c] Core only.

[d] References: 1. Readhead et al. (1983). 2. Romney et al. (1984b). 3. Pearson, T. J. and Seielstad, G. A. (1984, private communication). 4. Cohen and Unwin (1984). 5. Unwin et al. (1985). 6. Unwin and Biretta (1984). 7. Medd et al. (1972). 8. Pauliny-Toth et al. (1984). 9. Walker et al. (1984). 10. Benson et al. (1984). 11. Andrew et al. (1978). 12. Mutel (1984). 13. Unwin et al. (1983). 14. Bartel et al. (1984). 15. Biretta et al. (1983). 16. Moore et al. (1983). 17. Cohen et al. (1983a). 18. Cohen et al. (1983b). 19. Bååth et al. (1980). 20. Mutel and Phillips (1980). 21. Shaffer (1984). 22. Brown et al. (1981). 23. Lawrence et al. (1985). 24. Bååth et al. (1984). 25. Marscher and Broderick (1981). 26. Romney et al. (1984a). 27. Eckart et al. (1985). 28. Marscher and Broderick (1985). 29. Readhead et al. (1984). 30. Worrell (1984). 31. Bregman et al. (1984). 32. Kellermann et al. (1973). 33. Jones (1984). 34. Reid et al. (1984). 35. Simon et al. (1983). 36. Simon et al. (1984). 37. Preuss et al. (1984). 38. Seielstad et al. (1983). 39. Porcas (1984).

The second interpretation is preferred because then 4C 39.25 would be similar to the other superluminal sources.

Superluminal motion was predicted by Marscher and Broderick (1981) in NRAO 140 because of its low X-ray intensity, and the prediction has been verified. NRAO 140 is the only source displaying all of the main characteristics that suggest bulk relativistic motion: low-frequency variability, weak X-rays, and superluminal motion (Marscher and Broderick 1985).

3C 147 has weak X-rays and also has compact radio components, but so far has not shown superluminal motion. This does not disprove the relativistic beam hypothesis for 3C 147. Superluminal motion requires both the beam and a condensation in it that shows proper motion, and the appearance of the condensation could be episodic or totally missing in some sources. 0735 + 178 has a particularly flat spectrum, produced by a superposition of several components each having a self-absorbed synchrotron spectrum. Its morphological similarity to 3C 273 and 3C 345 led to the suggestion that it would show superluminal motion (Marscher 1980), and this has proved to be the case. There are other examples, however, of sources that resemble 3C 345 but do not have superluminal motion.

3C 179 has been studied because it has the strongest core of a group of lobe-dominated quasars selected at 966 MHz. It was thought that this group would be randomly oriented and thus that the cores would not show superluminal motions, in most cases. In fact the core of 3C 179 did show superluminal motion, and this surprising result has led to VLBI studies of other similar sources. It will be several years yet before the superluminal sources in this group can be identified. The core of 3C 179 is substantially weaker than any of the others in Table 1, and it has been difficult to obtain good maps of it. Most of the flux comes from two components that have been steadily separating for 4 years.

(ii) Possibly Superluminal. 3C 454.3 is a strong highly variable source. It has the common VLBI core–jet structure, except that the jet components appear much weaker, relative to the core, than they do in 3C 345. The proper motion (Table 1) is based on comparisons of data taken 7.5 years apart; the procedures at the two epochs were different, and the expansion needs to be confirmed. The data for 3C 446 are limited but suggestive. They were obtained in 1978 and 1979, but since then there has been a major outburst, with the flux density doubling in 1983–1984. This source clearly deserves more attention.

0235 + 164 is an optically violently variable (OVV) source. Its radio structure has been different at every observing session and tracking the components will require observations at monthly intervals. This was the case also with 3C 120 and BL Lac where frequent monitoring did finally reveal steady superluminal motion. There is some evidence in 0235 + 164 for very strong jet curvature or for ejection at different position angles.

The data from two epochs suggest superluminal motion in 1928 + 73. This source is similar to 3C 120 and 3C 273 in having many components that are moving, but it is a better candidate for study because the latter two are at low declination where the VLBI maps are essentially one-dimensional. 3C 371 and 0711 + 356 are two other sources in the Pearson and Readhead (1984) survey showing structural variations that may be superluminal. The two-epoch data for 0711 + 356 show that it is a double which appears to be contracting. There is no doubt that this source is changing, but more observations are needed before the nature of the change can be certified.

(iii) Subluminal. 3C 84 is the only source that shows subluminal internal proper motions. This source is in the nearby Seyfert galaxy NGC 1275 and is core-dominated only at high frequencies. The morphology of the core is complex; it apparently contains a core–jet structure with other bright radio components within a few parsecs.

(iv) Stable. Table 1 contains three "stable" sources, those showing no structural changes in VLBI maps obtained at several epochs. However, they have variable flux density (except for the nucleus of M87), and the history of 4C 39.25 shows that they may become superluminal. NRAO 150 is the strongest extragalactic radio source without a measured redshift. At centimeter wavelengths it is a double whose components vary in intensity but have a fixed separation. At 18 cm a more distant component appears, and here, too, the data are consistent with the separation being constant over a 1-year interval. 2134 + 004 has been studied for more than a decade. At short wavelengths it is a stable double whose components vary slowly in intensity. The upper limit on proper motion is low, and even with the large redshift it cannot be superluminal.

M87 is a nearby giant elliptical galaxy containing an optical jet and a weak radio nucleus, 1228 + 126. This source has been of interest for many years and attempts to find variations in its structure at three wavelengths (3.8, 6, and 18 cm) have all failed.

(v) Others. 3C 147 is an especially interesting case because the combination of X-ray and VLBI data led Simon et al. (1983) to predict that superluminal motion would be found. Second-epoch observations at 6 cm wavelength did not show any structural changes, but observations at 18 cm taken 6 years apart do show changes in the compact core. OJ 287 is particularly compact and is often used as a calibration source in VLBI observations. At 22 GHz its effective diameter is 0.2 mas. NRAO 530 is extended on VLBI scales but does not have multi-epoch observations.

C. Relativistic Motion

The relativistic beam theory explains the following features of the superluminal sources.

(i) Superluminal Motion. A beam that has two or more bright spots can show superluminal motion if the values of γ are different. If one spot is stationary (the core), then Eq. (3) gives v/c.

(ii) Variability. All the superluminal sources have variable flux density, except possibly the weak core in 3C 179. In most cases the measured angular sizes are substantially bigger than $c\tau$, the light-travel distance for the observed variability time scale. For example, in 3C 345 $c\tau \sim 1$ pc which corresponds to 0.2 mas, but the measured sizes of the inner jet components, where the variations occur, are of order 1 mas (Unwin et al. 1983). As discussed in Section I, the best explanation for this discrepancy is relativistic motion, with $\delta/(1 + z) \sim 5$ for 3C 345.

(iii) Morphology. The well-studied superluminal sources all show a one-sided curved core–jet morphology whose features find a natural explanation in terms of a relativistic beam. The central engines are probably two-sided (because we see two outer lobes in many cases) but we see only the approaching one, which is Doppler-boosted. The core is the region where the optical depth is unity, near the base of the jet, and the components are shock waves or condensations in the jet (Begelman, Blandford, and Rees 1984). The core is smaller and denser than the jet components, so its frequency at the maximum of synchrotron emission is higher, and at centimeter wavelengths the core has spectral index $\alpha > 0$ while the jet components have $\alpha < 0$.

In some cases the inner jet curves through 45° or more, and it is difficult to see how a relativistic jet can be bent so much. However, if θ is small, the bend is amplified by about csc θ in projection, and the intrinsic bend is only a few degrees (Readhead et al. 1983).

(iv) Equal-Strength Components. It is remarkable that the core and inner jet components have such similar flux densities near the maximum of their spectra. This can be seen in Fig. 2, and it is true also for C_4 of

3C 345, and for all the other superluminal sources. The core and jet components must have similar values of γ, or the Doppler boosting would make them very different unless by coincidence the ratio of intrinsic strengths matched the Doppler boost. Near the synchrotron maximum the brightness temperature may be limited by the inverse Compton effort to about 10^{12} K, and if the solid angles and values of γ are the same, then the flux densities will be the same. As shown below, in 3C 345 the X-ray data give evidence that both the core and component C_3 are moving relativistically. But the core is stationary; so this is strong evidence for the inhomogeneous jet model, where the core is the stationary region of an expanding relativistic conical flow at optical depth near unity.

It should be noted that (i) and (ii) provide strong evidence for relativistic motion, but (iii) and (iv) merely show consistency. For example, (iv) could equally well be explained with no relativistic motion; the components have equal strength because they are limited to 10^{12} K by the inverse Compton effect, and they have equal solid angles.

D. Statistics

Table 1 is "nearly complete" for sources that had S_m (10.7) $\geq$ 7.0 Jy between 1966 and 1984, and $\delta \geq -13°$. S_m refers to the bright compact core, with dimensions less than 1 kpc. All the flux densities have a negative bias, since more complete monitoring of each source could give a higher but not a lower S_m. A more complete monitoring and a complete sky survey might have rearranged the list, and brought others in. Nonetheless, we expect that Table 1 contains most of the sources in the designated category.

In 1977 Cohen et al. (1977) showed a list of the top ten core-dominated sources at 8 GHz. They are the same as the top ten in Table 1. In 1977 four of the ten were superluminal; today six of the ten are known to be superluminal, and another is possibly superluminal. Nine of the top fourteen are in these categories. It appears that over half of the strong core-dominated sources are superluminal. Any theory for these sources must explain their common occurrence.

The flux density is a distance-dependent parameter, and it seems more natural to use the emitted luminosity, column 5 of Table 1, for statistical studies, rather than S_m. The luminosity function, however, is dominated by distant sources with low apparent flux density, and any list arrayed in luminosity is extremely incomplete. Table 1 merely shows that superluminal and nonsuperluminal sources cover similar ranges of luminosity. It is interesting, though, that 2134+004 is stable and is also the most luminous. If we use the beam theory and correct the luminosities for anisotropy, then 2134+004 would be stronger than the superluminal sources, by two to three orders of magnitude. This peculiarity disappears if we postulate that 2134+004 is like 4C 39.25; it contains a relativistic beam with two bright spots which for unknown reasons have not shown proper motion during the last decade.

The statistics of the occurrence of superluminal sources has been discussed by Scheuer and Readhead (1979). They showed that in a flux-density-limited sample of beamed objects at random orientation but with all having the same γ, about 70 percent of the sources will have $v/c > 0.5\gamma$. This is in crude agreement with the results in Table 1, if γ is of order 5 or 10. This will be discussed further in Section V, after the presentation of the X-ray results which restrict δ.

IV. SSC LIMITS

A. The Synchrotron Self-Compton (SSC) Model

It is believed that the radio radiation and some of the IR–UV radiation from quasars is due to incoherent synchrotron radiation from relativistic electrons gyrating in a magnetic field. In simple synchrotron models the compact components have magnetic field strength B from 10^{-3} to 10 gauss, and the high-energy particles have Lorentz factor γ_e from 10^2 to 10^4. Some of the radio or IR photons scatter off the electrons by the inverse Compton (IC) process and their energy is raised by about γ_e^2. This produces a spectrum of UV and X-ray photons that forms a lower limit to the overall high-energy photon flux. It is thought that in some sources the bulk of the X-ray flux does arise in this way.

SSC calculations for a compact source were presented by Jones, O'-Dell, and Stein (1974), and recent discussions are by Gould (1979) and Marscher (1983). In the usual application the source is assumed to be homogeneous, with an isotropic power law distribution of electrons. The angular diameter is measured with VLBI, or is estimated from the variability time scale. The radio or IR spectrum is measured, and the peak is assumed to be due to synchrotron self-absorption. The spectrum is then transformed into the comoving coordinate system (this brings in δ, the Doppler-shift factor), and the magnetic field strength, the photon density, and the X-ray spectrum are calculated. By this means the Doppler factor is expressed as a function of the observed radio and X-ray spectra. As was first emphasized by Burbidge, Jones, and O'Dell (1974), the result involves only the brightness (power per unit area per unit frequency per unit solid angle) and so is independent of distance and of lens effects. This eliminates a range of objections to the interpretations and makes the conclusion, that some sources are relativistic, very strong.

The Doppler boost (Section II) means that the calculated photon and electron densities, and the X-rays, are strong inverse functions of the Doppler factor. Thus the calculated δ is an inverse function of the observed X-ray flux, and weak X-rays are a sign of relativistic motion. Since other processes could also contribute to the observed X-rays, the calculated δ is a lower limit to the true value.

In terms of observable quantities, the limit to δ can be written (Marscher 1983; Unwin et al. 1983; Cohen 1985)

$$\delta_{\min} \sim S_m \nu_m^{-p} \phi^{-q} S_x^{-r} (1 + z)^s \tag{4}$$

where S_m, ν_m, and ϕ are radio quantities: the flux density and frequency at the peak of the synchrotron spectrum, and the angular diameter. S_x is the X-ray flux density in a band near 1 keV, and $p \approx 1.3$, $q \approx 1.6$, $r \approx 0.2$, and $s \approx 1.0$. S_m and S_x are variable but since r is small the calculation is insensitive to uncertainties in the X-ray flux. The frequency ν_m is usually poorly defined, but the biggest uncertainty comes from ϕ, since it is hard to determine accurately, and q is the largest of the exponents. The IC X-rays are strongly dependent on diameter; from Eq. (4), $S_x \sim \phi^{-8}$.

It is difficult enough to directly measure ϕ with VLBI, but it is worse when it has to be estimated from a variability time scale. The Hubble constant is needed to calculate the distance and $\delta_{\min}$ now varies approximately as $\delta_{\min} \sim (\tau H_0)^{-0.6} S_x^{-0.08}$. The calculation is no longer distance independent, and it uses the variability time scale τ, which is not known with any precision. Additional assumptions also are needed; a typical one is that changes propagate with the speed of light. A more dubious assumption sometimes made (for lack of other information) is that the X-ray variability is indicative of the radio size. This can lead to unreasonably small angular diameters, which probably means that in these cases the X-rays come from a tiny region that is different from the radio-emitting volume.

(i) Inhomogeneous Models. The homogeneous model is clearly inadequate in many cases. The self-absorbed (low-frequency) side of the spectrum from a homogeneous synchrotron source has $\alpha = 2.5$ but this is rarely observed; for example, the core of 3C 273 has $\alpha = 0.8$ and this is a sign of inhomogeneity (Unwin et al. 1985). In other cases a VLBI component may appear to have a smaller size at higher frequency; this may be an observational artifact due to low sensitivity, but if real it denotes a central condensation.

Marscher (1977) and Gould (1979) discuss spherically symmetric inhomogeneous models that produce the preceding characteristics, but they have had only limited use. Königl (1981) has modeled the relativistic beam with a conical flow of high-energy particles, with the particle density and magnetic field decreasing in a power law from the apex. The observer is outside the cone, which is narrow. This inhomogeneous jet model may be more realistic than a sphere but unfortunately it has many parameters and cannot be simply or uniquely constrained. But in the few cases where it has been applied it roughly agrees with the sphere model in demanding bulk relativistic motion when the X-rays are particularly weak.

Another variety of inhomogeneity would result if the source were subdivided into n equal, randomly spaced subcomponents. Consider first the extremely dilute case with no shielding, or equivalently with filling factor $f \ll n^{-1/2}$. Straightforward application of the synchrotron and X-ray formulas shows that the total X-ray flux increases faster than $(f^2n)^{-1} \gg 1$, because the angular size of a subcomponent goes down rapidly with n. But the problem is that the observed X-ray flux is too small; thus this subdivided model increases the bulk Lorentz factor of the source. As f is increased, the Lorentz factor approaches the value appropriate to a homogeneous source. If the source is centrally condensed, its effective diameter is smaller and this increases the required bulk velocity. If the center were hollow, the opposite would be true, but this case seems unlikely.

B. Results

Table 2 shows fifteen of the sources from Table 1, now arrayed in right ascension. Values of $(v/c)h$ (column 2) are repeated from Table 1. Column 3 gives $\delta_{\min}$ for eight sources; the values have been estimated by Eq. (4) or by another method, described below. The remaining columns will be discussed in the next section. A number of other cases exist in the literature, but most of them do not use reliable diameters directly measured with VLBI. Worrall (1984) treats a group of five sources (including two in Table 1) and concludes that they all require relativistic beaming. Schwartz, Madjeski, and Ku (1982) discussed 16 sources that had limited VLBI data (three of them are in Table 1), and concluded that half of them required relativistic beaming.

We now give a brief description of those sources in Table 2 for which

TABLE 2.

Source	$(v/c)h$	δ_{min}	$H_0 = 100$		$H_0 = 50$	
			θ_{max} (deg)	γ_{min}	θ_{max} (deg)	γ_{min}
0316 + 413	0.20	1	—	—	—	—
0333 + 321	6.2	4	13	6	8.5	12
0430 + 052	3	—	37*	3	19*	6
0538 + 498	—	7	8	4**	8	4**
0711 + 356	(−2.5)	—	(44)*	(3)	(23)*	(5)
0723 + 679	5.7	—	20*	6	10*	11
0735 + 178	3	4	14	3	13	6
0932 + 392	(−2.6)	—	(42)*	(3)	(22)*	(5)
1226 + 023	5.5	2	18	6	10	11
1253 − 055	3.5	—	32*	4	16*	7
1641 + 399	8	7	8	8	6	16
1730 − 130	—	(1)	—	—	—	—
1807 + 698	—	(10)	(6)	(5)**	(6)	(5)**
2200 + 420	2.4	—	45*	3	24*	5
2251 + 158	(9)	—	(13)*	(9)	(6)*	(18)

* Assumes $\gamma \to \infty$.
** Assumes $v/c = 0$.

an estimate of δ_{min} has been made. The reader not interested in the details of how the entries were arrived at can skip directly to Section C.

(i) **0316+413 = 3C 84.** The radio nucleus of 3C 84 is complex but at a high enough frequency (22 GHz) a kinked core–jet structure is seen (Readhead et al. 1983). The core has angular size ~0.3 mas, which is small, but the frequency of the synchrotron maximum is high, near 30 GHz, and the IC X-ray flux (calculated with $\delta = 1$) is comparable with the observed value. The jet contains components that show proper motion relative to the core, but at a subluminal rate (Table 1). There is no evidence that this nucleus contains bulk relativistic motion, although Readhead et al. discuss the one-sided morphology in terms of relativistic beaming.

(ii) **0333+321 = NRAO 140.** Marscher and Broderick (1985) have made VLBI observations of NRAO 140 at 1.6, 5.0, and 10.7 GHz. This source has a core and three jet components within 9 mas (50 h^{-1} csc θ pc), but the spectra of only the core and the first jet component can be estimated with confidence. An SSC calculation for the first jet component, based on the homogeneous sphere model, gives $\delta_{min} = 3.7$. Marscher and

Broderick discuss several varieties of inhomogeneities and conclude that their effect would be to increase δ_{min}.

(iii) 0538+498 = 3C 147. 3C 147 is a compact steep-spectrum source that has the complex structure characteristic of that class (Wilkinson et al. 1984). VLBI observations at 329 MHz, and X-ray data, allowed Simon et al. (1983) to calculate $\delta_{min} \sim 3.6$. The model could be further constrained by using a diameter limit based on the radio variability, and this gave $\delta_{min} \sim 7$. These large values of δ_{min} led to a prediction of superluminal motion, but so far none has been found (Table 1).

(iv) 0735+178. The OVV object 0735+178 has been studied by Bregman et al. (1984), who obtained spectra at four epochs between 1979 and 1981. The IR–UV fluxes varied rapidly while the radio and X-ray fluxes varied slowly, and because of this Bregman et al. suggest that the X-rays are IC radiation from the large radio region. This may or may not be the case, but the observed value for the X-ray flux is certainly an upper limit to the IC radiation, and limits can be calculated for the relativistic motion. Bregman et al. estimate that the bulk Lorentz factor is less than 5 in the IR–UV emitting region, but this is based on the variability time scale and is uncertain. Another limit can be calculated from the X-ray observations of Bregman et al. and the earlier VLBI data reported by Cotton et al. (1980). If Cotton's values for component C (S_m = 1.2 Jy, ν_m = 6 GHz, ϕ = 0.48 mas, α = 0.2) are used, Eq. (5) gives δ_{min} = 4. However, the models of Cotton et al. came from a small amount of data, and the SSC limit needs to be refined by making good multifrequency VLBI maps.

(v) 1226+023 = 3C 273. SSC models for 3C 273 have been discussed by Unwin et al. (1985), who found that the first jet component has δ_{min} = 2.4. For this component $\gamma > 1.4$, which is only mildly relativistic; but since $v/c \approx 5.5\ h^{-1}$ it is likely that $\delta > \delta_{min}$ and thus that the X-rays are not due to the IC process. This conclusion disagrees with that of Jones (1979), who ascribed the X-rays to first-order Compton scattering and the γ-rays to second-order Compton scattering. Königl (1981) reached a conclusion similar to that of Unwin et al.: it is difficult to fit the radio and the X-rays with the inhomogeneous jet model. It is possible to reconcile a low δ with a large v/c, by assuming large values of γ and θ (see Fig. 2). However, this would greatly increase the intrinsic luminosity, and add to the statistical peculiarity already associated with 3C 273 (see Section V). It is preferable to assume that $\delta \approx 6$ and that the X-rays arise in some other way, possible in a process directly associated with the accretion disk.

SSC studies of the core of 3C 273 do not give any useful limit to the relativistic motion (Unwin et al. 1985). The result of using both the homogeneous sphere and the inhomogeneous jet models is that the IC flux from the core is less than the observed total X-ray flux.

(vi) 1641+399 = 3C 345. Unwin et al. (1983) applied homogeneous SSC models to 3C 345 and found $\delta_{min} = 18$ and 8 for the core (D) and inner jet component (C_3), respectively. Considerations of a possible change in effective diameter with wavelength led Cohen (1985) to the weaker but more secure limits $\delta_{min} = 6$ and 7 for D and C_3.

Component D, the core, is stationary or nearly so, and $\delta_{min} = 6$ means that the emitting material is flowing relativistically through a stationary volume. But this is inconsistent with the quasistatic approximation used in the sphere calculation, and a model that integrates through a relativistic flow should be used. The Königl (1981) expanding jet model does this, and can be applied by assuming that the parameters measured for C_3 ($v/c = 8\,h^{-1}$, $\delta_{min} = 7$) are also applicable to the flow through the core D. In this case the geometry of the jet is restricted by the observed X-ray flux and by various angular sizes. The chief result is that $\theta \leq 8°$ if $H_0 = 100$, but $\theta \leq 4°$ if $H_0 = 50$. It seems likely that in 3C 345 the bulk of the X-rays are indeed due to the IC process.

(vii) 1730−130 = NRAO 530. Limited VLBI observations at 10.7 GHz enabled Marscher and Broderick (1981) to model NRAO 530 as a close double. SSC models then show that the expected IC flux is comparable to the observed value, and relativistic motion is unnecessary. This result is tentative. NRAO 530 is clearly elongated with several components (see also Romney et al. 1984) and is a good candidate for detailed study.

(viii) 1807+698 = 3C 371. Worrall et al. (1984) use SSC models to explain the IR to X-ray spectrum of the BL Lac object 3C 371. To apply the homogeneous model they add the assumption that the electrons are not reaccelerated after injection (this gives the size of the source region) and are then able to calculate $\delta_{min} \sim 10$. This model is self-absorbed in the IR and is unable to explain the radio radiation. Similarly, the inhomogeneous jet model is able to explain the IR to X-ray spectrum with $\delta \approx 10$ but does not explain the radio. 3C 371 has several VLBI components at radio wavelengths, and it is likely that a detailed radio to X-ray comparison, using measured values of angular size, would yield a value for δ_{min} that is on a more secure basis.

C. Constraints

Four sources in Table 2 (0333+321, 0735+178, 1226+023, 1641+399) have useful values of both (v/c)h and δ_{min}. These are plotted in Fig. 2 and the corresponding values of θ_{max} and γ_{min} are shown in columns 4–7 of Table 2 for both $H_0 = 100$ and 50 km sec^{-1} Mpc^{-1}. No attempt is made to assess errors but they must be substantial. The viewing angle θ decreases monotonically as δ increases (for fixed v/c) so θ_{max} is found at δ_{min}, but γ_{min} may occur at $\delta > \delta_{min}$.

Eight sources have a tabulated (v/c)h, but with $\delta \leq 1$ or with no value for δ. The corresponding values of θ_{max} are calculated for $\delta = 0$ (i.e., $\gamma \rightarrow \infty$), and are rigorous limits but unrealistically large. Two sources have a tabulated $\delta_{min} > 1$ but no $(v/c)h$; the limits for γ_{min} are calculated for $v/c = 0$ and are unrealistically small.

For $H_0 = 100$, most values of γ_{min} are between 3 and 6, and for $H_0 = 50$ they are about a factor of 2 bigger. In most cases θ_{max} and γ_{min} do not occur at the same point in Fig. 2, and only a few values of θ_{max} are unstarred and have realistic limits. These limits are from 6 to 18° for $H_0 = 100$, and smaller for $H_0 = 50$.

Table 2 shows that the observational data are consistent with the narrow relativistic beaming model, with minimum Lorentz factors between 5 and 10, and maximum viewing angles 5 to 20°. No other model is so economical in explaining the independent phenomena of superluminal motion, weak X-rays, and rapid variability, while at the same time explaining the one-sidedness, greatly reducing the curvature problem, and providing a convenient mechanism for transporting energy and momentum to the outer radio lobes. The nature of the moving blobs and the physics of the jet are not understood in detail, and our knowledge of the collimation processes and of the ultimate energy source is even scantier (Begelman, Blandford, and Rees 1984). But at present there is no reason to expect that new physics will be required to explain superluminal sources. Our lack of detailed understanding does not vitiate the kinematic success of the relativistic beam.

V. STATISTICS OF BEAMED SOURCES

The measured properties of beamed sources should have certain statistical distributions that can be calculated using the theory described above. Comparisons of measured and theoretical distributions thus can be used to test the theory. Unfortunately, we only measure limits to δ, so tests involving δ will be weak. The data also have unknown errors and the number of sources is small, so that in fact these tests at present are merely suggestive. In this section we sketch some aspects of these statistical considerations.

A. Flux-Density Limited Sample

A flux-density limited sample contains every source of a specified type, with flux density stronger than some limit, in a certain section of the sky. Table 1 is an attempt to construct such a sample for core-dominated objects, but as discussed in Section III it may be incomplete. We hypothesize that these objects are pointed nearly at us and we see Doppler-boosted radiation. The ones we see form a small subset of a large population that is oriented at random.

It can be shown that the sample will preferentially contain objects with $(v/c) < \gamma < \delta < \csc\theta$, in spite of the small solid angle associated with such small values of θ. This comes about because of the strong Doppler boost; sources with small θ can be seen to great distances and the increasing volume overcomes the small solid angle. In Fig. 2 the objects in the sample should mainly lie below the diagonal $v/c = \delta$ and above the line $\gamma\theta_{deg} = 10$. The plotted points represent the four sources where there are independent estimates of $(v/c)h$ and δ_{min}. No valid conclusions can be generated from only four points, but they do not seem to lie in the expected part of the diagram. They possibly suggest that $\delta > \delta_{min}$, on average. Note that the location of the points depends on the Hubble constant. With a larger and more accurate set of data we might be able to place constraints on the Hubble constant, although since the scheme outlined above only gives δ_{min} and not δ, there always will be ambiguity.

Lynden-Bell (1977) was the first to discuss the Hubble constant in this context, although his model is no longer considered valid. Marscher and Broderick (1982) discussed the derivation of the Hubble constant from combined superluminal and X-ray data. Detailed calculation of the statistical distributions for beamed sources was made by Scheuer and Readhead (1979).

B. Orientation-Independent Sample

The sources in Table 1 are mainly pointed at us and we cannot study their statistics without a beaming theory. A more powerful procedure would be to define an orientation-independent sample and then, for example, we could study the distribution of (v/c), which would be characteristic of sources with a known (random) distribution of orientations. The sample has to be picked on the basis of some quantity that is independent of the bright central source. It could be the strength of the optical spectral lines, or of the outer radio lobes. However, hidden correlations may exist between these quantities and the central jet, so that any discussion will require great care.

This problem is being attacked in various ways, and some of the work is reported by Porcas (1984) and by Antonucci and Ulvestad (1985). One interesting preliminary result comes from radio studies of "triple" quasars. A triple source is one that has large strong outer lobes and a weak central core and it often has faint outer jets going toward the lobes. The core may be relativistic but this has little effect on the selection of a sample because the (nonrelativistic) lobes give most of the flux density. In a number of well-studied cases the inner and outer jets are one-sided and roughly colinear, or at least on the same side (Owen and Puschell 1984; Porcas 1985). This also is true of all the superluminal sources. As Scheuer (1984) points out, this means that if the inner jet (10 pc scale) is relativistic, the outer jet (100 kpc scale) should be relativistic also. This follows from a realization that the inner jet is one-sided because it is Doppler-boosted, and not because that side is intrinsically stronger. The stronger outer jet is on the same side, and if it were not Doppler-boosted we would expect it to be on the opposite side half the time, on average. It therefore appears that these beams can maintain a relativistic flow with $\gamma = 5$ or 10 for hundreds of kiloparsecs, and that the ejection direction can be constant for at least a million years! Owen and Puschell (1984), however, argue that $\gamma < 2$, on average, for the outer jets, because their distribution in intensity does not match the one expected for more highly relativistic cases.

C. The Parent Population

The strong core-dominated radio sources can be understood in terms of Doppler-boosted radiation from a beam with Lorentz factor γ and oriented at angle θ to the line of sight. This means that they form part of a larger group that is oriented randomly in the sky, so that there are about $8\gamma^2$ times as many "misdirected" beams as the strong ones we see coming toward us. (This assumes that the beams are bidirectional and that a typical core-dominated source has $\theta \approx (2\gamma)^{-1}$). The nine superluminal sources must have about 1800 weak counterparts if $\gamma = 5$. What are these myriad sources?

Two different populations have been suggested for the misdirected sources: the "radio-quiet" quasars, and the quasars with steep-spectrum extended radio lobes, the classical "radio doubles." Although quasars were originally discovered through their radio properties, they are defined by their optical properties and in fact the great majority of them are very weak radio emitters. The weak ones are called "radio-quiet" for historical reasons: they were below the limit of detection until the Very Large Array (VLA), two years ago, became able to measure their flux density. Scheuer and Readhead (1979) calculated the expected ratio of radio-quiet to radio-loud quasars, using the relativistic beam model and a power law distribution of intrinsic strengths. They found the ratio to be a few times γ^2, in agreement with the crude estimate in the paragraph above, and in rough agreement with the observations. By using the VLA, Kellermann et al. (1983) were able to measure the flux density of many weak quasars. They found that the overall range in quasar flux density is greater than 10^3 and that the distribution is consistent with the beaming model if $\gamma \leq 10$. This is consistent with the superluminal and the X-ray data, but there is no independent evidence to suggest that these optically bright quasars actually contain relativistic beams. They may simply be a subset of the overall radio source population, which has similar statistics.

More attention is now devoted to the possibility that superluminal sources can be unified with the classical radio doubles. The major impetus for this is the discovery that nearly all the superluminal sources are surrounded by large, low-surface-brightness halos, which could be the extended radio lobes seen end-on (Browne et al. 1982; Schilizzi and de Bruyn 1983; de Bruyn and Schilizzi, 1984; Antonucci and Ulvestad, 1985). These extended halos in fact can produce a substantial flux density S_{ext}, and in several cases S_{ext} is big enough that the sources would be in the flux-limited catalogues even if the central cores were missing. Dynamical arguments show that the outer lobes are not relativistic and therefore radiate isotropically; and if the beam model is correct, we must believe that the catalogues contain many misdirected beamed sources that appear as extended doubles with weak cores. This is particularly true of the 3C catalogue, originally made at 159 MHz, where the outer lobes dominate the radiation in most sources and there is little bias for sources aimed at us. Seven of the superluminal (or possibly superluminal) sources in Table 1 are in the original 3C catalogue (Edge et al. 1959) and three of the superluminals, 3C 179, 3C 273, and 3C 279, have outer structure that by itself qualifies for inclusion in 3C. 3C 273, however, is a special case. Its outer structure consists exclusively of a one-sided jet that may be relativistic, and if it were turned sideways it probably would become very weak and would not be in 3C. Thus two superluminal sources would be in the catalog even if their cores were misdirected, and so with $\gamma \approx 5$ we expect that roughly 400 of the 471 entries in 3C are extended sources with weak misdirected cores. This fits, since (1) the statistical error associated with two objects is enormous, and (2) it is known that most of the 3C sources are extragalactic and indeed are extended with weak cores. Their study is of great interest. If as many as half a dozen of the extended 3C objects show superluminal motion, then the theory will be strained.

The unification of superluminal sources with the steep-spectrum extended sources was originally suggested by Blandford and Königl (1979) and has been discussed by a number of other authors; see the reviews by Begelman, Blandford, and Rees (1984) and by Scheuer (1984). The number of objects is small, but so far the picture generally is self-consistent with $\gamma \approx 5$ (Orr and Browne 1982). This is roughly consistent with the X-ray data and with the superluminal data if H_0 is of order 100 km $\sec^{-1}$ Mpc^{-1}. However, some difficulties remain. Schilizzi and de Bruyn (1983) pointed out that deprojection by a factor γ makes the superluminal sources the largest objects for their redshift class. This is unacceptable if they are merely average sources accidentally pointing at us. However, this may not be a problem, since some superluminal sources have a curved beam, and this might destroy the colinearity of the pc- and hundred kpc-scale components. The one-sidedness also may be a problem—the generally double nature of the extended structure cells for a bidirectional beam, but the high-luminosity VLBI beams are nearly all one-sided. Doppler boosting can explain this but not if the beam is close to the plane of the sky, as it should be in some cases. For relevant discussions concerning the famous radio galaxy Cygnus A, which is known to lie close to the plane of the sky, see Linfield (1982); Saikia and Wiita (1982); and Perley et al. (1984).

Finally, we mention 3C 273, which presents a unique problem. It is close and is much the brightest quasar; but simultaneously it shows superluminal motion giving $\theta < 18°$, and it is the only superluminal source for which the outer structure (a long linear jet) is entirely one-sided. Are these characteristics all coincidental? Apart from proximity, they can be explained if θ is small and the inner and outer jets are both relativistic, although the absence of low-surface-brightness outer lobes is peculiar. The quasar continuum light itself should be unboosted, since the lines are not blue-shifted, and the optical line-to-continuum ratio is normal. Therefore there should be other objects like 3C 273, close and bright optically but with misdirected beams and no visible radio or optical jets. But there are no other bright close quasars. Thus 3C 273 appears to be a special case, accidentally placed and oriented to tantalize us with its interesting properties.

ACKNOWLEDGMENTS

I am grateful to T. J. Pearson and G. A. Seielstad for unpublished flux densities from the Owens Valley Radio Observatory, and to J. A. Biretta for a critical reading of this manuscript and useful comments on the statistics of beamed radio sources. This work was partially supported by the National Science Foundation.

REFERENCES

Andrew, B. H., MacLeod, J. M., Harvey, G. A., and Medd, W. J., 1978, *Astron. J.*, **83,** 863.

Antonucci, R., and Ulvestad, J. S., 1985, *Ap. J.*, **294,** 158.

Bååth, L. B., Cotton, W. D., Counselman, C. C., Shapiro, I. I., Wittels, J. J., Hinteregger, H. F., Knight, C. A., Rogers, A. E. E., Whitney, A. R., Clark, T. A., Hutton, L. K., and Niell, A. E., 1980, *Astr. Ap.* **86,** 364.

Bååth, L. B., 1984, in *IAU Symposium 110, VLBI and Compact Radio Sources*, R. Fanti, K. I. Kellermann, and G. Setti (eds.), Reidel, Dordrecht, p. 127.

Bartel, N., Ratner, M. I., Shapiro, I. I., and Herring, T. A., 1984, in *IAU Symposium 110, VLBI and Compact Radio Sources*, R. Fanti, K. I. Kellermann, and G. Setti (eds.), Reidel, Dordrecht, p. 113.

Begelman, M. C., Blandford, R. D., and Rees, M. J., 1984, *Rev. Mod. Phys.*, **56,** 255.

Benson, J. M., Walker, R. C., Seielstad, G. A. and Unwin, S. C., 1984, in *IAU Symposium 110, VLBI and Compact Radio Sources*, R. Fanti, K. I. Kellermann, and G. Setti (eds.), Reidel, Dordrecht, p. 125.

Biretta, J. A., 1985, Ph. D. thesis, California Institute of Technology.

Biretta, J. A., Cohen, M. H., Unwin, S. C., and Pauliny-Toth, I. I. K. 1983, *Nature*, **306,** 42.

Blandford, R. D., and Königl, A., 1979, *Ap. J.*, **232,** 34.

Blandford, R. D., and Rees, M. J., 1974, *M.N.R.A.S.*, **169,** 395.

Boroson, T. A., and Oke, J. B., 1984, *Ap. J.*, **281,** 535.

Bregman, J. N., Glassgold, A. E., Huggins, P. J., Aller, H. D., Aller, M. F., Hodge, P. E., Rieke, G. H., Lebofsky, M. J., Pollock, J. T., Pica, A. J., Leacock, R. J., Smith, A. G., Webb, J., Balonek, T. J., Dent, W. A., O'Dea, C. P., Ku, W. H.-M., Schwartz, D. A., Miller, J. S., Rudy, R. J., LeVan, P. D., 1984, *Ap. J.*, **276,** 454.

Bridle, A. H., and Perley, R. A., 1984, *Ann. Rev. Astron. Astrophys.*, **22,** 319.

Brown, R. L., Johnston, K. J., Briggs, F. H., Wolfe, A. M., Neff, S. G., Walker, R. C., 1981, *Astroph. Lett.*, **21,** 105.

Browne, I. W. A., Clark, R. R., Moore, P. K., Muxlow, T. W. B., Wilkinson, P. N., Cohen, M. H., and Porcas, R. W., 1982, *Nature*, **299,** 788.

Burbidge, G. R., Jones, T. W., and O'Dell, S. L., 1974, *Ap. J.*, **193,** 43.

Cohen, M. H., 1985, *Proceedings of Bangalore Winter School on Energetic Extragalactic Sources.*

Cohen, M. H., Kellermann, K. I., Shaffer, D. B., Linfield, R. P., Moffet, A. T., Romney, J. D., Seielstad, G. A., Pauliny-Toth, I. I. K., Preuss, E., Witzel, A., Schilizzi, R. T., Geldzahler, B. J., 1977, *Nature*, **268,** 405.

Cohen, M. H., Unwin, S. C., Lind, K. R., Moffet, A. T., Simon, R. S., Wilkinson, P. N., Spencer, R. E., Booth, R. S., Nicolson, G. D., Niell, A. E., Young, L. E., 1983b, *Ap. J.*, **272,** 383.

Cohen, M. H., Cannon, W., Purcell, G. H., Shaffer, D. B., Broderick, J. J., Kellermann, K. I., and Jauncey, D. L., 1971, *Ap. J.*, **170,** 207.

Cohen, M. H., and Unwin, S. C., 1984, in *IAU Symposium 110, VLBI and Compact Radio Sources*, R. Fanti, K. I. Kellermann, and G. Setti (eds.), Reidel, Dordrecht, p. 95.

Cohen, M. H., Unwin, S. C., Pearson, T. J., Seielstad, G. A., Simon, R. S., Linfield, R. P., and Walker, R. C., 1983a, *Ap. J.* (*Lett.*), **269,** L1.

Cotton, W. D., Wittels, J. J., Shapiro, I. I., Marcaide, J., Owen, F. N., Spangler, S. R., Rius, A., Angulo, C., Clark, T. A., and Knight, C. A., 1980, *Ap. J.* (*Lett.*), **238,** L123.

de Bruyn, A. G., and Schilizzi, R. T., 1984, in *IAU Symposium 110, VLBI and Compact Radio Sources*, R. Fanti, K. I. Kellermann, and G. Setti (eds.), Reidel, Dordrecht, p. 165.

Eckart, A., and Witzel, A., 1984, in *IAU Sympsoium 110, VLBI and Compact Radio Sources*, R. Fanti, K. I. Kellermann, and G. Setti (eds.), Reidel, Dordrecht, p. 65.

Eckart, A., Witzel, A., Biermann, P., Pearson, T. J., Readhead, A. C. S., and Johnston, K. J., 1985, *Ap. J.* (*Lett.*) **296,** L23.

Edge, D. O., Shakeshaft, J. R., McAdam, W. B., Baldwin, J. E., and Archer, S., 1959, *Mem. R.A.S.*, **68,** 37.

Gould, R. J. 1979, *Astr. Ap.*, **76,** 306.

Hoyle, F., Burbidge, G. R., and Sargent, W. L. W., 1966, *Nature*, **209,** 751.

Jones, D. L., 1984, *Ap. J.*, **287,** 33.

Jones, T. W., 1979, *Ap. J.*, **233,** 796.

Jones, T. W., O'Dell, S. L., and Stein, W. A., 1974, *Ap. J.*, **188,** 353.

Kellermann, K. I., Clark, B. G., Cohen, M. H., Shaffer, D. B., Broderick, J. J., and Jauncey, D. L., 1973, *Ap. J.* (*Lett.*), **179,** L141.

Kellermann, K. I., and Pauliny-Toth, I. I. K. 1969, *Ap. J.* (*Lett.*), **155,** L71.

Kellermann, K. I., Sramek, R., Shaffer, D., Schmidt, M., and Green, R., 1983, in *Quasars and Gravitational Lenses: Proc. No. 24 International Astrophysics Colloquium, Liege*, Université de Liege, Institut d'Astrophysique, p. 81.

Königl, A., 1981, *Ap. J.*, **243,** 700.

Lawrence, C. R., Readhead, A. C. S., Linfield, R. P., Payne, D. G., Preston, R. A., Schilizzi, R. T., Porcas, R. W., Booth, R. S., Burke, B. F., 1985, *Ap. J.*, **296,** 458.

Linfield, R., 1982, *Ap. J.*, **254,** 465.

Longair, M. S., Ryle, M., and Scheuer, P. A. G., 1973, *M.N.R.A.S.*, **164,** 243.

Lynden-Bell, D., 1977, *Nature*, **270,** 396.

Marscher, A. P., 1977, *Ap. J.*, **216,** 244.

Marscher, A. P., 1980, *Nature*, **288,** 12.

Marscher, A. P., 1983, *Ap. J.*, **264,** 296.

Marscher, A. P. and Broderick, J. J., 1981, *Ap. J.*, **249,** 406.

Marscher, A. P. and Broderick, J. J., 1982, *Ap. J.* (*Lett.*), **255,** L11.

Marscher, A. P. and Broderick, J. J., 1985, *Ap. J.* **290,** 735.

Medd, W. J., Andrew, B. H., Harvey, G. A. and Locke, J. L., 1972, *M.N.R.A.S.*, **77,** 109.

Moore, R. L., Readhead, A. C. S., and Bååth, L. B., 1983, *Nature*, **306,** 44.

Mutel, R. L. and Phillips, R. B., 1980, *Ap. J.* (*Lett.*), **241,** L73.

Mutel, R. L. 1984, in *IAU Symposium 110, VLBI and Compact Radio Sources*, R. Fanti, K. I. Kellermann, and G. Setti (eds.), Reidel, Dordrecht, p. 117.

Orr, M. J. L., and Browne, I. W. A., 1982, *M.N.R.A.S.*, **200,** 1067.

Owen, F. N. and Puschell, J. J., 1984, *Astron. J.*, **89,** 932.

Pauliny-Toth, I. I. K., Porcas, R. W., and Zensus, A. 1984, in *IAU Symposium 110, VLBI and Compact Radio Sources*, R. Fanti, K. I. Kellermann, and G. Setti (eds.), Reidel, Dordrecht, p. 149.

Pearson, T. J. and Readhead, A. C. S., 1984, in *IAU Symposium 110, VLBI and Compact Radio Sources*, R. Fanti, K. I. Kellermann, and G. Setti (eds.), Reidel, Dordrecht, p. 15.

Perley, R. A., 1984, in *IAU Sympsoium 110, VLBI and Compact Radio Sources*, R. Fanti, K. I. Kellermann, and G. Setti (eds.), Reidel, Dordrecht, p. 153.

Perley, R. A., Dreher, J. W., and Cowan, J. J. 1984, *Ap. J.* (*Lett.*), **285,** L35.

Porcas, R. W., 1984, in *IAU Symposium 110, VLBI and Compact Radio Sources*, R. Fanti, K. I. Kellermann, and G. Setti (eds.), Reidel, Dordrecht, p. 157.

Porcas, R. W., 1985, in *Active Galactic Nuclei*, J. Dyson (ed.) Manchester University Press, Manchester, p. 22.

Preuss, E., Alef, W., Whyborn, N., Wilkinson, P. N., and Kellermann, K. I. 1984, in *IAU Symposium 110, VLBI and Compact Radio Sources*, R. Fanti, K. I. Kellermann, and G. Setti (eds.), Reidel, Dordrecht, p. 29.

Readhead, A. C. S., Hough, D. H., Ewing, M. S., Walker, R. C., and Romney, J. D., 1983, *Ap. J.*, **265,** 107.

Readhead, A. C. S., Pearson, T. J., and Unwin, S. C., 1984, in *IAU Symposium 110, VLBI and Compact Radio Sources*, R. Fanti, K. I. Kellermann, and G. Setti (eds.), Reidel, Dordrecht, p. 131.

Rees, M. J., 1967, *M.N.R.A.S.*, **135,** 345.

Rees, M. J., 1971, *Nature*, **229,** 312.

Reid, M. J., Schmitt, H. M. M., Wilkinson, P. N., and Johnston, K. J., 1984, in *IAU Symposium 110, VLBI and Compact Radio Sources*, R. Fanti, K. I. Kellermann, and G. Setti (eds.), Reidel, Dordrecht, p. 145.

Rickett, B. J., Coles, W. A., and Bourgois, G., 1984, *Astron. Astrophys.*, **134,** 390.

Romney, J. D., Padrielli, L., Bartel, N., Weiler, K. W., Ficarra, A., Mantovani, F., Bååth, L. B., Kogan, L., Matveenko, L., Moiseev, I. G., and Nicholson, G., 1984a, *Astron. Astrophys.*, **135,** 289.

Romney, J. D., Alef, W., Pauliny-Toth, I. I. K., Preuss, E., and Kellermann, K. I. 1984b, in *IAU Symposium 110, VLBI and Compact Radio Sources*, R. Fanti, K. I. Kellermann, and G. Setti (eds.), Reidel, Dordrecht, p. 137.

Saikia, D. J., and Wiita, P. J., 1982, *M.N.R.A.S.*, **200,** 83.

Scheuer, P. A. G., 1984, in *IAU Symposium 110, VLBI and Compact Radio Sources*, R. Fanti, K. I. Kellermann, and G. Setti (eds.) Reidel, Dordrecht, p. 197.

Scheuer, P. A. G., and Readhead, A. C. S., 1979, *Nature*, **277,** 182.

Scheuer, P. A. G., and Williams, P. J. S., 1968, *Ann. Rev. Astron. Astrophys.*, **6,** 321.

Schilizzi, R. T., and de Bruyn, A. G., 1983, *Nature*, **303,** 26.

Schmidt, M. 1972, *Ap. J.*, **176,** 273.

Schwartz, D. A., Madejski, G., and Ku, W. H.-M., 1982, in *IAU Symposium 97, Extragalactic Radio Sources*, D. S. Heeschen and C. M. Wade (eds.), Reidel, Dordrecht, p. 383.

Seielstad, G. A., Pearson, T. J., and Readhead, A. C. S., 1983, *PASP*, **95,** 842.

Shaffer, D. B., 1984, in *IAU Symposium 110, VLBI and Compact Radio Sources*, R. Fanti, K. I. Kellermann, and G. Setti (eds.), Reidel, Dordrecht, p. 135.

Simon, R. S., Readhead, A. C. S., Moffet, A. T., Wilkinson, P. N., Allen, B., and Burke, B. F., 1983, *Nature*, **302,** 487.

Simon, R. S., Readhead, A. C. S., Wilkinson, P. N., 1984, in *IAU Symposium 110, VLBI and Compact Radio Sources*, R. Fanti, K. I. Kellermann, and G. Setti (eds.), Reidel, Dordrecht, p. 111.

Unwin, S. C., Cohen, M. H., Biretta, J. A., Pearson, T. J., Seielstad, G. A., Walker, R. C., Simon, R. S., and Linfield, R. P., 1985, *Ap. J.*, **289,** 109.

Unwin, S. C. and Biretta, J. A., 1984, in *IAU Symposium 110, VLBI and Compact Radio Sources*, R. Fanti, K. I. Kellermann, and G. Setti (eds.), Reidel, Dordrecht, p. 105.

Unwin, S. C., Cohen, M. H., Pearson, T. J., Seielstad, G. A., Simon R. S., Linfield, R. P. and Walker, R. C., 1983, *Ap. J.*, **271,** 536.

Walker, R. C., Benson, J. M., Seielstad, G. A., and Unwin, S. C., 1984, in *IAU Symposium 110, VLBI and Compact Radio Sources*, R. Fanti, K. I. Kellermann, and G. Setti (eds.), Reidel, Dordrecht, p. 121.

Whitney, A. R., Shapiro, I. I., Rogers, A. E. E., Robertson, D. S., Knight, C. A., Clark, T. A., Goldstein, R. M., Marandino, G. E., Vandenberg, N. R., 1971, *Science*, **173,** 225.

Wilkinson, P. N., Spencer, R. E., Readhead, A. C. S., Pearson, T. J., and Simon, R. S., 1984, in *IAU Symposium 110, VLBI and Compact Radio Sources*, R. Fanti, K. I. Kellermann, and G. Setti (eds.), Reidel, Dordrecht, p. 25.

Worrall, D. M., Puschell, J. J., Bruhweiler, F. C., Miller, H. R., Rudy, R. J., Ku, W. H.-M., Aller, M. F., Aller, H. D., Hodge, P. E., Matthews, K., Neugebauer, G., Soifer, B. T., Webb, J. R., Pica, A. J., Pollock, J. T., Smith, A. G., and Leacock, R. J., 1984, *Ap. J.*, **278,** 521.

Worrall, D. M., 1984, *IAU Symposium 110, VLBI and Compact Radio Sources*, R. Fanti, K. I. Kellermann, and G. Setti (eds.), Reidel, Dordrecht, p. 187.

11.

Space Telescope and Cosmology

RICCARDO GIACCONI

Riccardo Giacconi is the Director of the Space Telescope Science Institute at The Johns Hopkins University in Baltimore, Maryland. One of the pioneers of the field of X-ray astronomy, he led the extremely successful UHURU and Einstein Observatory X-ray satellite projects. He has been at the forefront of the design and launch of detectors that have opened up new windows on the universe.

I. INTRODUCTION

When Space Telescope begins operations in 1986, it will provide a major improvement in observational capabilities for optical astronomy. It will, in some sense, represent the first qualitative improvement in telescope capabilities in the optical domain since the completion in 1948 of the 200-in. Hale Telescope at Palomar Mountain.

The unique capabilities of Space Telescope derive from the fact that the Observatory will be placed into orbit around the earth, so that its "seeing" will be unaffected by the obscuring and distorting effects of the earth's atmosphere. This will allow the attainment of an angular resolution, a sensitivity, and a wavelength coverage unachievable from the ground.

Space Telescope (ST) is a joint National Aeronautics and Space Administration (NASA) and European Space Agency (ESA) project to launch a 2.5-m optical telescope in near earth orbit (Fig. 1). ST is provided with a diverse complement of focal plane instruments, including cameras and spectrometers, which can be changed and improved over time. It will be used as a general-purpose observatory available to astronomers from all over the world to carry out studies impinging on all areas of current astronomical interest.

The launch, currently scheduled for late 1986, will occur by means of

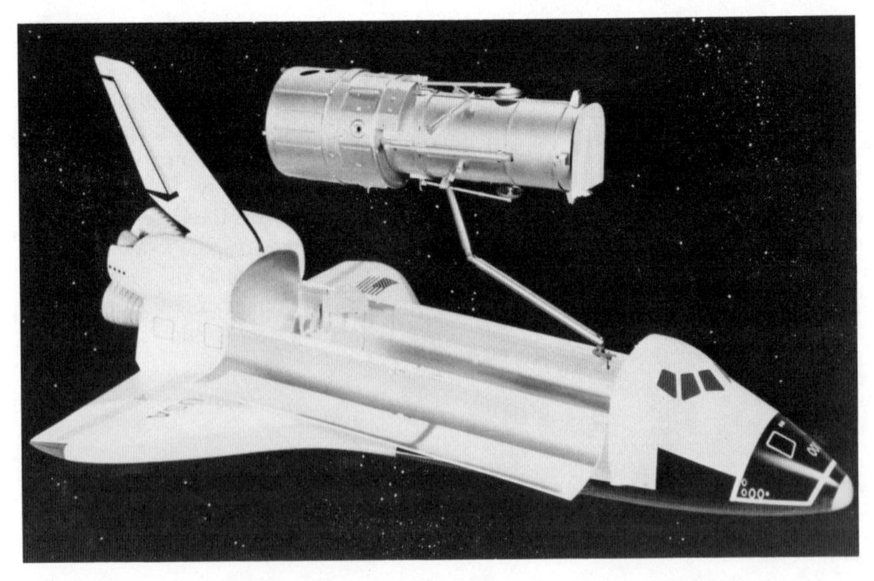

FIGURE 1. Artist's conception of the Space Telescope Observatory being deployed by the shuttle.

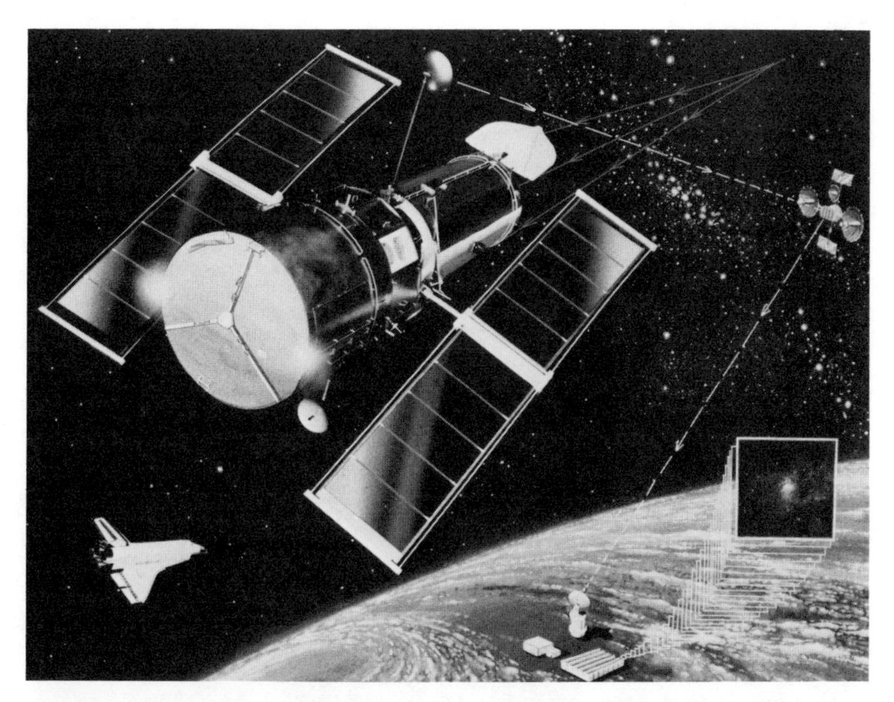

FIGURE 2. Data from and commands to Space Telescope are transmitted through the tracking and data relay satellite system.

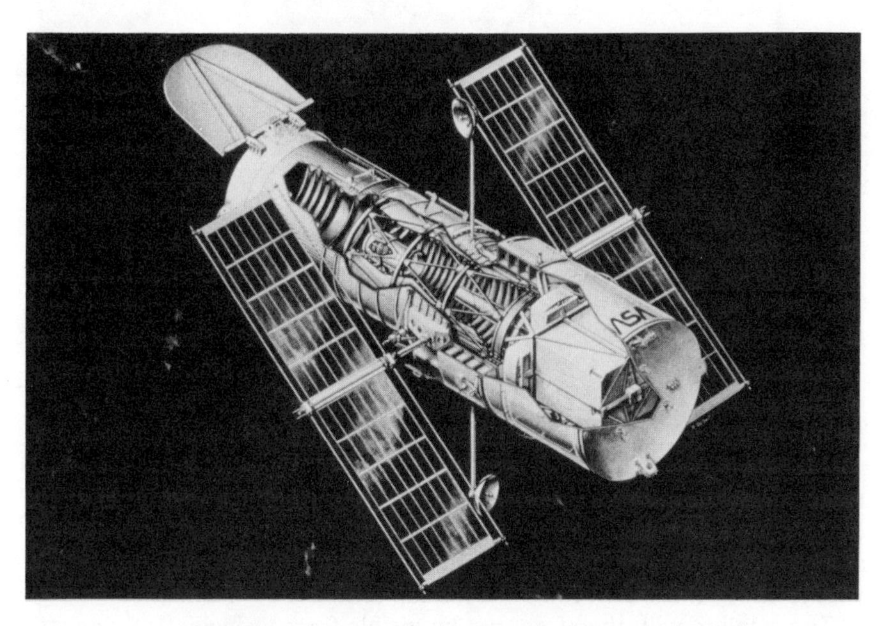

FIGURE 3. Cutout view of ST major components.

FIGURE 4. ST spacecraft outer shell and support module.

the Space Shuttle Orbiter. The shuttle system will be used not only to carry the observatory aloft but to service it at regular intervals to achieve a lifetime of at least 15 years. In order to be compatible with current shuttle capabilities the orbit is at 500 km altitude, 30° inclination.

The satellite will, therefore, not be visible from a single receiving station on Earth for much of the orbit. To maintain continuous accessibility

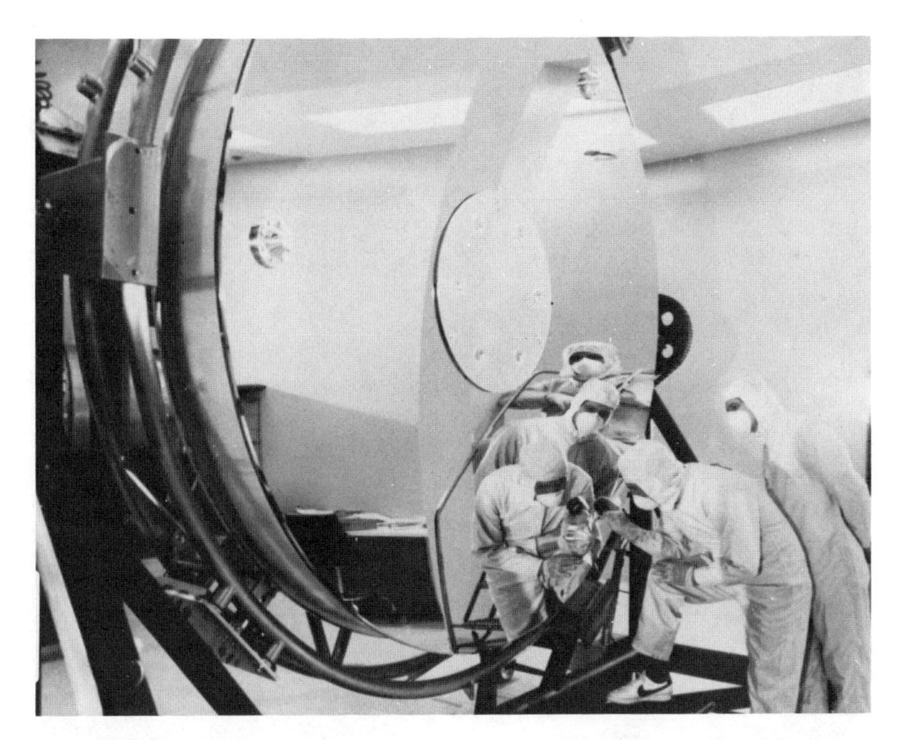

FIGURE 5. The ST mirror is inspected at Perkin Elmer after vacuum coating.

for uplink and downlink of commands and data the satellite will utilize the Tracking and Data Relay Satellite System (TDRSS), a system of three geosynchronous orbit communication satellites (Fig. 2).

The ST project is one of the most complex scientific missions yet undertaken by NASA. Observing time on ST will be heavily subscribed and is a precious commodity. We ultimately expect that the archives of the ST data will be as widely distributed and used as the 1950 Palomar Observatory Sky Survey.

An idea of the physical size of the ST major components can be gained from Fig. 3. The spacecraft's outer shell and support module is on schedule at Lockheed (Fig. 4); the mirror fabrication and vacuum coating is completed (Fig. 5); the Optical Telescope Assembly (OTA) (Fig. 6) has been integrated at Lockheed with the spacecraft.

II. INSTRUMENTATION

Description of Space Telescope and its instruments is contained in articles by several authors in *The Space Telescope Observatory* (Hall 1982). Lon-

FIGURE 6. The optical bench constructed of graphite apoxy is ready for integration into the optical telescope assembly.

gair and Warner (1979) and Macchetto et al. (1979) contain contributions by a number of authors on the uses of ST for specific astronomical applications.

The focal plane configuration of the telescope is shown in Fig. 7. Light collected by the 2.4-m-diameter hyperbolic primary mirror is directed onto a 0.3-m-diameter secondary mirror (also hyperbolic) located 4.9 m away. The secondary forms a beam that comes to a focus 1.5 m behind a central hole in the primary mirror. The on-axis beam is directed by a pickoff mirror to feed the 2.7 arcminute square field of the Wide Field Planetary Camera. The axial bay instruments, the Faint Object Camera, the Faint Object Spectrograph, the High Resolution Spectrograph, and the High Speed Photometer occupy the four quadrants about the optical

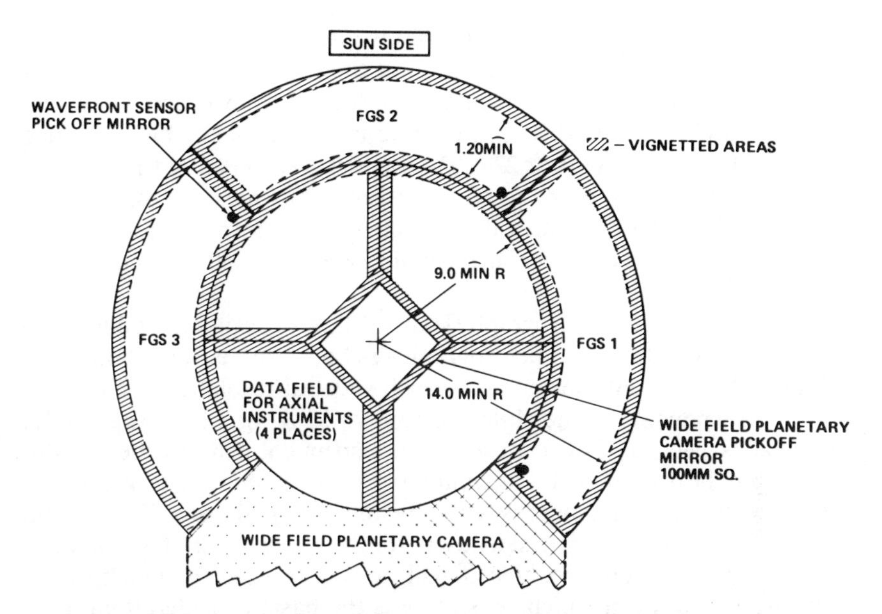

FIGURE 7. Map of the focal plane showing the segments used for scientific instruments and fine guidance sensors.

axis. Three Fine Guidance Sensors, each with an associated optical control subsystem, occupy quadrant "pickles" at the perimeter of the field.

The primary and secondary mirrors of the Space Telescope have a reflecting layer of aluminum overcoated with MgF_2. This combination provides broad wavelength coverage from below Lyman α (the measured reflectivity at 121.6 nm is $\gtrsim$ 70%) to submillimeter wavelengths with good reflectivity at the optical wavelengths ($\gtrsim$ 85% at 632.8 nm).

The following paragraphs contain brief descriptions of the scientific instruments.

The Wide Field Planetary Camera can be operated in two modes (Hall 1982). The first gives a 2.7×2.7 arcmin^2 field (planetary). Each pixel of the detector corresponds to 0.1 arcsec in the WF mode (wide-field) and to 0.043 arcsec in the PC mode (planetary camera). The WF mode yields the largest field of view available on ST at the expense of some resolution; the PC mode attains nearly the full resolution of the optics. The camera is sensitive in the range 115.0–1100.0 nm. Recent measurements of extremely low noise characteristics insure that the instrument can achieve sensitivity of better than apparent visual magnitude $m_v = 28$, for point sources in 1-hr. exposure. The camera will certainly be one of the most versatile and widely used instruments on ST.

The Faint Object Camera (Hall 1982) is intended as a complementary instrument to the WF/PC and will exploit the full performance of ST by attaining the highest angular resolution and sensitivity in a narrower wavelength band (120–500 nm). It can operate in two modes, which produce images of 11×11 arcsec^2 or 22×22 arcsec^2. Each pixel of the detector in the two modes will correspond to 0.02 and 0.04 arcsec, respectively. The instrument records individual photons and, as a special feature, the instrument can achieve very high angular resolution (0.007 arcsec) in very narrow fields.

The Faint Object Spectrograph (Hall 1982) is a sensitive, medium-resolution spectrometer. The spectrograph will be sensitive from about 115–800 nm. The different modes can furnish spectral resolution $\lambda/\Delta\lambda$ of 10^2 or 10^3 corresponding to the ability to resolve 30 Å, or 3 Å, respectively. The smallest slit size is 0.1 arcsec. The instrument can also be used to study the polarization of the incoming light and to study time variability of the spectra of bright sources with a resolution of 10 msec. The system is designed to have adequate sensitivity for the study of very faint objects. It is anticipated, therefore, that it will be widely used for much of the spectroscopic work in which sensitivity is the basic consideration.

The High Resolution Spectrograph (Hall 1982) is primarily intended for use in the 110–320 nm region with resolution of $\lambda/\Delta\lambda$ 2×10^4 or 1.0×10^5. The highest resolution corresponds to 0.05 Å, and is obtainable only for relatively bright objects. Sensitivity of this instrument will be some 100 times better than that of the International Ultraviolet Explorer (IUE), with finer spectroscopic resolution and photometric accuracy.

TABLE 1. Scientific Instruments

Instrument	Field	Resolution	Band (Å)	Limit
Wide Field/ Planetary Camera	2.7′;1.2′□	0.1″;0.04″	1150–11,000 Å	~28^m
Faint Object Camera	11″;22″□	0.02″;0.04″	1200–6000	28^m
Faint Object Spectrograph	0.1–4.3″	3 Å;30 Å	1150–7000	22;26^m
High Resolution Spectrograph	0.25–2″	0.03;0.15;1.5 Å	1100–3200	11;14;17^m
High Speed Photometer	0.4″,1″,10″	16 μsec	1200–8000	24^m
Fine Guidance System	69□′	0.003″	4670–7000	17^m

The High Speed Photometer/Polarimeter (Hall 1982) is designed to carry out high time resolution studies of sources over a wide band of wavelengths, from 120–700 nm, over intervals as short as 10 μsec for the brightest sources. A number of filters can be used for broad-band spectroscopy. Apertures from 0.1 to 2.8 arcsec can be used. Polarimetry of the focused radiation can be carried out in the near ultraviolet. The instrument is designed to be simple, precise, and rugged. It can be used to study, for instance, time variability of some of the optical counterparts of collapsed objects such as stellar mass black holes or neutron stars. More generally, it will be used to establish photometric standards and time variability for a number of stellar objects.

The capabilities of the instruments are summarized in Table 1.

III. IMPACT OF ST ON ASTRONOMY

The broad wavelength coverage, angular resolution and sensitivity of the ST observatory will allow us to perform a number of observations currently unfeasible with ground-based instruments. Several scientific symposia and workshops have addressed the potential scientific returns to be expected from ST in each subdiscipline of astronomy. A particularly effective summary by Longair is contained in (Hall 1982).

ST will make a fundamental contribution to the determination of the distance scale and age of the universe, by providing an extension of a factor of 10 in the distance over which we can measure the properties of the standard candles, used in the logical ladder leading to the determination of the Hubble constant.

Study of clusters of galaxies and individual galaxies at larger redshifts than possible from the ground will provide significant tests of cosmological models and cosmic evolution.

The extension of morphological classification of galaxies to redshifts of order unity and the extension of stellar population studies to very faint levels will provide important clues on the formation, dynamics, and evolution of galaxies.

Study of the central regions of active galactic nuclei and quasars (QSOs) can provide the basic data necessary to understand the physics of the nuclear regions of galaxies, and the source of the tremendous energies released there, possibly by accretion onto giant collapsed objects.

ST promises the extension of stellar studies in our own galaxy to the faint population of stars that populate the corona. Studies of physical conditions and chemical composition of the intersteller medium and studies of the injection of material through stellar winds, evolution of outer atmospheres of stars, coronal winds, circumstellar shells, planetary nebulae and supernova remnants can be extended with ST to the UV range of wavelengths.

An area of great potential for ST observations relates to the study of the birth and evolution of single and binary stars and globular star clusters. Future extensions of ST capabilities to longer wavelengths in the infrared will further improve the capabilities for this type of research.

Astrometric measurements with ST promise also an improvement in precision of a factor 10. This will have a dramatic effect on many fundamental problems of astronomy. Parallaxes can be obtained with the same precision for objects ten times more distant than possible from the ground. Visual orbits of some spectroscopic binaries can be obtained, leading to a substantial improvement in our knowledge of stellar masses. A search can be carried out for planetary systems of other stars. Finally, a link can be forged between optical, radio, and dynamical reference frames, and the inertial reference frame can be tied down by measurement of star motions relative to QSOs.

In planetary astronomy ST will play a major role in the coming decade because of the possibility of performing synoptic studies of planetary atmospheres and environments, high resolution and UV studies of comets, and further inventory of the solar system by searching for faint satellites and rings. ST will also permit, either through direct imaging or through astrometric measurements, the search for planetary systems of several hundred nearby stars.

It is clear that the use of ST will have a predictable and substantial impact on all subdisciplines of astronomy. Even more exciting is the potential offered by such a major improvement in observational capabilities for new and unsuspected discoveries.

Although I can not deal in detail with all of the aspects of astronomical research with ST, I have chosen to discuss in some greater detail the impact of ST observations on cosmology.

IV. SOME ASPECTS OF COSMOLOGICAL RESEARCH WITH ST

To set the problems in some perspective, I will adopt the generally accepted view that we live in a Friedmann universe (Big Bang model). The universe started in a singularity that cannot be studied directly by observation of photons (electromagnetic radiation). In fact, it is only some 10^6 years after the origin that the original fireball, matter and radiation, had sufficiently expanded so that radiation could escape. The radiation first escaping from the furthest horizon we can "see" has been detected.

TABLE 2. Use of ST for Cosmology

Relics of the Hot Big Bang
H, D, ^{4}He, ^{7}Li
Baryonic mass density, isotropy, primordial fluctuations
The Cosmological Model
Test of world models
Determination of H_0, q_0, Ω, Λ
The mass problem
Evolution
Evolution of astrophysical objects over time $\tau \sim \frac{1}{H_0}$
Evolution of galaxies, QSOs, AGNs, radio and X-ray galaxies
Formation of Structure in the Universe
Cross correlation functions
Formation and dynamic evolution of galaxies and clusters
Origin of the Big Bang
??

It is the 3 K microwave background radiation discovered in 1965 by Arno Penzias and Robert Wilson. This discovery and the concordance of the predicted cosmological production of light elements with the observed abundances are taken as strong evidence for a hot big bang. In addition Hawking, Ellis, and Penrose have proved that in general relativity theory the existence of the 3 K background implies that the universe began from a hot big bang.

Table 2 summarizes some aspects of cosmological research to which ST can contribute.

V. RELICS OF THE HOT BIG BANG

Advances in elementary particle physics, including the development of Grand Unified Theories, make it possible to pursue some of the most fundamental problems of cosmology such as the origin of the isotropy of the universe, the origin of baryon asymmetry, the origin of the elements and the origin of the fluctuations from which structures were later formed. Until recently these properties had to be taken as given initial conditions from which the universe evolved. We are now in the position to show how these conditions could have come about in the very early epochs of the universe through evolution from an initial singularity (a Hot Big Bang) according to the laws of physics as we know them from laboratory experiments. The number and interactions of the particles that were present at times close to the singularity, when the temperature and density of the universe were very large, are crucial to the evolution of the universe. They determine quantities that are observable today and that can be used to extract information about the early universe.

The fossil relics include the 3-degree microwave background, the existence of structure (galaxies and clusters), the abundances of ^{4}He, D, ^{3}He and ^{7}Li, and the ratio of matter to radiation. The 3 K radiation reflects the state of the universe 10^6 years after the Big Bang, when matter and radiation decoupled. The existence of galaxies and clusters of galaxies implies the existence of small deviations from homogeneity and isotropy in the early universe. The mass spectrum and spatial distribution of galaxies reflect the nature of these fluctuations.

When the universe was about 100 seconds old, light elements were synthesized from primordial neutrons and protons. In the high-temperature soup of the initial universe the number of neutrons and protons are in equilibrium through rapid exchange processes involving electrons, positrons, neutrinos and antineutrinos.

In the successive phase of the expansion the fusion of deuterium can occur and through a series of further thermonuclear reactions ^{4}He is formed.

The amount of D that is left is very sensitive to the matter density during nucleosynthesis and can be used to infer the current value of the mass density of the universe. The process of D destruction and He formation depends on the density of baryons at a given temperature and therefore depends on the ratio T^3/δ, where T is the temperature at the time of element formation and δ is the density of normal matter. If T^3/δ had been small, all deuterons would have been transformed into He, which would result in 33% ^{4}He and 67% H.

In Fig. 8 the relative distribution of chemical elements synthesized in the initial expansion is plotted as a function of the current density of normal matter in the universe, (^{2}H $\equiv$ D and ^{3}He is due to tritium decay, ^{7}Li to ^{7}B decay).

As mentioned earlier, the fraction of ^{4}He produced ($\sim\frac{1}{3}$) is sensitive to the expansion rate of the universe at the time of nucleosynthesis, which depends on the energy density; in turn, this is dependent on the number of species of particles present.

Thus if we knew the relative abundance of chemical species in the universe at the initial conditions, we could derive the density and temperature at the time of nucleosynthesis. It is clear, however, that when we measure the relative abundances today we cannot compare these num-

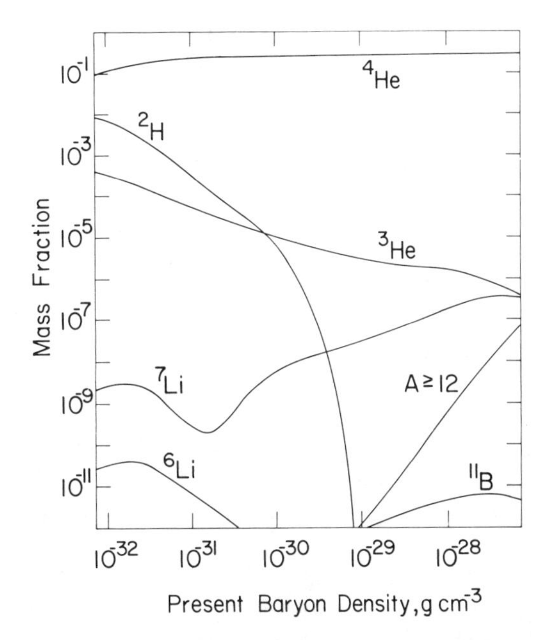

FIGURE 8. Relative distribution of chemical elements created during the initial expansion. The parameter δ_0 is the current value of matter density. ^{2}H indicates deuterium.

bers directly with those predicted by different cosmologies since elements can be synthesized as well as destroyed in the course of stellar evolution.

For this reason we must study the problem of chemical composition in a variety of evolutionary settings. ^{4}He abundance can be obtained from current data for the interior of young stars, stellar atmospheres, and the interstellar gas. In most instances the ratio $Y = {}^4\text{He/H}$ is found to be between 0.26 and 0.32 and the best estimate of the He content in prestellar matter is about 0.3, very close to the maximum value of 0.33 that has been inferred.

On the other hand, deuterium cannot be formed in the interior of stars because it is rapidly transformed into ^{3}He. Therefore any deuterium we find must be due to nucleosynthesis at the initial epoch of expansion. Recent measurement in interstellar gas and in planetary systems yield $D/H = 5 \times 10^{-5} - 1.5 \times 10^{-4}$. If this deuterium is really primordial, then we must conclude that the density of the universe must be rather low, of order $\delta_0 \approx 3 \times 10^{-31}$ g/cm^2 and $\Omega = 0.1$, insufficient for closure.

ST will contribute to this problem by permitting much more sensitive and accurate measurements of chemical abundances in planetary systems, in comets, in stars, and the interstellar medium.

For instance it has been estimated that the D/H (if $>10^{-4}$) ratio could be measured for comet Halley at the time of closest encounter. The Lyman-alpha line of deuterium at 1215.340 Å is about one-third of an angstrom short of the hydrogen Lyman-alpha wavelength or five resolution elements in the HRS medium resolution mode of ST (Brandt 1982). Current estimates of sensitivity show that this measurement could be obtained with a signal-to-noise ratio of $\sim$10 in 3000 sec. Although the delay of the ST program will not permit this measurement to be carried out, the estimate can be applied to other comets that will be observable during the ST lifetime.

Bahcall (1979) has pointed out that for $z_{abs} \geq 0.15$ several Lyman lines, through Ly-δ, are redshifted into the region $\lambda \geq 1100$ Å accessible to ST spectrographs. It will be possible therefore, in principle, to observe with ST the D/H ratio in absorption, with brighter sources and less crowded spectra than are feasible from the ground. Thus one could hope to determine D/H ratios for intergalactic clouds, halos of galaxies, and cluster gas.

Bahcall also has pointed out that helium resonance lines from distant quasars will fall in the range of wavelengths to which ST is sensitive. For $z \geq 1.1$, these lines include He I lines at 584.3, 537.0, and 522.2 Å, while for $z \geq 2.6$ there is He II at 303.9 Å. Studies of systems with $z_{abs} \gtrsim 2.6$ may also be extremely interesting since one could measure the strengths (or upper limits) of lines from H I, He II, O I, O II, and O III.

ST may therefore open up a number of new observational possibilities with which to determine the relative abundance of light elements in objects at cosmologically interesting distances.

Also, ST will permit better studies of galaxies and cluster dynamics. These measurements can be used as a probe of the properties of the dark matter we believe to be present in these objects.

VI. THE COSMOLOGICAL MODEL

Clearly, however, the main contribution of Space Telescope to cosmology will be in the exploration of times much greater than 10^6 sec. Our knowledge of the age of the universe $\tau = 1/H_0$ depends on the determination of parameters (H_0 and q_0, H_0 the Hubble constant and q_0 the deceleration parameter) we know only approximately and so, of course, does the relation between z (the redshift) and τ. Our predictions about the future of the universe, whether it will expand indefinitely or will ultimately collapse, also depend on our knowledge of q_0 or of Ω, the ratio of matter density to the critical density for the universe to be closed.

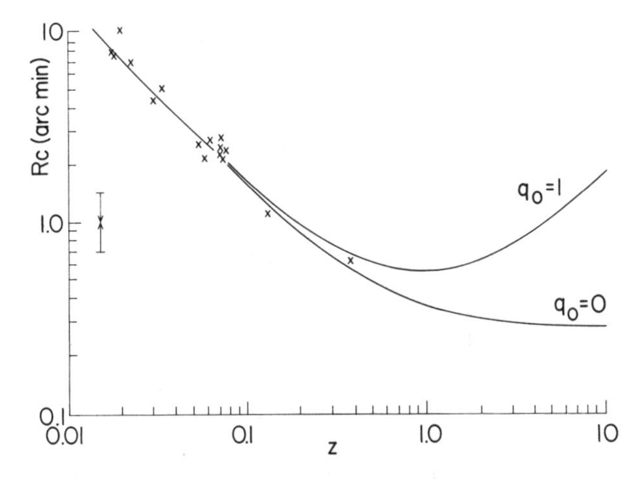

FIGURE 9. Core radius of clusters as a function of redshift z.

We can use ST to carry out investigations that will help us establish more firmly the Friedmann cosmology and eliminate other possible world models.

For instance we can test whether the cosmological redshift $\Delta\lambda/\lambda$ is independent of λ, or whether the surface brightness goes as $(1 + z)^{-4}$ as predicted by the theory. As an example of how ST would do this, consider the observation of giant eliptical galaxies, which will be extended to $z = 1$. Fifty percent of the light would fall in an angular diameter of less than 1 arcsec. We could obtain the profile of brightness m_r to 3% with the WF/PC or FOC cameras within an hour observation.

Many other cosmological tests will become possible such as:

1. Count-magnitude relation to $z = 1$.
2. Count angular size relation to $z = 0.5$–1.
3. Magnitude-redshift relation for giant galaxies to $z \sim 0.5$–1.
4. Redshift count relation for great clusters to $z \sim 0.5$–1.
5. Redshift distribution of galaxies by apparent magnitude.

A specific test proposed by N. Bahcall (see Longair and Warner 1979) for the angular size–redshift relation of the core radius of great clusters is shown in Fig. 9. This test could also be used to bound the values of q_0.

But of course the glory of ST would be the determination of the values of the H_0 and q_0 parameters of the Friedmann model.

Determination of H_0 and q_0 completely determines the cosmological model, in the assumption of isotropy and homogeneity (and cosmological constant equal to zero). The best determination of these quantities can be obtained by measuring the relation between apparent magnitudes and redshift for distant sources ($m_{bol} - \log z$). The model predicts a relationship that expands in powers of z yields $m_{bol} = 5 \log(1/H_0) + 1.086(1 - q_0) z + \ldots + M_{bol} + 25$.

In principle, it is necessary only to measure the observed apparent magnitude m_{bol} for two sources for which we know M_{bol} (absolute luminosity measured at the time the signal is emitted). But the basic difficulty in establishing the values of the H_0 and q_0 is precisely the definition of M_{bol} for a distant source. We do not have such a standard candle and in its absence we have to use a number of sources whose luminosity is at least approximately known to us. For example, if one postulates the well-tested hypothesis that the brightest galaxies in clusters have approximately the same absolute magnitude, one can construct the type of diagram shown in Fig. 10 (Sandage 1975). Note that in the diagram m_v rather than m_{bol} is shown. This is because the actual detectors measure not m_{bol} but m_v, the apparent brightness in a fixed band (the visible). At different redshifts the detector samples different spectral portions of the emitted source spectrum and this must be compensated for by applying a correction (the so-called k correction). Other corrections are required to take into account absorption in our Galaxy and several other effects.

Clearly, the clustering of galaxies about the straight line and the absence of very large deviations shows that, at least for this range of values of z, the working hypothesis of constant M_{bol} for the brightest galaxies in a cluster is satisfied and, moreover, that evolutionary effects are small.

But to obtain H_0 from the observations, we must also *know* what absolute value to assign M_{bol}. To do this, one attempts to determine M_{bol} for brightest galaxies in some of the nearest clusters, for instance Virgo. We find that we must follow a complex chain of derivations in which we must establish a continuous set of standard candles progressing in distance from nearby objects stars at few light years to Virgo (at ~10 Mpc (Fig. 11).

First we determine absolute magnitudes for the nearest stellar clusters (typically the Hyades) from statistical parallax methods. Then we can determine absolute magnitudes for main sequence stars and construct a Hertzsprung–Russel diagram. Given this relation, we now know absolute magnitude for all main sequence stars as a function of color.

Using this information, we can derive absolute magnitudes for stars in all the clusters within the galaxy and, in particular, the absolute magnitude of variable stars within them. A particular class of these variable stars (the Cepheids) exhibits a strict correspondence between the period P of

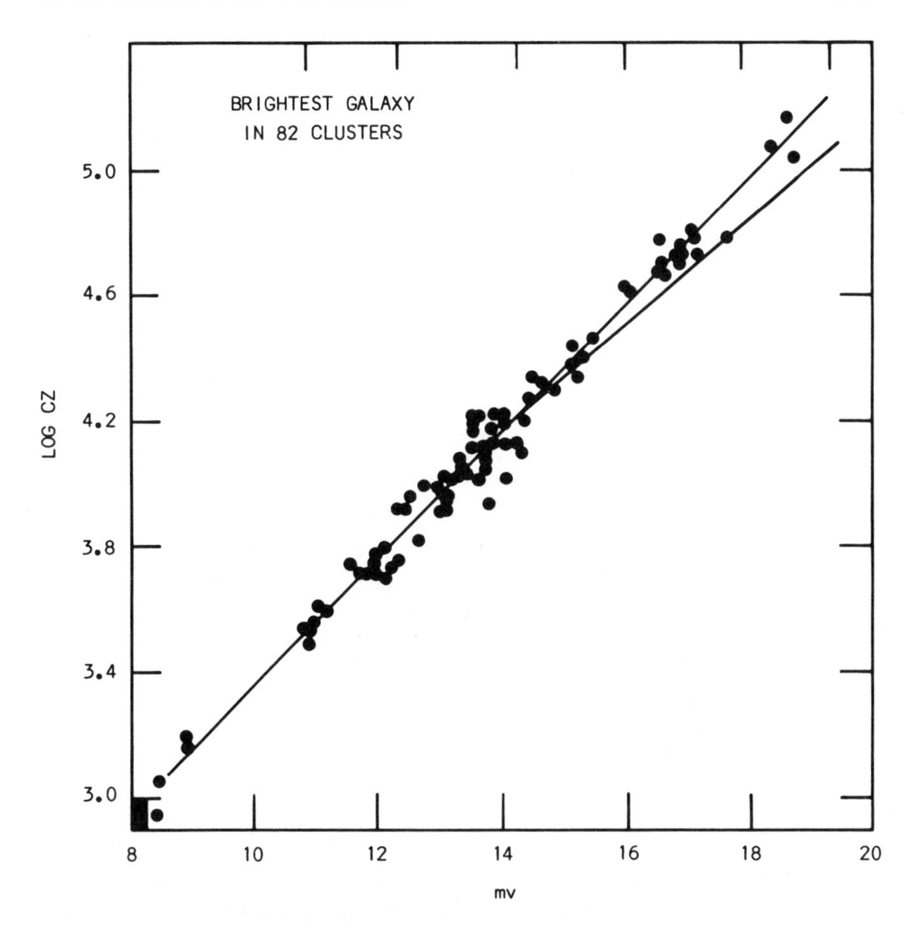

FIGURE 10. Apparent magnitude as a function of redshift for the brightest galaxies in clusters.

the variability and the absolute magnitude *M* and is very luminous. Knowing the absolute magnitude of the Cepheids, we can establish the zero point of the *P–M* relation.

Cepheid variables are visible at very great distances such as the Andromeda galaxy (M31). We can, therefore, measure the distance to Andromeda (~0.5 Mpc). Unfortunately, however, we can not take this galaxy as our standard candle, since it is a sample of one and is not in a cluster. On the other hand, we can not observe Cepheids in the Virgo cluster. We need another intermediate step. Observation of globular clusters in Andromeda yields absolute magnitudes for this new class of brigh-

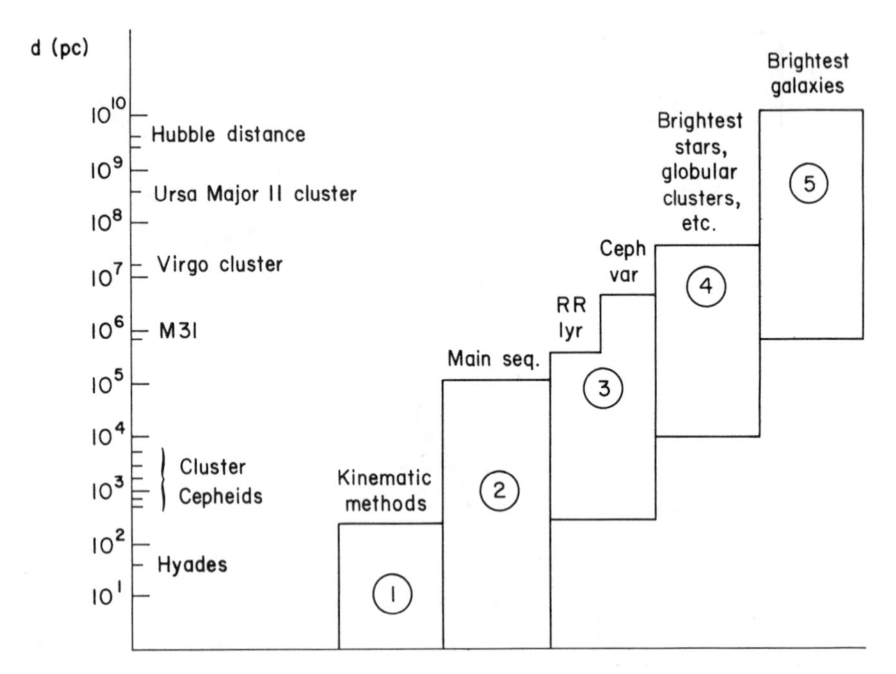

FIGURE 11. The distance scale ladder indicates the range of distances over which various types of distance indicators are currently used.

ter standard candle. Globular clusters containing 10^5 stars can be observed also in the galaxies in Virgo (and in particular in the great elliptical M87). By observing their apparent magnitude in M87 and knowing their absolute magnitude in M31, we can deduce the distance to Virgo. We can then sample the luminosity of the brightest galaxies in Virgo and assign to them M_{bol}.

Clearly, the errors that can accumulate in the construction of this elaborate procedure are quite large and it is no wonder that after 30 years of work, H_0 is only known to within a factor of 2 with accepted values ranging from 50 to 100 km sec^{-1} Mpc^{-1}. This is equivalent to the statement that we do not know the age of the universe ($\tau = 1/H_0$) to within a factor of 2.

One of the major contributions that we expect from Space Telescope is the observations of Cepheid variables in the thousands of galaxies in the Virgo cluster. Bart Bok and others calculated the expected apparent magnitude of Cepheids in Virgo, under the assumption of one or the other of the extreme values of H_0 (Bok 1983). The Cepheids would have apparent magnitude $m_v = 26$ or $m_v = 27$ for $M_v = -4.5$. Since in either case, the Cepheids would be observable from ST in times of order of one-half hour, he concluded that one day of ST observing time, spread judiciously, should be sufficient to measure H_0 to 10%. Further, observations of globular clusters could be extended from Virgo to the Coma cluster. They could then be used, as described previously, to extend the distance scale to the Coma cluster, which is some six times further away than Virgo (~60 Mpc). In principle, this would require observing times not much longer than one day as well. In practice more detailed evaluations show that the measurement of H_0 can be done using some 500 hr of ST observing time.

Tammann (1979) has taken a more skeptical view of the potential of this simple approach. Of course, there is agreement that ST will extend the detection of primary indicators to 10 times greater distances. Tammann himself emphasizes how profitable it would prove to study RR Lyz stars in M31 and Cepheids in M101, though he concludes that by these methods H_0 could not be determined to better than 10–15%.

Since there appears still to be a factor of 2 uncertainty in the value of H_0, between 50 and 100 km sec^{-1} Mpc^{-1}, I would conclude that a determination even if only at the 10% level would be extremely important.

Sandage and Tammann (1974) and Humphreys (see Longair and Warner 1979) have emphasized the use of other distance indicators such as the brightest M supergiants in late-type spirals to map the local velocity field and enable us to disentangle the local Virgocentric motion from the unperturbed Hubble flow. With $M_v \cong -8.0$, M supergiants would appear at $V_0 = 3250$ km sec^{-1}, $m_v = 26$ and $m_b = 28$, clearly good objects for ST observations.

Through their study one could hope to place a 10% limit on the random velocities of individual galaxies and determine the local Virgocentric motion and the Virgocentric flow model and measure the local density parameter Ω.

Another method advocated by several authors (see, for instance, references in Tammann 1979) is the use of Type I supernovas with emission lines (SNe I) as standard candles. Although SNe I are fainter at maximum by three magnitudes than the brightest cluster galaxies and therefore more difficult to detect, the hope is that their luminosity may be less affected by evolution. From the ground the faintest observed SNe I are at $m = 20$. At $z = 0.5$ SNe I would appear at $m = 23$. ST therefore would be ideal to study such objects. If a sufficient number could be found in ST fields, then one could hope to determine both H_0 and q_0 from the Hubble diagram to $z = 0.5$. Unfortunately, given the small field of view of the ST camera, such serendipitous discoveries may be extremely rare.

An alternative approach to the measurement of H_0 and q_0 based on the

Baade–Wesselink method has been suggested by several authors (see, for instance, discussion by Oke and by Wagoner in Longair and Warner 1979 and also see Chapter 7). The measurement consists of the simultaneous observation of apparent magnitude, spectrum, and line profiles in supernovae. With the assumption that the photosphere is understood in terms of atmosphere models, that it is symmetric and sharp, one can derive the (proper motion) distance to the SN in a single measurement. The importance of the method is that the SN need not be a standard candle.

A combination of ground-based observations at $0.01 \lesssim z \lesssim 0.1$ to de-

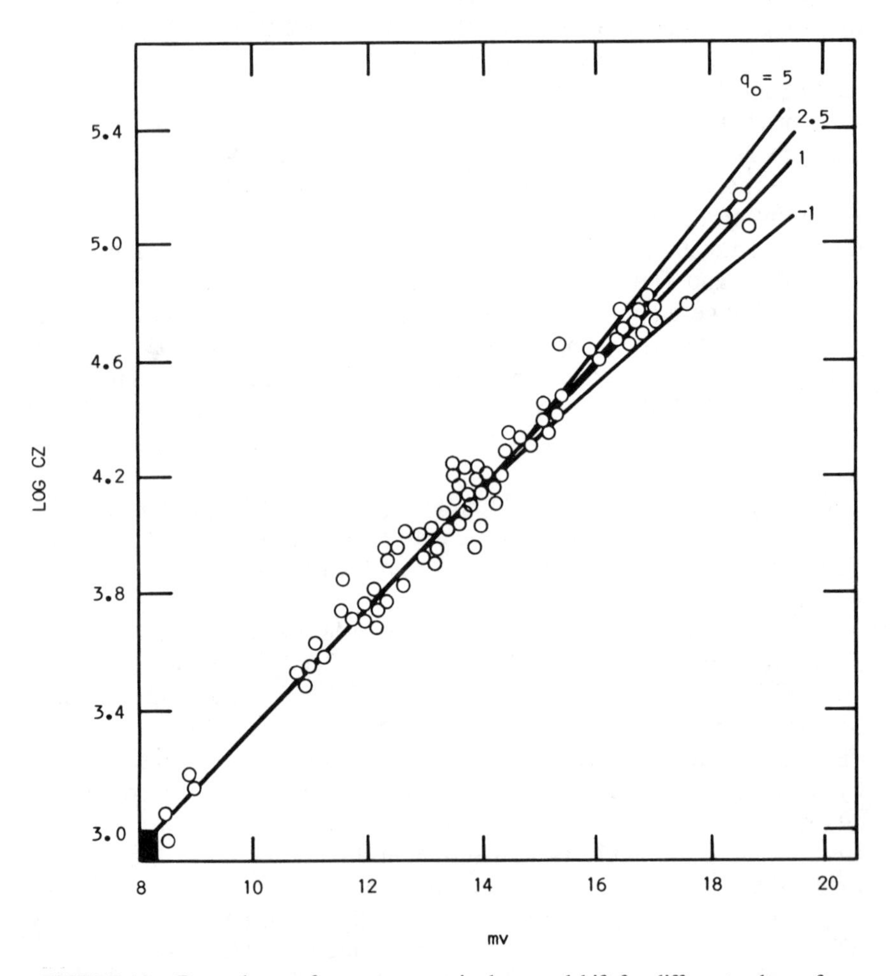

FIGURE 12. Dependence of apparent magnitude on redshift for different values of q_0.

termine H_0 and of ST observations at $z \gtrsim 0.1$ may permit determination of q_0. (In principle q_0 could be determined in a single observation.) Problems arise in the choice of SNs to use and in the confidence we may place in the detailed atmospheric models now available. Wagoner points out that SNe II may be better understood than SNe I, and therefore preferable for use of the Baade–Wesselink method. Macchetto, on the other hand, emphasizes that shell expansion velocities derived from UV and visible light observations are different, indicating that we still cannot adequately model SNe II atmospheres.

By one or the other of these methods or most probably by a combination of all of them, Space Telescope should give us a much firmer value for H_0. The deceleration parameter q_0 measures the quantity of matter in the universe (in a Friedmann universe $\Omega = 2q_0$). Thus q_0 can be, in principle, directly determined from the curvature of the log $z - m_{bol}$ relation even without precise knowledge of M_{bol}. In practice, however, the deviation from linearity in the range of observed magnitudes (or z's) is quite small (Sandage 1975). As shown in Fig. 12, moreover, the extension of this analysis to larger z's is hampered by the fact that significant evolutionary effects on M_{bol} can not be disentangled from the measurement of q_0 without a much more precise knowledge of galaxy evolution.

VII. EVOLUTION AND FORMATION OF STRUCTURE

The extension, which ST will make possible, of the study of galaxies to distances where evolutionary effects may become important promises significant contributions to the understanding of their evolution and dynamics. It can be confidently predicted that morphological identification of galaxies can be extended from $z = 0.2$ to $z = 1.2$. Figure 13 shows a vivid simulation by John Bahcall of the difference in appearance of a galaxy from a ground-based telescope and ST. ST will permit us to study the evolution of stellar populations, to examine the onset of cannibalism, and to determine disk to bulge ratios for these distant objects.

Furthermore, galaxies in the very early, high-luminosity stages of their evolution may become detectable at very large distances ($z \sim 5$). This can be deduced from rough estimates of the expected luminosity of such galaxies and from our knowledge of the expected sensitivity of ST. Figure 14 illustrates a simulation by J. Kristian of ST response to stars of different magnitudes in a 3000-sec exposure, based on measured values of the detector's noise. The brightest star in the field is of $m_v = 22$ and the faintest of $m_v = 28$. Tinsley (1979) has emphasized the power of ST in searching for primeval galaxies. Of course, our expectation of observing such ob-

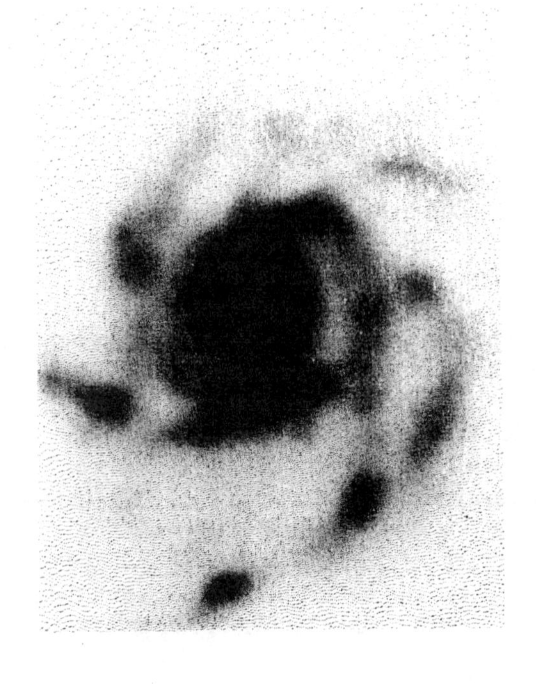

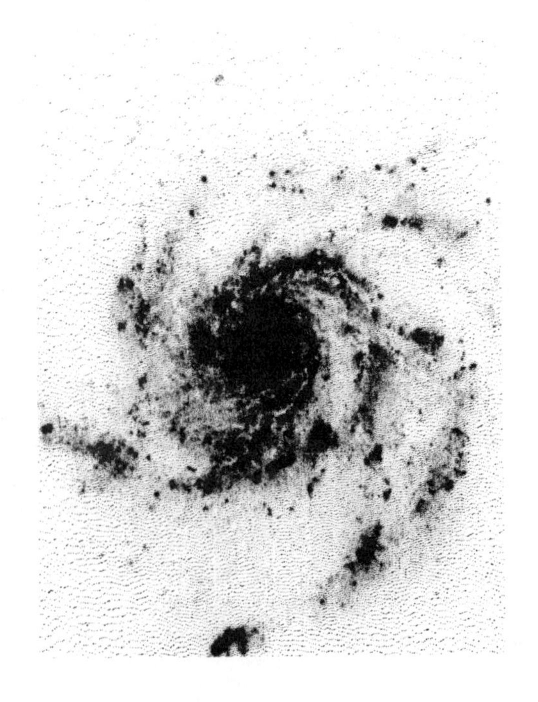

FIGURE 14. Sensitivity of a simulated 3000-sec exposure of WF/PC with ST.

jects depends on the detailed prediction of different theories. Partridge and Peebles (1967) predict a formation epoch for galaxies at $z \sim 10$–30 with an angular extent of $\theta \gtrsim 5''$. Such objects would be undetectable with ST. On the other hand, Sunyaev et al. (1978) predict z of 2–20 and $\theta \sim 1''$ and Toomre (1977) (and Tinsley and Larson) predict mergers at $z \sim 2$–10 with $\theta \sim 1''$. If the UV radiation from these objects were not absorbed by dust, ST could detect such primeval galaxies at $z \sim 5$ if $R \lesssim 10$ kpc.

It is clear that evolution of quasars and active galactic nuclei will form an important field of research to which the capabilities of ST are particularly suited. Recently, several authors (Bahcall 1979, Boksenberg et al. 1979, Weymann and Williams 1978) have pointed out that the study of absorption line systems from quasars furnishes a powerful probe of the density and composition of galaxy halos and intergalactic clouds.

Cluster evolution and galaxy content of clusters can also be studied

←

FIGURE 13. The improvement in angular resolution of ST with respect to ground-based telescopes is shown by comparison between the two pictures.

with ST to greater redshift than is currently feasible with ground observations. Clusters and superclusters distributions can also be used for specific tests of galaxy formation and evolution scenarios.

In a general sense by extending the study of galaxy evolution to such distances, it may become possible to distinguish between the quite different predictions of the two main theories on the formation of early structure in the universe that are currently receiving greatest attention.

These theories seek to describe the existence of the condensed structures we see at the current epoch, such as galaxies and clusters, as a consequence of the growth of random fluctuations from homogeneity in the still hot dense universe prior to the decoupling of radiation and matter ($\tau < 10^6$ sec). In the adiabatic model, proposed primarily by Zel'dovich and his school in the USSR, all fine-scale structure is washed out by damping mechanisms at $z \sim 1500$ so that large-scale structures form first. Masses of 10^{15}–3×10^{13} M will separate from the Hubble flow at $z \sim 5$ (Doroshkevich et al. 1973). Such massive objects (pancakes) would correspond to supercluster structures. Clusters, galaxies, and stars would later form from fragmentation of these massive objects. Galaxies would have formed in the rather recent past a z between a few and 10. The other main theory (isothermal), which has been due mainly to Peebles and his school in Princeton, would have the smaller structures form first and the larger by hierarchical clustering (Peebles 1971). Galaxy formation would occur in this theory much earlier, at $z = 10$–100. Space Telescope observations of protogalaxies at relatively small look-back time could immediately decide this question.

A qualitative treatment such as the one above gives an impression of the impact of ST on cosmological studies that is simultaneously much too optimistic and too pessimistic. Much too optimistic in that it sets the framework for questions in simple either-or terms that can be solved by direct observations. In reality, experience shows that the physical universe tends to reveal itself always to be more complex than described by the simplified descriptions of theory. It is quite probable that the 30-year quest for a precise value of q_0, which would reveal to us whether the universe will expand indefinitely or recollapse, may not end with Space Telescope. Typically, the problem will appear to us to have richer implications and to be more complex and different after ST than before. On the other hand, the view exposed above is too pessimistic in that we can hope for unexpected findings which may reveal clearly new aspects of the universe currently not understood, leading very rapidly to new and important insights.

REFERENCES

Bahcall, J. 1979, "Absorption Lines in the Spectra of Distant Objects," in *Scientific Research with the Space Telescope*, M. S. Longair and J. W. Warner (eds.), IAU Colloquium No. 54, NASA CP-2111, Princeton, N.J., pp. 215–240.

Bok, B. J., 1983, "The Promise of the Space Telescope," *J. Astr. Soc. Pacific*, Vol. XII, No. 2, pp. 66–75.

Boksenberg, A. Carswell, R. F., and Sargent, W. L. W. 1979, *Ap. J.*, **227,** 370.

Brandt, J. C. 1982, "The High Resolution Spectrograph for the Space Telescope," in *The Space Telescope Observatory: Proceedings of the Special Session of Commission 44*, D. N. B. Hall (ed.), IAU 18th General Assembly, Patras, Greece, pp. 76–105.

Doroshkevich, A. G., Sunyaev, R. A., and Zeldovich, Ya. B. 1973, *Proceedings of IAU Symposium No. 63*, Poland.

Hall, D. N. B. (ed.), 1982, *The Space Telescope Observatory: Proceedings of the Special Session of Commission 44, IAU 18th General Assembly*, Patras, Greece.

Longair, M. S., and Warner, J. W. 1979, (eds.), *Scientific Research with the Space Telescope*, IAU Colloquium No. 54 (NASA CP-2111), Princeton, N. J.

Macchetto, F. D., Pacini, F., and Tarenghi, M., 1979, *ESA/ESO Workshop on Astronomical Uses of the Space Telescope*, Geneva, Swizerland.

Partridge, R. B., and Peebles, P. J. E., 1967, *Ap. J.*, **147,** 868.

Peebles, P. J. E., 1971, *Physical Cosmology*, Princeton University Press, Princeton, N.J.

Sandage, A., and Tammann, G. A. 1974, *Ap. J.*, **191,** 603.

Sandage, A. R. 1975, in *Galaxies and the Universe*, A. Sandage, M. Sandage, J. Kristian (eds.), The University of Chicago Press, Chicago, p. 781.

Sunyaev, R. A., Tinsley, B. M., and Meier, D. L. (1978), *Comments on Astrophysics*, **7,** 183.

Tammann, G. A. (1979), "Precise Determination of the Distances of Galaxies," in *Scientific Research with the Space Telescope*, M. S. Longair and J. W. Warner, IAU Colloquium No. 54, NASA CP-2111, Princeton, N.J., pp. 263–293.

Tinsley, B. M. (1979), "Galactic Evolution with the Space Telescope," in *Scientific Research with the Space Telescope*, M. S. Longair and J. W. Warner, IAU Colloquium No. 54, NASA CP-2111, Princeton, N.J., pp. 263–293.

Toomre, A. (1977), in *The Evolution of Galaxies and Stellar Populations*, B. M. Tinsley and R. B. Larsen (eds.), Yale University Observatory, New Haven, CT., p. 401.

Weymann, R. J., and Williams, R. E. (1978), *Phys. Scripta*, **17,** 217.

PERSPECTIVES

12.

What is Astronomy?

PHILIP MORRISON

Philip Morrison is Institute Professor of Physics at M.I.T. His early work focused on nuclear physics, and he was a participant in the Manhattan Project at Los Alamos. More recently his contributions have been in theoretical astrophysics including the areas of quasars, cosmic rays, and X-ray astronomy. He has been an articulate spokesman for arms control and is also widely known as the omnivorous book reviewer for *Scientific American*.

Everyone knows, as the most commonplace part of any discussion in astronomy, with the man on the street or even in the horoscope next to the funny papers, that astronomy is in some version or other one of the oldest of explicit scientific efforts to understand the world. It offers the oldest sign of order in our universe; day and night entrain us all. Of course, the rhythm of the seasons adds more to that; and nothing except the intimacy of heartbeat and breath can be closer to us than the rhythms of day and night and summer and winter. It must go back a long ways in our species—and I venture to say in our genus—as long as we had any way of formulating ideas, that we recognized that there is something outside there that forms order.

When Isaac Newton at the end of that great century, the seventeenth century, when modern science began, formulated the first really powerful mathematically and experimentally complete "system of the world," he drew heavily on the results of astronomy. The apple was the connection between the moon and the earth; it was an interpolating device. But the Keplerian results on planets and satellites, the values of periods and radii, were the main facts on which Newton rested his system of the world. The effort to describe the tides, and the pendulum swinging in the laboratory, were important; they went toward universality, but they couldn't quite get there. The real impression was made by the understanding of the events in the sky, then of course kinematically extremely well known and studied for thousands of years. Even with the telescope there were results that had been well established for 60 years or so.

Even 50 years after Newton, the main support for Newtonian physics was astronomical, or at least large-scale. It was Maskelyne, the Royal Astronomer in the middle of the eighteenth century, who first measured the gravitational constant. He did not set up lead balls in the laboratory—that was to come later. He used a mountain in Scotland. By observing the zenith from two sides of that mountain he could see what the mountain did to his local definition of the vertical. So it was large-scale physics (not laboratory physics) that determined the most important results the physicist had in those days. It wasn't till Cavendish, 30 years later, that the physicists left the outdoors world and started to build their laboratories and their tunnels and put large lead balls in place, to do the physics that of course by now has risen to extraordinary power and maturity. Everyone will agree that it was celestial mechanics that gave rise to the remarkable mathematical structures that made contact between the particle physics of Newtonian mechanics and the partial differential equations of continuum theory that dominated nineteenth-century physics, ending in Maxwell.

By the time the 1930s came along, there had already been 70 years of spectroscopy, most of it done with very little understanding of atomic physics. The great success of the old quantum theory, culminating in the work of Saha in the 1920s, finally allowed the great pioneers of modern astronomy, Hertzsprung and Russell and Eddington and Milne and Payne, to do the atomic physics that made it for the first time possible to analyze the stars quantitatively. They discovered that the world, the whole visible cosmos, is made of hydrogen and all the rest we so much count upon.

Then, as everyone knows, came nuclear physics. It has its slow beginnings in World War I, and broke onto the world with an air of omen and power in World War II. In between it acquired that maturity which in the hands of such masters of the subject as Hans Bethe was able to answer a long-voiced query. Nobody who had the slightest feeling for the nature of the world could have failed to ask,"what makes the stars and the sun shine?" At last this could be answered.

Today we see astronomy as the application of physics to an inaccessible laboratory. It demands the best and the most complex physical understanding that we have. But it is without the parallel support and drive of an important industrial technology that demands laboratory answers for its applications, as, for example, aerodynamics or plasma physics. There we know the physicist is going into those particular interesting applications because they're enormously needed, and well paid for. Because of our ancient human connection with astronomy, and because it is so far yet from any abilities to control, it remains a special case. It is applied physics, where application, yes, is to special systems, not given to us to choose but only to observe from afar.

Other beautiful branches of applied physics have grown up in such a way that they mostly take away a field right out of the physics profession. There are now not very many physicists, only a few, who know a lot about aerodynamics, because aerodynamicists have become a world of their own; they have so much power and so much need and so much desire to enrich one particular subject. So it has gone for quite a few specialties, but not yet in astronomy. Astronomers did that for a while, from the time when the photographic plate came in through a time when the spectroscope was the astronomer's real monopoly.

The new beginning was presaged by Karl Jansky (the pioneer of radio astronomy) before World War II. But it was in the postwar world that physicists were enormously attracted to astronomy. It happened on two sides. On the side of experiment (probably the more important), new channels have been opened into the astronomical world by the techniques of the physicist. It also happens, of course, on the theoretical side, because we now have a pretty powerful command of classical and relativistic mechanics, of plasma, of nuclei, of electromagnetic theory, of kinetic theory, and the rest. These enable us to try to cope with the complex systems and bizarre parameters domains the cosmos is pressing upon us. We look theoretically only because some clever experimental investigation has shown us a new signal: for heaven's sake, what can that be? We try hard and then, like Franco Pacini and Tommy Gold, we may come up with a fundamental answer, like a spinning neutron star. That's about as far as we can get with certainty. We're still struggling. We are of course enriching, deepening, and making more quantitative our whole understanding.

Now, I'd like to draw some conclusion from this. Probably everyone agrees there's a dual investigation, the investigation of new channels—in my view absolutely indispensable—and the investigation with new understanding of complex problems, also requisite. It would be impossible to view modern astronomy without both of these invigorating entries, almost rude door-breakings, from the side of the physics laboratories. That is not to say that the classical discipline of astronomy is not a monument to the effort of the astronomers. Witness, say, the magnificent instrument 50 years old now which is Palomar. Theirs was the social effort and the intellectual effort, to make something that would last and be worthy of the sense of frontier that astronomers have felt for a long time. They represent the human mind struggling with its origins and fate, in the best way we know, by looking at the world outside our own domain, where we look so far and so long ago.

Now what we have found, in the last 30 years or so, is that we live in a dual cosmos. We should have realized that earlier, but for a clear reason we didn't. I'll try to say what that reason is. The reason is very simple.

But let me first sketch what the duality is, as I see it. In the first place, the cosmos, the telescopic view is dominated, like the time on telescopes and the papers in the *Astrophysical Journal*, everything we do is dominated by the stars. That's as it should be, for we are creatures of the stars. We are adapted to a star; we were evolved under the light of a star; we can see stars; we can see deep into their interiors with neutrinos, even if only in a murky and puzzling fashion at the moment, but at least we try! Now the star is an example of something very familiar to humankind, familiar probably in the Rift Valley six or eight hundred thousand years ago, if not earlier, when the first campfires were kindled under the dark sky by those who were our forebears. The hearth is the characteristic word I'd associate with that ancient experience.

If you ask, what is the nature of this as a physical phenomenon, I think it is quite plain to say: The issue in the course of evolution of the astronomical world, dominated on the large scale by gravitation as it is, is the struggle between gravity seeking to crush and anything else that tries to maintain. What maintains the stars, of course, just as what maintains terrestrial objects against the crush of gravity, is the resistance of random motion, whether it be thermal motion, or zero-point energy, the fluctuations of the electrons in the van der Waals forces between atoms. Whatever way you look at it, speaking broadly, I think it is fair to say that it is a randomized energy in which many motions in many directions take part. That's what we're used to. Accompanying this are certain characteristic features: the emission of radiation is unpolarized, because no direction is singled out. Thus the characteristic shape of this astronomical world is a sphere. The sun is a sphere, and the earth is a sphere, and Jupiter is a sphere, and Titan is a sphere, and the stars are spheres. Again, this is a sign of the lack of particular direction, the lack of particular motion; the sense of symmetry is characteristic.

Now we understood pretty soon that this is not all there is in our astronomical surrounding. There is something else unchanging; it is like the laws of Medes and Persians. We see it both in the solar system and in the galaxies. It is not only random motion that fights gravity, it is also ordered circular motion. The planets circle around the sun, and in the galaxies, the stars, like the Sun, circle in the galaxy. It's not now a question of randomness; in fact, we see very well that the solar system and the spiral galaxies are characterized by an axis, by flatness. So we're quite right in saying it's not all spheres, as I described above. But look how stable the orbit is! Look how long we've been circling this little earth, ever since the sun was made, in the same orbit, with hardly any changes. There are only slight changes that our savants now tell us might make the Ice Ages, but so small that you can hardly believe it; it's not quite visible. The energy content of the motion of the Sun around the galaxy is about the same as the internal thermal energy of the Sun, and just as important. Yet nothing changes. Why worry about it? It's not a phenomenon of interest; it was made so in the beginning, and "as it was in the beginning, so it will be to the end," until 10^{25} years from now—gravitational radiation will take its ultimate toll. But we can't worry about that. That's the state of affairs in the universe we know and the universe we were born to. But what we have discovered is that beside the glowing hearth, a second completely different world exists in the universe. It is a world in which polarization, lack of sphericity, sudden change, explosion and expulsion play an important role. It is no hearth; rather it is the whirlpool and the fountain. This is the characteristic side of the new cosmos we are now beginning to investigate.

Why did we not see it in the past? Again the answer is very clear: we didn't see it because we looked with the eye. The eye was evolved for the gentle light of Planck radiation at 6000 K, the big broad curve that gives us white light. Everything else is a small departure from that—the dyestuff on the tablecloth or what you will. But what we recognize is that sunlight, radiant sunlight, is what we study. When the astronomers work, they study starlight; maybe they push a little bit to infrared or they go a little down near ultraviolet. If you look hard at the stars and have good eyes, you can see that there is some kind of color in Rigel or Betelgeuse, but it isn't very saturated; it's not the kind of vivid color we are talking about. There are hints, of course. One of those hints is held in the aurora that are visible at high latitudes. The aurora is not white; the aurora is not steady; the aurora is not in thermal equilibrium; the aurora has no simple temperature; the aurora is full of mystery, so we hardly understand it yet.

In fact, a similar hint is what led me directly into astronomy from the side of physics. The cosmic-ray physicists were also studying something a little more randomized, but again not temperaturelike, no hearthlike equilibrium. It has that characteristic power-law spectrum that goes with turbulence, and not the absolutely necessary Boltzmann distribution that goes with thermal equilibrium. We all recognized that, so we followed the cosmic rays, first into the laboratories. They took particle physics away from cosmic rays, since they could make mesons at 10 million times the rate of the cosmic-ray beam. But they left us the chance to follow the cosmic rays back to their origins, which we're still looking for far into the sky!

One other thing I want to mention, of course, the comets. Halley's comet has just come and gone. The comets represent a kind of gravitational instability not present in the rest of the solar system. Comets come in, we don't know when and we really don't know whence; they are not in the plane; they are not uniform; they are not combed out, like the planets, into circles in a plane. Slowly a great light dawns upon us—it is blazoned big among the clusters of galaxies—that gravitation, celestial mechanics, is not the quiet subject we all think it is. The almanac is there, and the experts in celestial mechanics, what do they do? They push the numbers one digit farther, to add another hundred years to prediction. That's the view I grew up with. I laughed at the perturbation theory that they taught us in dynamics, because, while it's very nice, elegant, and good for mathematicians (and no doubt osculating orbits are worthwhile, as well as Poisson brackets), there's nothing in this subject: no novelty, no romance, nothing new under the sun, just the sun and the circulating planets with their slow orbits. Well, the comets don't quite fit that, and nowadays we learn to our astonishment that it may be that the evolution of life itself owes a great deal to perturbations of comets that, half-expected, bring comets hurtling into the central solar system, to cross all the orbits, destroy regularities, break down the stabilities: no more circles, but plenty of collisions in the solar system. That, of course, is a sign of real trouble.

If you think about that, you realize that the reason we see light with the eye, not X-rays and infrared, and millimeter radiation, the reason we see circular orbits logarithmically spaced out into the solar system in a flat plane is that we come late upon the cosmic scene. We depend for our origins on the very systems we're studying. Of course, gravitation appears to us the rock-bottom safety of circular orbit. But that's not the way it really is. That's only the way it's been combed out to be after five billion years of circling around the sun. What it really is, is galaxies in collision and accretion disks, fierce fountains bubbling up along the axes of spinning disks in a way that we can barely begin to understand, on the largest scale and on small scale too.

Gravitation is an explosive phenomenon, not an orderly one. It depends on what conditions you have opened it to. The Laplacian notion that you can predict the future with precision, that if only we improve our measurements, we can project predictions further off, depends entirely on dealing with the two-body approximation to the circulating planets. As soon as you start putting in multiple bodies, reduce the degree of symmetry of the system, and ask for the evolution, then you find from Monte Carlo calculations the evaporation of clusters and the tails behind interacting galaxies and a hundred other phenomena, too many to evoke here. In fact, there really isn't any physical system of interest that you can predict with Newtonian precision far into the future. It's only because we have looked hard for those few systems that do work that way that we have the view that classical physics is a simple determinate mechanical structure. That's where we live: at a hearth. Of course, we teach that to the students, because otherwise they can't solve the problems. So for a long, long time separable differential equations and the two-body problem solved by the constants of the motion are everything in the textbooks. The fact is that if you go to three bodies only and generalize, you don't know what is going to happen next. You can study it forever; there is nothing but a finite circle of convergence. No matter how well you measure it, there will come a time when you can't say whether a system will hold together or not. That is even true probably for our solar system. A young colleague of mine at MIT has now built a grand digital orrery with chips to try to push the orbits of Jupiter and the other planets five or ten billion years into the future, just according to the laws of Newton, to see if in fact they will blow up one day. The stability theorem that governs the solar system is simply not adequate to answer this question.

So that is the way it stands now: there are two worlds, the world of the hearth and the world of the whirlpool. We recognize the universe contains both of them; we recognize that we belong to the one, while the other is alien. That comes from history alone, from the fact of our heritage and legacy, our adaptation to this world, the hearth-world of sun and planets, where life has lived for four billion years, one way or another.

I think we have come to the most important juncture in physical sciences, certainly since the glorious days near 1930 of the discovery of quantum mechanics, which makes possible much of what we do. We find ourselves surrounded by the radiation everyone knows about, the 3^0 radiation, in which we are clearly not central. Our motion shows very strongly in the map of the sky, so it isn't that. Copernicus was right; we are not central. But there is something strange about it, because we know of no other such perfection on the large scale as we see in the uniformity (to $\pm 1/1000$) of the incoming blackbody radiation. There is no other case where any macroscopic ball has that degree of accuracy, and for reasons that we don't understand. This makes one queasy.

The current proposal to explain this has remarkable implications. Those implications, taken together with what else we know, suggest that we live in a universe here in which we have not only found the whirlpool component we had neglected, but will find one day soon that most of the matter is our universe is alien to us! It is matter we have never seen, matter we have never examined. I'm not trying to say this is a sure prediction, but it is fair to say that that is the most popular and the most intriguing proposal of the day. The mass that holds the orbiting stars of galaxies in a tight grasp, even though the starlight within is not adequate to explain its presence were they normal stars, might be a kind of matter not known to us at all, a kind of matter left over from a previous regime of matter when the protons were broken up and free quarks reigned in the world. This would be remarkable if it were true. It's a challenge, after four centuries, to confront for the first time a failure of that great principle of Copernicus: as here below, so above. The world is really somehow one; it is up to the physicists, the astronomers, to show how. Maybe we're going to find that it is more dual than I ever expected; maybe we will not, but perhaps we will know that soon.

It seems to me it's a glorious time. We have had in the last 35 years a fitful peace, a happy, breathless time, we physicists and we astronomers. I hope we can continue in peace in this troubled world, to try to puzzle out the nature of the cosmos into which we were born.

13.

Quick Is Beautiful!

FREEMAN DYSON

Freeman J. Dyson is a Professor at the Institute for Advanced Study in Princeton. A pioneer in quantum electrodynamics, he was instrumental in formulating the theory of field renormalization. He has contributed to many diverse subjects, bringing to bear the tools of the theoretical physicist. Winner of numerous prizes and awards, including the Heineman Prize of the American Institute of Physics, he has also written several widely acclaimed books and popular articles.

I. WHAT IS THE PROBLEM?

I was once a professor at Cornell. It is remarkable, if you look back on the old days at Cornell, how quickly everything used to get done in science. Bob Wilson was in charge of experimental high-energy physics in those times. He would give the order for a new accelerator one day and in a couple of years it would be finished. That's how we used to do physics in those days. It's a shame that it's not done that way anymore. However, I understand it's still done that way at Cornell. I once asked Tommy Gold how long it took to build Arecibo, the marvelous radiotelescope in Arecibo which was a Cornell project. He said, "Three years." I also asked Ken Wilson how long it took to build CESR, the Cornell Electron Synchrotron Ring, and he said, "Three years."

If you do the thing fast, it doesn't cost so much, it generally works better, and it answers the questions you had in mind when you designed it. That's something that hasn't been happening recently in the space science business. And that's the main theme of my chapter.

Why is it that we've got into this slow mode of operation in space science, that we talk about instruments for years and years, and it always seems to be ten or twelve years before you have something you can use? This is a terrible way of doing science. It's bad from many points of view, primarily because it's discouraging to students. No student is interested in a project that's not going to fly for the next ten years, and if students aren't excited about it, then the thing is dead. It's a terrible situation.

How did we get into that situation, and what can we do about it? I'm going to blame the Field Committee to some extent. The Field Committee[1] undoubtedly did the best job it could; it made a thoughtful and well-considered study of the situation in astronomy. But the results have been disastrous, and I would like to understand why. Probably the best way it could have been done would have been to announce at the start that only graduate students were eligible to serve on the Field Committee. Then we might have got something moving!

There's a terrible mismatch in time scale between science and space missions. The cutting edge of science moves rapidly; new discoveries and new ideas often turn whole fields of science upside down in a few years. The discovery of pulsars in 1967, for example, just burst onto the astronomical scene. So did the discovery of quasars four years before, the discovery of X-ray sources; those things transformed within a year or two the way we think about late stages of stellar evolution and active

[1] Editor's Note: The National Academy of Sciences commissioned George Field to chair a prestigious committee to assess priorities for astronomy and space sciences for the 1980s.

galactic nuclei. The effect of such discoveries is to change the priorities, to change the questions we want to have answered.

Every young scientist would like to be able to say what the nineteen-year-old mathematical genius, Evariste Galois, said in 1830: "I have carried out researches which will halt many savants in theirs." Science must always be ready to halt and switch its objectives at short notice. But to make this possible the tools of science should be versatile and flexible, and that's what we seem to be missing in space missions.

In the space program, plans for missions and designs of instruments tend to be frozen years in advance. Intervals of eight or ten or twelve years are not uncommon. In most of the major missions the instruments were designed to answer questions that seemed important to scientists a decade or more earlier. And the bigger and the more ambitious the missions become, the more difficult it is to reconcile the time scale of the missions with the time scale of science. Space science begins to look like a two-horse shay with a cart horse and a race horse harnessed together.

The main message I'm preaching is that economies of scale are usually false economies, especially if they're bought at the expense of speed. Quick is beautiful! I don't say necessarily that small is beautiful. You don't have to be small—Arecibo is not small, but it is quick. If we want to do good science in space, the most important requirement is to have available a wide variety of missions and instruments, so that we can jump quickly to take advantage of unexpected opportunities. The most important discoveries are those that could not be planned in advance.

II. LESSONS OF THE PAST

What should the space program do to recapture its lost youth? I'll try, at the end of the chapter, to give specific suggestions of things we might do in the next 25 years to rejuvenate space science. But first, I'd like to go back to the past, to look at the history of the last 25 years and see if we can learn from the mistakes of the past, and perhaps also learn from the successes of the past how not to do worse in the future.

It's convenient to divide the 25 years since Sputnik into two periods, Apollo and post-Apollo. The Apollo period ends with the departure of Harrison Schmitt and Eugene Cernan from the moon in December 1972. The Apollo period is particularly instructive, because we can see clearly with the benefit of hindsight which parts of the space enterprise were scientifically the most productive.

In the Apollo period there was a strong negative correlation between budgetary input and scientific output. This negative correlation was not planned; it just happened because science is unpredictable. The most expensive missions produced the least significant science, and the cheapest missions produced the most exciting science. I'll just summarize what we learned scientifically from some of the representative missions of those days.

The Apollo period included three main types of exploratory mission. First, there were the manned missions, culminating with Apollo; second, unmanned planetary missions, culminating with Mars Mariner and Mercury Mariner; and third, the series of X-ray sounding rocket missions, culminating with the launch of the first X-ray satellite, Uhuru. The costs of the Apollo missions, the Mariner missions, and the X-ray missions were roughly in the ratio of a hundred to ten to one. You can't attach numerical values to the scientific results. The relative values of different types of scientific information are a matter of personal taste. For me, however, the beginning of X-ray astronomy, which opened up a new window into the universe and revealed the existence of several new classes of astronomical objects, has been the most important single scientific fruit of the whole space program up till now.

The newly discovered X-ray sources gave an entirely fresh picture of the universe, dominated by violent events, explosions, shocks, rapidly varying dynamical processes. The X-ray observations finally demolished the ancient Aristotelian view of the celestial sphere as a serene region populated by perfect objects moving in eternal peace and quietness. The old quiescent universe of Aristotle, which had survived essentially intact the intellectual revolutions associated with the names of Copernicus, Newton, and Einstein, disappeared forever as soon as the X-ray telescopes went to work. And that new universe of collapsed objects and cataclysmic violence originated in the cheap little sounding rocket of the 1960s, popping up out of the Earth's atmosphere and observing the X-ray sky for only a few minutes before it fell back down. The most brilliant achievement of the sounding rocket era was Herb Friedman's 1964 measurement of the angular size of the X-ray source in the Crab Nebula, using the moon as an occulting disk.

The cost of X-ray astronomy in that period was less than 1 percent of the total budget for space. The manned missions that absorbed the bulk of the budget in those days yielded a harvest of solid scientific information about the moon: samples of various types of moon rock were brought home and analyzed and dated; the stratigraphy of the moon was clarified and its early history elucidated; its seismic and magnetic characteristics were measured. All of this was good science, but it was not great science.

For science to be great it must involve surprises; it must bring discoveries of things nobody had expected or imagined. There were no surprises on the moon comparable with the X-ray burst sources or with the X-ray binary sources that gave us the first evidence of the actual existence of black holes in our Galaxy. Everything discovered on the moon could be explained in terms of conventional physics and chemistry.

However, God played us a joke that made all of this scientifically worthwhile. Because NASA was interested in bringing back rocks from the moon, NASA was also interested in funding meteoritic science in a rather lavish fashion. You had to have good instruments in order to analyze the moon rocks. So a whole generation of meteoritic chemists was supplied with the necessary instruments, good mass spectrometers and microchemical apparatus of various sorts, and there they were with this beautiful apparatus, waiting for the rocks to come back from the moon. And at just that happy moment, God decided to play a hand and threw down in Mexico a piece of rock that was more interesting than all the moon rocks combined, namely, the Allende meteorite. It was not only more interesting, it was also bigger. It weighed more than twice as much as all the moon rocks put together, and that meteorite contained a wealth of wonderful things, isotopic anomalies that gave evidence of presolar composition of the meteorites, evidence for things you can actually hold in your hand that are older than the Sun, evidence of events that took place before the solar system was formed. All these things came from the chemical analysis of microinclusions in the Allende meteorite. All that we got more or less for free.

The unmanned planetary missions of the Apollo period were intermediate, both in cost and in scientific importance, between the manned missions and the sounding rockets. They were less costly than Apollo and less exciting scientifically than X-ray astronomy. The most exciting aspect of the planetary missions was their technical brilliance. They gave us celestial billiards, a game which Giuseppe Colombo invented, with Mariner 10 bouncing around in the inner solar system, encountering Mercury repeatedly. It was a spectacular demonstration of skill. Those missions also gave us some beautiful scientific surprises; we found high temperatures and pressures and absence of water in the atmosphere of Venus, giant volcanos and canyons, ancient craters and the absence of canals on Mars. But the surprises were not of such a magnitude as to cause a scientific revolution. The newly discovered features of Mars and Venus were mysterious but not wholly unintelligible.

So much for the Apollo period. If space exploration had stopped at the end of 1972, we might have deduced from the history that I have described that a simple mathematical law governs the scientific effectiveness of missions in space, namely, that scientific output varies inversely with financial input. If that law held universally, the administration of space science programs would be a simple matter. Just cut the budget and watch the science improve. Unfortunately, that simple managerial method does not always work as it should. The history of space science in the post-Apollo period shows a more complicated pattern. In the 1970s we again had three programs continuing the work begun in the 1960s, Skylab and shuttle missions taking over from Apollo, Viking and Voyager missions taking over from the Mariners, and the Einstein X-ray Observatory taking over from Uhuru.

It was still true that the X-ray observations were first in scientific importance. The Einstein Observatory during its sadly short life poured out a steady stream of revolutionary discoveries, including the discovery of X-ray variability of quasars on a time scale of hours. This rapid variation of quasars implies that we have in some of these objects a switch that can turn the energy output of ten billion suns on and off within a hundred minutes. Some switch! The X-ray telescope allowed us for the first time to penetrate close to the central core of the mysterious engines that drive those most violent objects in the universe.

It was still true in the 1970s that the X-ray discoveries were of greater fundamental importance than the planetary discoveries, even though the Viking and Voyager missions gave us a wealth of new scientific information, as well as pictures of incomparable beauty. It was still true in the 1970s that the planetary missions outstripped Skylab in scientific value, but in the 1970s unlike the 1960s, there was no longer a factor of ten difference in cost between the three types of mission. All three types of mission had become comparably expensive. Einstein was a little cheaper, but not enormously cheaper, than Voyager. Voyager was a little cheaper, but not enormously cheaper, than Skylab.

What have we done since then? We have had two magnificent missions in the last six years: the IUE, International Ultraviolet Explorer, and IRAS, the Infrared Astronomy Satellite. Those two missions have been magnificent in their scientific output and intermediate in their level of cost. I wish we'd had more like that. We desperately need more like that.

If you go and look at IUE at Goddard, that is, at the console where the astronomer actually sits and points the telescope and takes the observations, it is wonderfully direct. She sits there at the console and she points the telescope and takes the data. It is all done in real time; she can decide what to do next, and if the sky is not particularly good, then she can turn to something else. It's a beautiful system, and it's small enough and informal enough so that people can get access to the telescope rather easily. It works!

And it's also, I'm told, the most productive telescope in existence; if you measure productivity as the bureaucrats like to do—it's a terrible way to do it, but I don't know how to do it better—by pages published per year in the *Astrophysical Journal*, this little telescope in the sky, the IUE, is twice as productive as any other telescope in the world. IRAS is dead, as an instrument, but its data are still pouring out; we haven't analyzed more than a small fraction of it. It's still very much alive as far as the processing of data is concerned. And that's also an amazingly productive mission.

I'm not saying that the big expensive missions are unnecessary or undesirable; I'm not trying to kill the big missions. The Voyager and the Einstein missions were great scientific achievements, and we must be grateful to the taxpayers of the United States for the generous funding that makes these things possible. The big expensive missions do have an essential role to play; but a space science program needs to put a substantial fraction of its effort into small missions if it is to keep pace with the shifting needs of science. And that is why I find the present situation so bad.

Large missions have two outstanding defects that are apt to lead to scientific trouble. The first defect is the long lead times that make large missions unable to respond to new ideas. The second defect is the tendency of big missions to become one of a kind. That is related to the political climate within which they have to be sold to the government and the public.

In order to get funds for a large scientific mission the proponents are forced to talk about all the great scientific problems that mission by itself will solve. And then in order to stay honest they are forced to conform the design of the mission to their promises. So in the end it inevitably becomes a one-shot affair, designed and announced to the public as the "final answer" to some big scientific question. That was the unhappy fate of the Viking mission, which was forced by the political circumstances of its origin to accept the impossible scientific task of deciding all by itself whether life exists on Mars or not. If one looks in detail at the Viking experiments, it is difficult to imagine any combination of results that would have definitely proved or disproved the existence of life on Mars, unless we had been lucky enough to find a cactus bush or an armadillo, sitting immediately in front of the television camera.

One-shot missions are not a good way to do science. If we wanted to investigate seriously the question of life on Mars, the best way would be to plan a regular series of Mars missions, each one far less ambitious and less elaborate than Viking, so we could learn from the results of one mission the right questions to ask from the next. We could also learn from the mistakes of one mission how to avoid mistakes on the next. In almost any field of space science, whether we are exploring planets or galaxies or the earth, a series of modest missions is more likely than a big single spectacular mission to produce important discoveries.

There's a similar problem with X-ray astronomy today. The Einstein X-ray Observatory was a one-shot mission. We desperately need more X-ray missions. There are a few, but most of them are not American. There is a similar problem in infrared astronomy and there will be in optical astronomy. We will have this wonderful Space Telescope (see Chapter 8), but I would be much happier if we had four or five 1-m telescopes in orbit, carrying various specialized instruments, to which more people could get access. It would be wonderful to have the instrumentation up there and see how the sky looks at a tenth of a second arc resolution, even if this is not quite as good as what the Space Telescope will give. There is a tremendous lot you could do with pictures of a tenth second of arc resolution from smaller orbiting telescopes. We should have those in addition to the Space Telescope, and we could have had them long ago if the program had been pushed in that direction.

The Space Telescope is only one item in the plans for space science in the coming decade. The plans are subject to great uncertainties; a number of missions have been proposed and recommended by various committees of distinguished scientists, in particular by the Field Committee, and we're suffering from the disease of "committeeitis." I have nothing against this particular committee and certainly nothing against George Field, but these committees do harm merely by existing. It's the same whether you're talking about astronomy or high-energy physics or computing or nuclear power or plasma physics or anything else. What happens when you have a committee? Inevitably it concentrates its attention on the big items. That's the way committees go. The agenda is dominated by the biggest and most costly projects. No matter what the intent, those things receive the major attention and end up being emphasized in the final report.

The Field Committee says that we need small missions, and they give lists of small missions, but actually the big missions have far more space devoted to them. They appear first, and they are the ones that have got all the attention. That's the way the world seems to work. The big missions get overemphasized, and when the experts go and testify in front of congressional committees and ask for funds, all the conversation is about the big items. In the old times, when missions were more numerous, when you had smaller and more frequent flights, you didn't have so much political trouble because things weren't so visible. We should take a lesson from the Japanese, who are now in this happy state. They're flying one science mission a year. It's still possible to do that, and the science they're doing is pretty good.

III. THE NEXT DECADE

Let me talk now about the plans for the future. The three missions I'll talk about, which are going up in the next few years, are Space Telescope, Galileo, and Hipparcos. Space Telescope and Galileo are shuttle missions, both stretching the limits of budgetary and political feasibility. Both have been subject to long delays and technical uncertainties resulting from difficulties in the development of the shuttle. I don't blame George Field and his friends for that. The shuttle has been a terrible setback for science from almost every point of view. But the Space Telescope will be fine once it goes up. It will explore the fine details of selected objects, mostly very dim and distant objects that cannot be looked at from the ground, or even from any other instrument in space.

Galileo is a planetary mission that will do for Jupiter what Viking did for Mars, sending probes deep into Jupiter's atmosphere and providing fairly complete photographic coverage of Jupiter and its satellites. But Galileo and Space Telescope are both one-of-a-kind missions. No further large missions to Jupiter are planned before the end of the century, and no further large optical telescopes. If, as is likely, Galileo raises important new questions, we will have to wait a long time for the answers.

Hipparcos is a bird of an entirely different color. In the first place it is not a NASA project; it belongs to the European Space Agency, having been invented and originally proposed by Professor Lacroute of Dijon, France. In the second place it is independent of the shuttle, being small enough to be put into geostationary orbit by the French Ariane launch system. In the third place it is cheap enough to be the first of a series. If the first Hipparcos mission works well, it will be easy to launch follow-on missions to give us higher precision or more extensive coverage. If the first Hipparcos fails, it will not be a major disaster.

Hipparcos is an astrometric satellite, designed to do nothing else than measure very accurately angular positions of stars in the sky. It will get positions about ten times as accurate as those measured by ground-based instruments. That sounds like a modest and not very revolutionary objective. But in fact the improvement of positional accuracy is of central importance to astronomy.

If we can improve the accuracy of angular position measurement by a factor of ten, we increase by a factor of ten the distance out to which we can determine distances of stars by the method of parallaxes, and so we increase by a factor of a thousand the sample of stars whose distances we can reliably measure. The tenfold extension of our range of stereoscopic vision will have qualitative as well as quantitative importance. The few hundred stars whose distances we now measure accurately are a random sample of stars that happen to lie close to the earth. Almost all of them are dwarf main-sequence stars of the commonest type, giving little information about interesting phases of stellar evolution. When we have a sample of several hundred thousand, the sample will include rare types, for example, variable stars of various kinds that we are observing at crucial phases of their lives. So in this and many other ways the data from Hipparcos will give a wealth of new information about the constitution and evolution of stars and about the dynamical behavior of the galaxy.

The Hipparcos mission includes a completely automated data processing system on the ground, so that star positions will not be measured laboriously one at a time, but will be computed wholesale in batches of several thousand. The data processing system will be a more revolutionary improvement of the state of the art than the satellite itself.

I hope I am not exaggerating the virtues of Hipparcos, but two facts about it seem to be of fundamental importance. First, it is the first time since Sputnik that a major new development in space science has come from outside the United States. And second, it is the first time since the days of Uhuru that a major new development comes from a small mission, a mission that can be repeated and further developed without putting excessive strain on budgets. Those facts seem to me a good augury for the future. And Hipparcos is probably only the first of many good ideas that will come from the space science programs of Europe and Japan, giving us the competitive stimulus the Soviet space science program once promised, and perhaps will again provide, now that Raoul Sagdeev is in charge.

IV. IDEAS FOR THE NEXT 25 YEARS

I went recently to a meeting of the Space Science Board futures committee, which is set up to study space science in the future, and it was a caricature of the way NASA does things. They don't allow you to talk about anything before 1995, because that might compete with the Space Station. We were only allowed to talk about things that could be launched after 1995. That's the way they operate. They would hate to have anything revolutionary come up in the period when the Space Station is still being built, because it might make it obsolete or undesirable. So we talked about space missions for the first decade of the twenty-first century, that is, from 2000 to 2010. I was disappointed, because we talked only about huge missions. There were some nice ideas; there was an idea from Peter Bender at Boulder about a new gravitational-wave instrument that would measure the strain of the metric of space-time to one part in 10^{21}. Unfortunately, it's in the billion-dollar class, so it's one of those things that will probably hang around for 25 years and be obsolete long before it flies. That's the way things tend to go at these sessions. People don't seem to want to think about things you might do in a couple of years. It's a shame.

Let me then stick my neck out and talk about things that I think we ought to do. What should we do in the next 25 years? My emphasis is on things that could be done quickly and cheaply as far as that's feasible. The first thing is gallium. It's crazy and scandalous and absurd that we didn't yet fund the gallium experiment (see Chapter 4). That is number one on the list. The gallium neutrino detector is of fundamental importance, it is cheap and simple, and we know how to do it.

Second, the small optical space telescope: let's have lots of them. I don't know how much they would cost if you did them in a cheap and dirty style. After all, a 1-m space telescope gets you down to twenty-sixth magnitude, and the Space Telescope is only 28; it's not such a tremendous difference. Twenty-sixth magnitude would be nice.

The extensions of Hipparcos are number three on the list. An infrared version of Hipparcos could look astrometrically at small infrared sources. That would be extremely interesting. You'd see nearby high-velocity objects; you'd discover huge numbers of asteroids.

The fourth item on my list is to repeat a lot of things we have done. NASA has the idea that you should never do anything twice. This is crazy. If you're a scientist, you want to do things twice, maybe three times. There are many missions that we would like to do over again. And they would be cheaper if you did them several times over. Why don't we repeat IUE, why don't we repeat IRAS, why don't we repeat Einstein? All those things would make sense.

Another thing we could do is to put more money into the space science programs of the Europeans and the Japanese. We could have much more collaboration. We do some of that. IRAS was a collaboration. We could do a lot more. It would be much more cost effective than doing big missions by ourselves.

Fifth on the list is the Orbiting VLBI Observatory. This is not very expensive; it is just a radio telescope in orbit with a fairly small antenna hooked in with a very long baseline interferometry network on the ground. It gives you an enormous increase in efficiency of the VLBI system if you have just one antenna in orbit. It does not need to be very big and it could be rather cheap. It gives you larger coverage of the UV plane and it gives you more rapid coverage. It increases the output of the VLBI system by something like a factor of ten at modest cost.

The VLBI systems on the ground are also fantastically successful. One of the rules for supporting science that every bureaucrat ought to follow is to put your money on the winning horse. If a program has been doing well in the past you should go on supporting it, and the VLBI has been doing spectacularly well. The radio astronomers have done far better than the optical astronomers in resolution, so one should go on, and the orbiting antenna is the next thing to do in that business.

The orbiting VLBI antenna is the kind of mission that could easily ride piggyback into orbit with some higher-priority spacecraft. It is cheap enough to be repeated if it works well, so within ten years you could have a network of orbiting antennas in a variety of orbits, pushing the angular resolution of radio astronomy toward the ultimate limit set by the lumpiness of the interstellar plasma. And all that you could do on a learn-as-you-go basis, free from the rigidities of a one-shot mission.

What comes next? Sixth is optical interferometry. That's a little bit further ahead. There again, it's a question of being modest. In Woods Hole, at the futures meeting, we heard a delightful talk from Irwin Shapiro about a big optical interferometer in space. He would observe things to a precision of a microarcsecond, or 10^{-11} rad. To my mind, it is too grand, too ambitious. Why not be content with milliarcseconds for the time being, and do the thing quick and do the thing now, rather than waiting 25 years? There's a lot you can do with optical interferometry. It doesn't require big telescopes or large rigid structures. Early missions could be quite modest, with baselines of 3 m or 10 m, telescope apertures of a few inches. This would enable us to map the optical structure of bright objects with angular resolution ten times better than Space Telescope. After that, we could develop the technology further so as to reach faint objects and get still higher resolution. It took the radio astronomers 20 years to learn the art of very long baseline interferometry; it will probably take about as long for the practitioners of the art of optical interferometry to catch up.

The seventh item on my list is active optics. That means the use of optical interferometry to bring light, nonrigid telescopes into exact focus. It is a simple idea in principle, difficult in practice. The different parts of a telescope mirror surface are to be held exactly in the right positions by servo controls, using feedback signals from interferometric sensors. We have been trying to do this on the ground with moderate success. That's a far more difficult problem.

If you try to use active optics on the ground to get clear images through the atmosphere, you have to compete with the time scale of atmospheric turbulence, which is a few milliseconds. You have to do all your servoing and correcting of the focus within a few milliseconds and you rapidly run out of photons. It's a technique that works up to a point, but you can only do it if you have a bright object in the field of view, and the field of view is small, because the patch of atmosphere that you can correct is small, a few seconds of arc, typically. This thing has a limited utility on the ground.

But in space it would be wonderful. You have then to deal not with turbulent fluctuations of the atmosphere but with mechanical errors and distortions of your optical surfaces, which have a time scale of minutes rather than milliseconds. You have tens of thousands more photons to run the sensors so you can use much fainter objects to focus on, and also you no longer have the problem of the small patch. You can correct a wide field all at the same time. The thing would be far more effective in space than on the ground. That's a technology that ought to be going ahead. We should be building small space telescopes that way. It would be a poor man's Space Telescope.

All the suggestions I've made so far have been somewhat modest, somewhat conventional. I'd like to end with a few more unconventional things. Now I'm talking about things that probably are 20 years away. But you never know. Sometimes things happen faster than you expect if you're willing to jump and not wait for some committee to approve it! So I end with four longer-range technological initiatives.

The first of these is something that has been talked about at the Jet Propulsion Lab. It is called the microspacecraft. The enormous advances in data processing during the last 20 years have been mainly a result of miniaturization of circuitry. The idea of the microspacecraft is to miniaturize everything, the sensors, the navigational instruments, the communication systems, the antennas, and everything else, so that the whole apparatus is reduced in weight like a modern pocket calculator. You still have to have antennas of a certain size, but they can be very thin. The whole thing could be very light. There doesn't seem to be any law of physics that says a high-performance exploratory spacecraft like Voyager has to weigh a ton. We might be able to build vehicles to do that sort of a job in the 1-kg weight class.

The second of the technological initiatives I like to think about is a serious effort to exploit solar sails as a cheap and convenient means of transportation around the solar system, at least in the zone of the inner planets and the asteroids.The main reason why solar sailing never seemed practical to the managers of NASA is that the sails required to carry out interesting missions were too big. Roughly speaking, a 1-ton payload requires a square kilometer of sail to drive it, and a square kilometer is an uncomfortably large size for the first experiments in deployment. Nobody wants to be the first astronaut to get tangled up in a square kilometer of sail. But if you think in terms of microspacecraft, then you might think of solar sails of a more manageable size. A 1-kg microspacecraft would go nicely with a 30-m square sail, and a 30-m square is a reasonable size to experiment with. The problems of sail management and payload miniaturization will be solved more easily together than separately.

Third on this list of more far-out projects is a thing I call minilaser propulsion. I've long been an advocate of laser propulsion, which was invented by Arthur Kantrowitz. It is a system of launching rockets from the ground using a laser beam as the energy source, so that the rocket climbs up the laser beam into the sky. The propellant can be something convenient like water. It's heated up to a very high temperature by the laser beam, and it gives a specific impulse of a thousand seconds, which is convenient for getting to escape velocity with a single stage.

The classical laser propulsion system of Kantrowitz had a gigawatt laser to supply the power and would launch payloads of 1 ton, which is a reasonable payload for scientific missions. It needed for that only about 1 ton of propellant, so the vehicle could be small. It would be extremely cheap as compared with a conventional three-stage chemical rocket. The trouble was that you had to have your gigawatt laser first. That's a very expensive item. The whole thing suffers from this grandiosity that makes it, from a practical point of view, useless. It will inevitably be ages before something like that gets built, and by that time all the interesting science will have changed.

However, it might be different if you scale the whole thing down. I've been looking at a "minilaser propulsion system" that uses lasers of a few megawatts. It's a comfortable sort of laser; we know how to build it. You've got to have a focusing system and a steering and guidance system and various other things—it's not going to be cheap, but it's in the hundred million dollar class, not in the billion dollar class. The motor doesn't yet exist—nobody's yet designed the motor to run this thing with; it's at the moment more or less in the stage of Mr. Goddard's chemical rockets in the year 1928. But with luck, the motor will work. You could launch these little spacecraft, then, wonderfully cheap; if you calculate how much electricity it takes to get them into orbit, or even into escape, it's about a hundred kilowatt-hours per launch. The energy costs are really low. Of course you've got to have a high volume of traffic in order to make the thing economic. But conceivably, if you had a system like this, people would enjoy using it. You could build your spacecraft cheaply. The rocket that goes with it would weigh a few pounds and would be a mass-production item.

The last of my projects is the "space butterfly." That's a way of exploiting for the purposes of space science the biological technology that allows a humble caterpillar to wrap itself up in a chrysalis and emerge three weeks later transformed into a shimmering beauty of legs and antennae and wings. I don't know whether you've ever watched a Monarch butterfly climbing into its cocoon and then afterwards climbing out again. It's an awe-inspiring sight. Sooner or later, probably fairly soon, we'll understand how that's done. Somehow it's programmed into the DNA and we should pretty soon learn how to do it. It's a good bet, within the next 25 years, this technology will be fully understood and available for us to copy. So it's reasonable to think of the microspacecraft of the year 2010, not a structure of metal and glass and silicon, but a living creature, fed on Earth like a caterpillar, launched into space like a chrysalis, with one of these little lasers, and metamorphosing itself in space like a butterfly. Once it's out there in space, it'll sprout wings in the shape of solar sails, thus neatly solving the sail deployment problem. It'll grow telescopic eyes to see where it's going, gossamer-fine antennae for receiving and transmitting radio signals, long springy legs for landing and walking on the smaller asteroids, chemical sensors for tasting the asteroidal minerals and the solar wind, electric-current-generating organs for orienting its wings in the interplanetary magnetic field, and a high-quality brain will enable it to coordinate its activities, navigate to its destination, and report its observations back to Earth.

I don't know whether we'll actually have that, but it's a good bet we shall have something equally new and strange, if only we turn our backs to the past and keep our eyes open for the opportunities that are beckoning in the twenty-first century. When I compare the biological world with the world of mechanical industry, I'm impressed with the enormous superiority of biological processes, in speed, in economy, and in flexibility. To give an example: a skunk dies in a forest. Within a few days, an army of ants and beetles and bacteria goes to work, and after a few weeks, barely a bone remains. An automobile dies and is taken to a junkyard. After ten years, it is still there.

If you consider anything that our industrial machines can do, whether it's mining, or chemical refining, or materials processing, or building, or scavenging, biological processes in the natural world do the same thing more efficiently, more quietly, and usually more quickly. That's the fundamental reason why genetic engineering must, in the long run, be beneficial to science and also profitable. It offers us the chance to imitate nature's speed and flexibility in industrial and in scientific operations.

It's difficult to speak of specific examples of things genetic engineering can do for us. Specific examples always sound too much like stories out of *Astounding Science Fiction*, but I'll give you four possible examples: the "energy tree," programmed to convert the products of photosynthesis into conveniently harvested liquid fuels instead of into cellulose; the "mining worm," a creature like an earthworm, programmed to dig into clay or metalliferous ore, and bring to the surface the desired chemicals in purified form; third, the "scavenger turtle," with diamond-tipped teeth, a creature programmed to deal in a similar fashion with human refuse and derelict automobiles; and fourth, the "space butterfly," acting as our scout in the exploration and eventually in the bringing of life to the solar system. None of these creatures performs a task essentially more difficult than the task of the honeybee with which we are all familiar.

But it's a sound instinct that leads us to be distrustful of such grand ideas. If we pursue long-range objectives of this kind, we're likely to find ourselves involved in a 20-year development program, with all the inertia and inflexibility of a Space Telescope or a nuclear power program. The whole advantage of biological technology will be lost if we let it become rigid and slow.

I don't think the theoretically possible dangers of genetic engineering will turn out to be real; I think that the benefits will be large and important, so I wish luck and joy to the young scientists who are now beginning their careers as genetic engineers. I hope they will move into astronomy and space exploration in the next 20 years, just as the radio and electrical engineers moved into astronomy during the last 40 years. The genetic engineers will find, as the radio astronomers found, that the astronomical community is a bit conservative and a bit resistant to new ideas, but after the initial barriers are broken down, we will in the end welcome them as colleagues in the job of exploring the universe.

Index